FUZZY MODELING AND CONTROL
Selected Works of M. Sugeno

FUZZY MODELING AND CONTROL
Selected Works of M. Sugeno

Edited by

Hung T. Nguyen
Nadipuram R. Prasad

CRC Press
Boca Raton   London   New York   Washington, D.C.

Library of Congress Cataloging-in-Publication Data

Sugeno, Michio, 1940–
    Fuzzy modeling and control : selected works of M. Sugeno / edited by Hung T. Nguyen, Nadipuram R. Prasad.
       p.   cm.
    Includes bibliographical references and index.
    ISBN 0-8493-2884-5 (alk. paper)
    1. Automatic control. 2. Fuzzy systems. I. Nguyen, Hung T., 1944– II. Prasad, Nadipuram R. III. Title.
TJ213.S798 1999
629.8—dc21   99-19201
                                                                                                                      CIP

    This book contains information obtained from authentic and highly regarded sources. Reprinted material is quoted with permission, and sources are indicated. A wide variety of references are listed. Reasonable efforts have been made to publish reliable data and information, but the author and the publisher cannot assume responsibility for the validity of all materials or for the consequences of their use.

    Neither this book nor any part may be reproduced or transmitted in any form or by any means, electronic or mechanical, including photocopying, microfilming, and recording, or by any information storage or retrieval system, without prior permission in writing from the publisher.

    All rights reserved. Authorization to photocopy items for internal or personal use, or the personal or internal use of specific clients, may be granted by CRC Press LLC, provided that $.50 per page photocopied is paid directly to Copyright Clearance Center, 222 Rosewood Drive, Danvers, MA 01923 USA. The fee code for users of the Transactional Reporting Service is ISBN 0-8493-2884-5/99/$0.00+$.50. The fee is subject to change without notice. For organizations that have been granted a photocopy license by the CCC, a separate system of payment has been arranged.

    CRC Press LLC's consent does not extend to copying for general distribution, for promotion, for creating new works, or for resale. Specific permission must be obtained in writing from CRC Press LLC for such copying.

    Direct all inquiries to CRC Press LLC, 2000 Corporate Blvd., N.W., Boca Raton, Florida 33431.

    Trademark Notice: Product or cororate names may be trademarks or registered trademarks, and are used only for identification and explanation, without intent to infringe.

© 1999 by CRC Press LLC

No claim to original U.S. Government works
International Standard Book Number 0-8493-2884-5
Library of Congress Card Number 99-19201
Printed in the United States of America 1 2 3 4 5 6 7 8 9 0
Printed on acid-free paper

The publication of Fuzzy Modeling and Control -- Selected Works of M. Sugeno is an important event -- an event which highlights the deep and wide-ranging impact of his work and ideas. Unique among the contributors to fuzzy set theory and its applications, Professor Sugeno is equally at ease with his pen and his soldering iron. His fundamental work on what we call today the Sugeno measure and Sugeno integral is among the most mathematically sophisticated contributions to fuzzy set theory. On the other extreme, his building and subsequent demonstrating at the 1987 IFSA Congress in Tokyo of a model car that could park itself, showed his remarkable ability to translate abstract ideas into working devices and systems -- systems that exhibit a high degree of machine intelligence. This unusual ability is even more in evidence in Professor Sugeno's path-breaking work on helicopter control, which is described in this volume.

I met Professor Sugeno in 1974, when he came to Berkeley to participate in the USA-Japan Symposium on Fuzzy Sets and Their Applications. He was a student of Professor Toshiro Terano at the Tokyo Institute of Technology and had just completed his doctoral thesis -- a seminal work which laid the foundation for a theory of fuzzy measures and fuzzy integrals. Professor Terano, Professor Kyoji Asai and the late Professor Kokichi Tanaka of Osaka University played key roles in launching the fuzzy boom in Japan. Today, most of the leading contributors to fuzzy set theory and fuzzy logic in Japan are students of students of Professors Terano, Asai and Tanaka.

Professor Sugeno's work on fuzzy measures and fuzzy integrals opened a new direction in fuzzy set theory which led to many important results and applications, especially in the realm of decision analysis. Very significant contributions by Murofushi and Grabisch to the theory and applications of the Sugeno measure and Sugeno integral are included in this volume.

During the past decade, fuzzy control has emerged as by far the most visible and most active area of research in fuzzy-logic-based systems. Professor Sugeno has played and is continuing to play a leading role in the development of fuzzy control on both theoretical and practical levels. What has come to be known as the Takagi-Sugeno-Kang model has proved to be highly effective in many applications. In this model, the fuzzy rule consequents are assumed to be linear combinations of the input variables, and the output is a convex combination of the consequents, with coefficients that are the grades of membership of the inputs in the antecedents. The inputs -- which are usually sensor outputs -- are scalar.

In a seminal 1985 joint paper with Tomohiro Takagi -- which is included in this volume -- Professor Sugeno has advanced a new approach to rule induction from observations. In the Sugeno-Takagi approach, the parameters of linguistic variables in the fuzzy if-then rules are determined by minimization of the quadratic error in approximation. The Sugeno-Takagi approach inspired Roger Jang, L. X. Wang, Lee and Lin and others to develop steepest descent techniques in the spirit of the back propagation algorithm and the radial basis function methods -- methods which were developed in the context of neural network theory.

An original thinker in the true sense of the term, Professor Sugeno has opened a number of new directions in his work. Recently, he became interested in the interplay between linguistics and systems analysis. We can expect, as usual, that his interest in linguistics will lead to novel ideas and results relating natural languages and control theory.

The papers in *Fuzzy Modeling and Control: Selected Works of M. Sugeno* are more than a tribute to Professor Sugeno's profound influence on the evolution of fuzzy set theory and its applications. They serve, above all, to provide access to some of the most important ideas and results within the theory of fuzzy sets and point to new directions, especially in the realms of control, systems and decision analysis.

The editors of this volume, Professors Hung Nguyen and Ram Prasad, have done a superb job of selecting and organizing the key papers authored by Professor Sugeno. We owe them our thanks and congratulations.

Lotfi A. Zadeh
Berkeley, California

Michio Sugeno
Professor, Tokyo Institute of Technology
Department of Computational Intelligence and Systems Science
Yokohama 227, Japan

Born on February 3, 1940 in Yokohama, Japan
- 1962   Graduated from Department of Physics, The University of Tokyo
- 1962   Researcher, Mitsubishi Atomic Power Industry
- 1965   Research Associate, Department of Control Engineering, Tokyo Institute of Technology
- 1974   Received the degree of Doctor of Engineering, Tokyo Institute of Technology
- 1976   Visiting Researcher, Queen Mary College, London, U.K.
- 1976-77 Visiting Researcher, Laboratoire d'Automatique et d'Analyse de Systemes, Toulouse, France
- 1977-87 Associate Professor, Department of Systems Science, Tokyo Institute of Technology
- 1987-96 Professor, Department of Systems Science, Tokyo Institute of Technology
- 1996-   Professor, Department of Computational Intelligence and Systems Science, Tokyo Institute of Technology

Academic Activities

1984-87	Vice President of the International Fuzzy Systems Association (WSA)
1989-90	Vice President of the Japan Society for Fuzzy Theory and Systems (SOFT)
1989-91	President of IFSA Japan Chapter
1991-92	President of SOFT
1989-95	Leading Advisor of the Laboratory for International Fuzzy Engineering Research (LIFE)
1989-94	Advisor of the Science and Technology Agency (STA), the Prime Minister's Office
1989-94	Director of Fuzzy Systems Research Project of STA
1994-96	Chairman of IEEE SMC Tokyo Chapter
1995-	President elect of IFSA

Editorial Board Member:

Fuzzy Sets and Systems
International Journal of Intelligent Systems
International Journal of Uncertainty, Fuzziness and Knowledge-Based Systems
Mathematical Modelling of Systems
Mathware and Soft Computing

Advisory Board Member:

International Journal of Approximate Reasoning
IEEE *Transactions on Fuzzy Systems*

Awards

1986	Sawaragi Paper Award, The Institute of Systems, Control and Information Engineers Control and Information Engineers, Japan
1988	Best Publication Award, Nikkan-Kogyo Press
1990	Best Paper Award, North American Fuzzy Information Processing Society
1992	Best Publication Award, Japan Society for Fuzzy Theory and Systems
1994	Best Paper Award, Japan Society for Fuzzy Theory and Systems
1994	International MOISIL Prize

Professional Societies (Member)

Japan Society of Fuzzy Theory and Systems
The Society for Instrument and Control Engineers in Japan
The Institute of Systems, Control and Information Engineers in Japan

Japanese Society for Artificial Intelligence
International Fuzzy Systems Association
North American Fuzzy Information Processing Society
Balkanic Union for Fuzzy Systems and Artificial Intelligence
The IEEE Systems, Man, and Cybernetics Society
American Helicopter Society
New York Academy of Science
Japan Association for Systemic Functional Linguistics
Japanese Neural Network Society

	Foreword	v
	Michio Sugeno	vii
	Preface	xiii
1	Introduction	1
2	Development of an Intelligent Unmanned Helicopter	13
3	Multi-Dimensional Fuzzy Reasoning	45
4	A New Approach to Design of Fuzzy Controller	59
5	Fuzzy Control of Model Car	67
6	Fuzzy Algorithmic Control of a Model Car by Oral Instructions	77
7	Application of Fuzzy Reasoning to the Water Purification Process	89
8	A Microprocessor Based Fuzzy Controller For Industrial Applications	105
9	Fuzzy Modeling and Control of Multilayer Incinerator	113
10	Fuzzy Identification of Systems and its Application to Modeling and Control	129

11	Structure Identification of Fuzzy Model	161
12	Successive Identification of Fuzzy Model and its Application to Prediction of a Complex System	179
13	A Fuzzy-Logic-Based Approach to Qualitative Modeling	199
14	Stability Analysis and Design of Fuzzy Control Systems	251
15	An Interpretation of Fuzzy Measures and the Choquet Integral as an Integral with Respect to a Fuzzy Measure	277
16	Fuzzy Measure of Fuzzy Events Defined by Fuzzy Integrals	303
17	Fuzzy Measure Analysis of Public Attitude Towards the Use of Nuclear Energy	329
18	Pseudo-Additive Measures and Integrals	363
19	A Model for Human Evaluation Process Using Fuzzy Measure	389
	Appendix	409
	Index	423

Preface

Since the mid-70's, Professor Michio Sugeno has ranked among the world's foremost researchers in advancing fuzzy technology to its present state. Starting from his seminal work in his doctoral dissertation on the Theory of Fuzzy Measures and Integrals, which forms the basic contributions to Fuzzy Theory and applications to subjective evaluations, he went on to devote his time to a different topic, namely, systems control. All of his research areas possess the same significance: they are pioneering works, basic, and have laid the foundation for many important applications addressed by other researchers. The now popular fuzzy methodology in engineering, especially in the development of intelligent systems, has its roots in the Theory of Fuzzy Sets founded by Professor Lotfi Zadeh. Indeed, on the theoretical side, while Professor Zadeh proposed the Theory of Fuzzy Sets to model uncertainty in knowledge expressed in natural language, Sugeno supplemented it with a Theory of Fuzzy Measures and Integrals. This is in line with the basic ingredients used by Kolmogorov to lay down the foundations for the modern Theory of Probability and Statistics, namely the Theory of Functions and Measure Theory. Sugeno's Fuzzy Measures have found applications in fields such as computer vision, subjective evaluations of patterns, etc.

On the engineering applications side, although the very first idea of Fuzzy Control was due to Mamdani, it was Sugeno, in 1985 with his fuzzy control of a model car, who really put Fuzzy Control on a firm basis for a wide range of applications. Indeed, his theory of Fuzzy Measures and Integrals is still a widely pursued topic and an active area of research. Applications of his ideas on this topic has spread over many fields. His pioneering work on fuzzy modeling, system identification techniques, stability analysis of fuzzy systems, and models bearing his name, have become fundamental tools in fuzzy technology today. He has put strong emphasizes on model-based fuzzy control. His current work on the unmanned helicopter fuzzy control, as described in his special contribution to this book (Chapter 2), clearly proves his exceptional engineering skills as well as his ability to generate highly innovative ideas for the design of complex control systems.

While there are numerous books currently on the market dealing with the fundamentals and applications of fuzzy logic, none have dealt entirely with the seminal contributions to this field made by Professor Michio Sugeno. Sugeno's works are known worldwide and are published in many journals. In the Fall of 1996, Professor Sugeno visited New Mexico State University and delivered a seminar on the helicopter project in The Klipsch School of Electrical and Computer Engineering. It was during this visit that we spontaneously came up with an idea to provide a book about his pioneering works. We thought it would be of immense value for researchers to have a volume handy that contained all his fundamental works as a reference for either applications or for further research. We proposed our idea to Professor Sugeno and suggested that the title of the book would be: "Fuzzy Modeling and Control - Collected Works of M. Sugeno". In a very humorous statement, his response was "I am still alive and hope to do some more work." The contents of this book are only a small selection of a very large number of topics that Professor Sugeno has addressed. Appropriately, therefore, we have called this book the *Selected Works of M. Sugeno*. Such a volume of his selected works is a great tribute from our part to a very special researcher and a close friend. We most certainly await more prodigious work from Professor Sugeno in the years to come.

The Introduction, which is Chapter 1 of this book, is devoted to providing a brief discussion of Professor Sugeno's works in the areas of Fuzzy Modeling, Fuzzy Control, and Fuzzy Measures and Integration. In this chapter, we will also address their significance in terms of their applications to real world problems. Chapter 2, as noted above, is a special chapter, in which Professor Sugeno describes the development and implementation of fuzzy control of an unmanned helicopter. The remainder of this book from Chapter 3 through Chapter 19 are selected works of Professor Sugeno that have been retyped verbatim from various journal publications.

We have taken extraordinary care in ensuring that the original contributions of Professor Sugeno are not altered in any way in the preparation of this edited volume. We have attempted to reorganize the text and graphical presentations to improve readability and to conform with the style of a textbook.

The book can be used both as a prescribed text in an engineering curriculum as well as a reference by practicing engineers and scientists. In the engineering curriculum, a course on "Fuzzy Control Systems" can be offered with a prerequisite of classical linear/nonlinear control systems. Although the prerequisite may be viewed as somewhat optional, background in classical methods may enhance the students' understanding and appreciation of the breadth of application of fuzzy logic based systems. Notions of controllability, observability, and stability may become more clear with appropriate background in classical techniques.

We are grateful to Professor Lotfi A. Zadeh for writing the Foreword for this book. We thank Mr. Robert Stern at CRC Press for his strong support and patience in publishing this book. We thank all publishing companies for allowing us to reproduce Professor Sugeno's papers from their journals so that all chapters of this book are uniform in style and appearance. Our sincere appreciation goes to Murali Siddaiah and Eric Nguyen for their deep commitment in the preparation of this book. Many thanks to Gary Anaaya for his help with the computer. The second author would like to thank Joe Fronczek, Stephen Hood, and his sons Abhishek and Aadarsh for their utmost patience and support throughout the preparation of this book. Finally, our very special thanks go to Professor Sugeno himself for contributing a chapter on his current research project on the unmanned helicopter.

Hung T. Nguyen
Nadipuram (Ram) R. Prasad,
Las Cruces, New Mexico.

Introduction

H. T. Nguyen and N. R. Prasad
New Mexico State University
Las Cruces, New Mexico

This chapter serves to introduce the basic topics addressed in the research papers of Professor Sugeno. We emphasize that the introductory material presented in this chapter is not a comprehensive treatment of fuzzy fundamentals in as much as a broad description of the topics needed to appreciate the depth of Professor Sugeno's contributions. For a complete treatise on fuzzy logic theory, the reader is urged to refer to *A First Course in Fuzzy Logic* by Hung T. Nguyen and Elbert Walker, CRC Press, 1996.

1. Generalities on Fuzzy Logic

Let X be a set. A *fuzzy subset* A of X is a map $A : X \to [0, 1]$. For $x \in X$, the value $A(x)$ is interpreted as the degree of membership of x in the underlying fuzzy concept represented by A. When the range of a map defined on X is reduced to $\{0, 1\}$, the map characterizes an ordinary (*crisp*) subset of X.

The application of fuzzy set theory to systems modeling and control is based upon concepts that this theory generalizes from ordinary set theory. These include operations among fuzzy sets and modes of reasoning with fuzzy concepts. The "real" logical aspects of fuzzy concepts are not used in full force.

In engineering applications, the reference set X is the real line $\mathbf{R}$, or an Euclidean space $\mathbf{R}^k$. In such a case, a fuzzy quantity A is said to be *normal* if $A(x) = 1$ for some $x \in X$. On $\mathbf{R}$, a *fuzzy quantity* A is said to be *convex* if its α-cuts are *convex*, i.e., the α-cuts are intervals.

The α-cuts A_α of A is defined to be:
$$A_\alpha = \{x \in X : A(x) \geq \alpha\}, \quad \alpha \in [0,1]$$

The set $\{x \in X : A(x) > \alpha\}$ is referred to as the *strong* α-cut of A at level α. The *membership function* A is recovered from its α-cuts via:

$$\forall x \in X, \ A(x) = \sup\{\alpha \geq 0 : x \in A_\alpha\}$$

A finite *fuzzy partition* of X is a finite set of normal fuzzy sets $\{A_1, A_2, \cdots, A_n\}$ of X such that $\sum_{i=1}^{n} A_i(x) = 1$ for all $x \in X$.

In using fuzzy sets to model linguistic labels, it is necessary to combine mappings, or functions, in various ways. For example, given two sets X and Y, and $f : X \times Y \to Z$, one can extend f to fuzzy sets as follows. Let A and B, represent fuzzy subsets of X and Y, respectively. Also, let "$\wedge$" and "$\vee$" denote *infimum* (minimum) and *supremum* (maximum), respectively. Then $f(A, B)$ is a fuzzy subset of Z with membership given by:

$$f(A, B)(z) = \bigvee_{\substack{(x,y) \\ f(x,y)=z}} (A(x) \wedge B(y))$$

The above equation is referred to as the *Extension Principle*. Note that if the function f is such that the right-hand side is achieved for each $z \in Z$, then,

$$[f(A, B)]_\alpha = f(A_\alpha, B_\alpha), \quad \alpha \in (0, 1]$$

A *fuzzy relation* on $X \times Y$ is a fuzzy subset of $X \times Y$. Once again, considering A and B to represent fuzzy subsets of X and Y, respectively, the *Cartesian product* $A \times B$ is the fuzzy subset of $X \times Y$ with:

$$(A \times B)(x, y) = A(x) \wedge B(y)$$

In particular, $A \times Y$ is called the *Cylindrical Extension* of A.

Operations among fuzzy subsets of a set X are generalizations of ordinary set operations.

Intersection of fuzzy sets is defined via the concept of *t-norms*.
A t-norm is a binary operation

$$\Delta : [0, 1] \times [0, 1] \to [0, 1]$$

such that

    i. $0 \triangledown x = x$

    ii. $x \Delta y = y \Delta x$

    iii. $x \Delta (y \Delta z) = (x \Delta y) \Delta z$

    iv.   If $u \leq x$ and $v \leq y$ then $u \Delta v \leq x \Delta y$

The minimum operator $\wedge$ is the only *idempotent* t-norm, i.e., $x \Delta x = x$ for all $x \in X$. A continuous t-norm Δ is classified as an *Archimedean t-norm* if $x \Delta x < x$ for all $x \in (0, 1)$. A t-norm Δ is said to be *nilpotent* if for $x \neq 1, x \Delta x \ldots \Delta x$ (n times) $= 0$ for some positive integer n where the n depends on x. For example, $x \Delta y = 0 \vee (x + y - 1)$ is a nilpotent *t-norm*.

For the study of sensitivity in systems modeling or control where t-norms (and other logical operations) are involved, isomorphism theorems are useful. Let $0 \leq a < 1$ and let $f : [0, 1] \to [a, 1]$ be an *order isomorphism*, i.e., f is one-to-one and onto, and $x \leq y$ if and only if $f(x) \leq f(y)$. It turns out that t-norms can be constructed from order isomorphisms. Specifically, if Δ is an Archimedean t-norm, then there exists an $a \in [0, 1]$ and an order isomorphism $f : [0, 1] \to [a, 1]$ such that; $x \Delta y = f^{-1}(f(x)f(y) \vee a)$. f is referred to as the multiplicative generator of Δ. If $f(0) > 0$, then Δ is nilpotent.

The complement A' of a fuzzy concept A is defined via the concept of *negation operators*. A negation operator is a map $\eta : [0, 1] \to [a, 1]$ such that $\eta(0) = 1, \eta(1) = 0$ and η is non-increasing. By a *strong negation operator*, we mean a negation operator η such that $\eta(\eta(x)) = x$, i.e., η is involutive. Such negations are of the form $f^{-1} \alpha f$, where $\alpha(x) = 1 - x$, and $f : [0, 1] \to [a, 1]$, is one-to-one and satisfies $f(x) \leq f(y)$ whenever $x \leq y$ (an automorphism of $([0, 1], \leq)$).

A negation of the form $(1 - x)/(1 - bx)$ with $b \in (-1, +\infty)$ is called a Sugeno negation.

Union of fuzzy sets is defined via the concept of *t-conorms*. This is a dual concept of t-norms.

A binary operation ∇ on $[0, 1]$ is a *t-conorm* if the following conditions hold:
    i. $0 \nabla x = x$
    ii. $x \nabla y = y \nabla x$
    iii. $u \leq x$
    iv.   If $u \leq x$ and $v \leq y$ then $u \nabla v \leq x \nabla y$.

We say that the t-norm Δ and the t-conorm ∇ are dual with respect to the negation η if:
$$x \nabla y = \eta(\eta(x) \Delta \eta(y))$$
In this case, (Δ, ∇, η) is called a DeMorgan system. Note that the space $\mathcal{F}(X)$ of all fuzzy subsets of X, equipped with $x \Delta y = x \wedge y$, $x \nabla y = x \vee y$ and $\eta(x) = 1 - x$, is a DeMorgan algebra.

Generalizing classical two-valued logic, fuzzy implications extend material implication. Thus, a fuzzy implication is a $[0, 1] \times [0, 1] \to [0, 1]$ with its truth table as in Fig. 1.

⇒	0	1
0	1	1
1	0	1

Fig. 1. *Truth table.*

For example,
$$x \Rightarrow y = \begin{cases} 1 & if \quad x \leq y \\ 0 & if \quad x > y \\ (1-x+y) \wedge 1 \\ y \vee (1-x) \end{cases}$$

Note that there are many ways to construct fuzzy implications.

The basic logical connectives are defined as operations among fuzzy sets. When the membership values of membership functions are imprecise, e.g., membership functions are interval-valued, (interval-valued fuzzy sets), the above logical connectives can be extended appropriately.

2. Fuzzy Modeling and Control

Traditionally, control laws of plants are derived from a knowledge of the plant models, e.g., dynamical differential equations. In general, plant models are not fully known: the form of the model may be known but the structure and parameters are not known. These unknown parameters have to be determined through a process known as parameter identification. Therefore, in general, some form of system identification needs to be performed before a control law can be derived.

Fuzzy control is an approach to systems control when the exact mathematical model(s) of the plant is unknown or the mathematical model is too complex to simulate or understand. Typically, in place of mathematical models, other weaker forms of knowledge about the plant or process behavior may be available. This knowledge may be in the form of:
1) Expert's knowledge about the control strategies,
2) Training examples from the plant where several operating scenarios are created and appropriate control actions are practiced.

In any case, the task of a knowledge engineer is to use as much of the non-conventional forms of knowledge to identify the input-output behavior of the process. Since most of the operating knowledge of plants is qualitative in nature, it becomes necessary to translate linguistic information into approximate mathematical expressions. In fact this is the fundamental basis for fuzzy logic-based systems. The idea then becomes one of reasoning with linguistic models rather than the use of a mathematical model. The theory of fuzzy sets and logic seems ideal for such a situation.

Directly or indirectly, the heuristic form of knowledge about a system is given by a collection of "*if-then*" rules. This collection forms the so-called rule base or the knowledge base of a process or system.

A rule base is of the form:

$$R_j : If \ x_1 \ is \ A_j^1 \ and \cdots x_n \ is \ A_j^n \ then \ y \ is \ B_j$$

where, $A_j^i, i = 1, 2, \cdots, n$; B_j are fuzzy subsets of $U_1, U_2, \cdots, U_n; V$, respectively, and $x = (x_1, x_2, \cdots, x_n), y$ are input and output variables of the plant. There are several aspects of reasoning with such a rule base that allow one to derive a control law.

The compositional rule of inference is an instance of approximate reasoning, which is useful for the interpolation of fuzzy "if-then" rules. The rule of inferencing is nothing more than a generalization of the pattern of reasoning known in classical two-valued logic as *modus ponens*. Specifically, let X and Y be two variables taken in U and V, respectively. Let A, A^*, and B be fuzzy subsets of appropriate spaces. Consider the following propositions:

1) p = "If X is A then Y is B"
2) q = "X is A^*"

The goal then is to find a reasonable fuzzy subset B^* in V, such that the proposition "Y is B^*" can be viewed as a logical conclusion. Now, the conditional proposition P is represented by a binary fuzzy relation R, i.e., a fuzzy subset of $U \times V$, and A^* is viewed as a unary fuzzy relation on U. As such B^* is obtained as $R \circ A^*$, where the operator '$\circ$' denotes the composition of relations, namely,

$$R \circ A^* = \bigvee_{u \in U} \left[A^*(u) \wedge R(u,v) \right]$$

Given a rule base $R_j, j = 1, 2, \cdots, r$, it is necessary to fuse the R_j's in order to produce an input-output map. Besides the logical connectives within each rule, one needs to find some appropriate logical connective that will allow combining of all the rules in the rule base. For this, the "OR" connective is used giving rise to what is known as the max-min compositional form of rules. Thus, a numerical value of the input $x = (x_1, x_2, \cdots, x_n)$ produces a fuzzy subset of V:

$$B_x(y) = \nabla \left(\Delta(A_j^i(x_i), i = 1, 2, \cdots, n; B_j(y)), j = 1, 2, \cdots, r \right)$$

where Δ and ∇ are some chosen *t-norm* and *t-conorm*, respectively.

This method of reasoning is referred to as the Mamdani method of fuzzy inferencing. In order to obtain a single output value that represents a crisp control signal to the plant or process, the fuzzy subset $B_x(\cdot)$ of V is summarized using some defuzzification procedure. The most common defuzzification technique that is used in control systems is the Centroid Method or the Center-of-Gravity method in which:

$$y(x) = \int_V B_x(y) y \, dy \bigg/ \int_V B_x(y) \, dy.$$

Another popular inferencing technique is that proposed by Takagi-Sugeno-Kang and is widely referred to as the TSK method. The TSK method is applied to rules of the form:

$$R_j : \text{If } x_1 \text{ is } A_j^1 \text{ and } \cdots x_n \text{ is } A_j^n \text{ then } y = f_j(x_1, x_2, \cdots, x_n)$$

where the right-hand side of each rule is a functional relationship of the inputs (antecedents). This has a unique structure in that it is similar to the output equation of a state-variable model in classical control theory. It should be interesting to recall that the output equation from a state-variable model is a function of all the states of the system.

The degree of firing of rule R_j is computed as:

$$\alpha_j(x) = \wedge \left(A_j^i(x_i), i = 1, 2, \cdots, n\right)$$

The overall output is then taken as:

$$y(x) = \sum_{j=1}^{r} \alpha_j(x) f_j(x) \Big/ \sum_{j=1}^{r} \alpha_j(x)$$

As stated previously, before a model can be identified or a control law can be obtained, problems such as representation of the rule base, method of reasoning, and the identification of appropriate design parameters need to be addressed. The problem of representing rules is in general dictated by available information about the plant under consideration, i.e., whether it is in the form of linguistic description or it is model-based information. Typically, the antecedent parameters are represented by parametric membership functions such as triangular, trapezoidal, Gaussian, piecewise polynomial, or spline functions. However, the important property of these functions is that they all be convex sets. Once the forms of the various membership functions are decided, tuning procedures are required to identify the appropriate system parameters. The identification is based upon some form of optimization of the consequent part of the rules. Optimization methods could include neural networks, genetic algorithms, least mean squares method, etc. While it is known that the universal approximation property of general fuzzy systems justifies the strategy in fuzzy control, the performance of fuzzy controllers depends on various factors such as the choice of logical connectives in the representation of rules. As a final note, the current model-based approach to fuzzy control is promising since it provides a framework in which stability issues can be addressed.

3. Fuzzy Measures and Integrals

Set-functions which are σ-additive are called measures in mathematics and its applications. Lebesgue measure on $\mathcal{R}^k$ measures volumes of Borel subsets of $\mathcal{R}^k$, whereas a probability measure on a measurable space measures the chances of occurrence of events. Non-additive set-functions are abundant in Analysis, such as Choquet capacities in Potential Theory, Hausdorff dimension on metric spaces.

In his dissertation (1974), Sugeno developed a mathematical framework to capture the concept of (human) subjectivity. This important aspect of human ability of judgement needs to be modeled and incorporated into intelligent systems. Sugeno's concept of "grade of fuzziness" aims at formalizing evaluations which depend heavily upon human subjectivity. He introduced the concept of fuzzy measures and stated "Fuzzy measures are interpreted as measures expressing grade of fuzziness and are compared with probability measures expressing grade of randomness." His theory of Fuzzy Measures and Integrals has applications in a variety of fields such as subjective evaluations, computer vision, etc. It has triggered a systematic study of non-additive set-functions.

Since Sugeno's work on fuzzy measures and integrals is less popular in the engineering community as compared to his seminal work in the area of modeling and control, we will take this opportunity to give a tutorial introduction to this novel contribution by Sugeno.

In his dissertation, Sugeno defined a fuzzy measure $\mathcal{M}$ as a map defined on a monotone family of subsets of a set Ω satisfying the following:
1) $\mathcal{M}(\emptyset) = 0$, $\mathcal{M}(\Omega) = 1$
2) $A \subseteq B \Rightarrow \mathcal{M}(A) \leq \mathcal{M}(B)$
3) If A_n is a monotone sequence, then $\lim_{n \to \infty} \mathcal{M}(A_n) = \mathcal{M}\left(\lim_{n \to \infty} A_n\right)$.

Later the range of a fuzzy measure is extended to $[0, +\infty]$, and the axiom (3) is dropped.

Definition: Let $\mathcal{A}$ be a family of subsets of a set Ω, with $\emptyset \in \mathcal{A}$. A map $\mathcal{M} : \mathcal{A} \to [0, \infty]$ is called a fuzzy measure if:
1) $\mathcal{M}(\emptyset) = 0$
2) If $A, B \in \mathcal{A}$ and $A \subseteq B$, then $\mathcal{M}(A) \leq \mathcal{M}(B)$.

As an example, we discuss the Sugeno λ-measure.

Let $(\Omega, \mathcal{A})$ be a measurable space. If $\mathcal{M} : \mathcal{A} \to [0, 1]$ is such that $\mathcal{M}(\Omega) = 1$, and $\mathcal{M}(A \cup B) = \mathcal{M}(A) + \mathcal{M}(B) + \lambda\, \mathcal{M}(A)\, \mathcal{M}(B)$, whenever $A \cap B = \emptyset$, for some $\lambda > -1$, then $\mathcal{M}$ is called a λ-measure. Note that a λ-measure $\mathcal{M}$ is of the form $f \circ P$ where P is a probability measure and,

$$f(x) = \begin{cases} x & \text{if } \lambda = 0 \\ \dfrac{1}{\lambda}\left[(1+\lambda)^x - 1\right] & \text{if } \lambda \neq 0 \end{cases}$$

As measure theory is in fact a theory of integration, let us consider the problem of integration of real-valued functions with respect to a fuzzy measure.

Originally Sugeno defined the integral of a function $f : \Omega \to [0, 1]$ with respect to a fuzzy measure $\mathcal{M}$ as:

$$I_{\mathcal{M}}(f) = \sup_{\alpha \in [0, 1]} \left\{\alpha \cap \mathcal{M}(f \geq \alpha)\right\}$$

Essentially, the operations involved in the definition of the Sugeno integral $I_{\mathcal{M}}(f)$ are $\vee$ and $\wedge$, replacing addition and multiplication in the Lebesgue integral. Note also that the construction of $I_{\mathcal{M}}(f)$ does not involve the monotone continuity of $\mathcal{M}$, as opposed to a stronger condition of ordinary measures, namely, the σ-additivity in the construction of the Lebesgue integral. Thus, even in the case where quantities like $I_{\mathcal{M}}(f)$ are to be taken as fuzzy integrals, the monotone continuity of $\mathcal{M}$ is not needed. Now because additive set-functions are special cases of fuzzy measures, it is natural to expect that fuzzy integrals generalize Lebesgue integrals. But, as it was shown in Sugeno's dissertation, $I_{\mathcal{M}}(f)$ is different from Lebesgue integral with respect to a measure $\mathcal{M}$. Thus, a new integral with respect to monotone set-functions (which need not be additive), generalizing Lebesgue integral is desirable.

Whatever approach is taken to define fuzzy integrals, we need to consider carefully the meaning of the quantities that fuzzy integrals are supposed to measure. For this purpose, let us consider a couple of situations.

a) Let X be a random variable taking values in the set $\Theta = \{\theta_1, \theta_2, \theta_3, \theta_4\} \subseteq \mathcal{R}^+$. Suppose that the density f_0 of X is only known to satisfy the following constraints:

$$f_0(\theta_1) \geq 0.4, \qquad f_0(\theta_2) \geq 0.2$$
$$f_0(\theta_3) \geq 0.2, \qquad f_0(\theta_4) \geq 0.1.$$

Thus, the expected value $E_{f_0}(X)$ cannot be computed exactly. However, we can obtain bounds on $E_{f_0}(X)$.

Indeed, let $\mathcal{F}$ be a class of densities on Θ satisfying the inequalities above. Obviously, we have:

$$\inf\{E_f(X): f \in \mathcal{F}\} \leq E_{f_0}(X) \leq \sup\{E_f(X): f \in \mathcal{F}\}$$

Let P_f denote the probability measure on the power set $P(\Theta)$ of Θ generated by f, i.e.,

$$A \subseteq \Theta, \quad P_f(A) = \sum_{\theta \in A} f(\theta)$$

Define $F: P(\Theta) \to [0, 1]$

$$F(A) = \inf\{P_f(A): f \in \mathcal{F}\}$$

Then clearly F is a fuzzy measure on Θ. Suppose $\theta_1 < \theta_2 < \theta_3 < \theta_4$. Define $g: \Theta \to [0, 1]$ as:

$$g(\theta_1) = F\{\theta_1, \theta_2, \theta_3, \theta_4\} - F\{\theta_2, \theta_3, \theta_4\}$$
$$g(\theta_2) = F\{\theta_2, \theta_3, \theta_4\} - F\{\theta_3, \theta_4\}$$
$$g(\theta_3) = F\{\theta_3, \theta_4\} - F\{\theta_4\}$$
$$g(\theta_4) = F\{\theta_4\}$$

Then $g \in \mathcal{F}$. Now,

$$E_g(x) = \int_\Theta x(\theta) dP_g(\theta) = \int_0^{+\infty} P_g\{\theta_i : \theta_i > t\} dt = \int_0^{+\infty} F\{\theta_i : \theta_i > t\} dt$$

The last integral is the Choquet integral of $X(\theta) = \theta$ with respect to the monotone increasing set-function F.

It can be checked that:

$$\inf\{E_f(X): f \in \mathcal{F}\} = E_g(X)$$

so that the Choquet integral above is the lower bound for expected values.

b) Here is another situation with imprecise information. Suppose that we perform a random experiment and face the following situation. The outcome of each experiment can only be located in some interval [a, b], i.e., $X(\omega) \in [a,b]$. Thus we have a mapping Γ defined on Ω with values in the class of non-empty closed subsets of $\mathcal{R}$ such that $X(\omega) \in \Gamma(\omega)$ for each $\omega \in \Omega$.

Let $g : \mathcal{R} \to \mathcal{R}^+$ be a measurable function. The random variable $X : (\Omega, \mathcal{A}, P) \to \mathcal{R}$ has probability law $P_X = P \circ X^{-1}$ on $\mathcal{R}$ and

$$E_g(X) = \int_\Omega g(X(\omega))\, dP(\omega) = \int_\mathcal{R} g(x)\, dP_X(x)$$

Observe that,

$$g_*(\omega) = \inf\{g(x) : x \in \Gamma(\omega)\} \leq g(x(\omega))$$
$$\leq \sup\{g(x) : x \in \Gamma(\omega)\} = g^*(\omega)$$

so that, $E(g_*) \leq E(g(X)) \leq E(g^*)$ provided g_* and g^* are measurable.

This is indeed the case when the multi-valued map Γ is measurable in the following sense:

Let $\mathcal{B}$ denote the Borel σ-field on $\mathcal{R}$. The map Γ is said to be strongly measurable if for any $B \in \mathcal{B}$

$$B_* = \{\omega : \Gamma(\omega) \subseteq B\} \in \mathcal{A}$$
$$B^* = \{\omega : \Gamma(\omega) \cap B \neq \varnothing\} \in \mathcal{A}$$

Now define $F_* : \mathcal{B} \to [0,1]$ by $F_*(B) = P\{w : \Gamma(\omega) \subseteq B\} = P(B_*)$

Then, $E(g_*) = \int_\Omega g_*(\omega)\, dP(\omega) = \int_0^{+\infty} P\{\omega : g_*(\omega) > t\}\, dt$

$$= \int_0^\infty P\{g_*^{-1}(t, \infty)\}\, dt = \int_0^\infty P[g^{-1}(t, \infty)]\, dt_*$$

$$= \int_0^\infty P\{w : \Gamma(w) \leq g^{-1}(t, \infty)\}\, dt$$

$$= \int_0^\infty F_*\left(g^{-1}(t, \infty)\right) dt$$

$$= \int_0^\infty F_*\{x : g(x) > t\}\, dt$$

Similarly,

$$E(g^*) = \int_0^\infty F^*\{x : g(x) > t\}\, dt$$

Thus, the bounds $E(g_*)$ and $E(g^*)$ for $E_g(X)$ are Choquet integrals of the function g with respect to fuzzy measures F_* and F^*, respectively.

We are now formalizing the concept of the fuzzy integral as a Choquet integral. To be specific, let $(\Omega, \mathcal{A})$ be a measurable space, and $\mathcal{M} : \mathcal{A} \to [0, \infty]$ a fuzzy measure.

Let $f : \Omega \to [-\infty, +\infty]$ be a measurable function. Then the fuzzy integral of f with respect to $\mathcal{M}$ is defined to be:

$$\int_\Omega f(\omega) d\mathcal{M}(\omega) = \int_0^{+\infty} \mathcal{M}(x : f(x) > t) dt + \int_{-\infty}^0 \left[\mathcal{M}(x : f(x) \geq t) - \mathcal{M}(\Omega) \right] dt$$

and for $A \in \mathcal{A}$,

$$\int_A f(\omega) d\mathcal{M}(\omega) = \int_0^{+\infty} \left[\mathcal{M}(x : f(x) > t) \cap A \right] dt$$

$$+ \int_{-\infty}^0 \left[\mathcal{M}\left((x : f(x) \geq t) \cap A\right) - \mathcal{M}(A) \right] dt$$

We say that f is $\mathcal{M}$ integrable on A when the right-hand side is finite.

The above fuzzy integral, as a Choquet integral, is monotone and positively homogeneous of degree one, that is $f \leq g$ implies $\int_\Omega f\, d\mathcal{M} \leq \int_\Omega g\, d\mathcal{M}$ and for $\lambda > 0$ $\int_\Omega \lambda f d\mathcal{M} = \lambda \int_\Omega f\, d\mathcal{M}$, but its additivity fails. For example, if $f = (1/4)I_A$ and $g = (1/2)I_B$ with $A \cap B = \emptyset$, then,

$$\int_\Omega (f+g) d\mathcal{M} \neq \int_\Omega f d\mathcal{M} + \int_\Omega g d\mathcal{M}$$

However, if $f = a\, I_A + b\, I_B$ with $A \cap B = \emptyset$, $0 \leq a \leq b$

$g = \alpha\, I_A + \beta\, I_B$ with $0 \leq \alpha \leq \beta$,

we do have $\int_\Omega (f+g) d\mathcal{M} = \int_\Omega f d\mathcal{M} + \int_\Omega g d\mathcal{M}$

The reason is that the pair (f, g) in the second case satisfies the inequality

$$\forall \omega, \omega' \in \Omega, \left[f(\omega) - f(\omega') \right]\left[g(\omega) - g(\omega') \right] \geq 0$$

which says that (f, g) are *comonotonic*. Thus, the fuzzy integral is comonotonic additive.

Finally, let us mention the concept of Radon-Nikodym derivatives of fuzzy measures.

Let $f : \mathcal{R}^k \to [0, 1]$ be an upper semi-continuous function.

Define $v, v_0 : B(\mathcal{R}^k) \to [0, 1]$ by

$$v(A) = \sup\{f(x) : x \in A\}$$

$$v_0(A) = \begin{cases} 1 & \text{if } A \neq \emptyset \\ 0 & \text{if } A = \emptyset \end{cases}$$

Then v and v_0 are non-additive, monotone increasing and zero on the empty sets, thus they are fuzzy measures.

Moreover, we have, $v(A) = \sup\{f(x) : x \in A\} = \int_0^\infty v_0\left[A \cap \{x : f(x) \geq t\}\right] dt$

Thus, by analogy with the situation in ordinary measure theory, f is a sort of "derivative" of v with respect to v_0. This leads to: If $\mathcal{M}$ and v are two fuzzy measures on $(\Omega, \mathcal{A})$ such that $v(A) = \int_A f d\mathcal{M}$, for any $A \in \mathcal{A}$, (where $\int_A f d\mathcal{M}$ is the Choquet integral of f on A with respect to the fuzzy measure $\mathcal{M}$), then f is called the Radon-Nikodym derivative of v with respect to $\mathcal{M}$, and is written as $f = \dfrac{dv}{d\mathcal{M}}$.

References

[1]. Diamond, P. and Kloeden, P. (1994), *Metric Spaces of Fuzzy Sets*, World Scientific, Singapore.
[2]. Driankov, D., Hellendoorn, H. and Reinfrank, M. (1996), *An Introduction to Fuzzy Control*, (Second Edition), Springer Verlag, Berlin.
[3]. Nguyen, H. T., Sugeno, M., Tong, R. and Yager, R. (Eds) (1995), *Theoretical Aspects of Fuzzy Control*, John Wiley & Sons, New York.
[4]. Nguyen, H. T., and Walker, E. A. (1997), *A First Course in Fuzzy Logic*, CRC Press, Boca Raton, Florida.
[5]. Nguyen, H. T. and Sugeno, M. (Eds) (1998), *Fuzzy Systems: Modeling and Control*, Kluwer Academic, Boston.
[6]. Palm, R., Driankov, D. and Hellendoorn, H. (1997), *Model-Based Fuzzy Control*, Springer-Verlag, Berlin.
[7]. Pedrycz, W. (1995), *Fuzzy Control and Fuzzy Systems*, John Wiley & Sons, New York.
[8]. Sugeno, M. (1974), *Theory of Fuzzy Integrals and Its Applications*, Ph.D. Thesis, Tokyo Institute of Technology, Japan.
[9]. Yager, R., Ovchinninkov, S., Tong, R. and Nguyen, H. T. (Eds) (1987), *Fuzzy Sets: Theory and Applications – Selected Papers by L. A. Zadeh*, John Wiley & Sons, New York.

Development of an Intelligent Unmanned Helicopter

M. Sugeno
Tokyo Institute of Technology
4259 Nagatsuta, Midori-ku, Yokohama 227, Japan

Abstract

This paper describes hardware and software systems for control of an unmanned helicopter. The helicopter control system developed by Sugeno Laboratory uses the knowledge and technique of an experienced pilot/engineer and puts them into a computer by using fuzzy control technology. An on-board fuzzy controller is installed to achieve intelligent control in the unmanned helicopter YAMAHA R-50. Using this fuzzy control system, the helicopter can be tele-controlled from the ground within the visual area by giving verbal commands such as "Fly forward" and "Hover". The helicopter can be navigated also in the invisible area by using GPS. Further it can track a moving object by an active image sensor.

1. Introduction

The helicopter has six degrees of freedom in its motions. There are four control inputs concerning its flight in addition to throttle control for the rotation of rotors. Coordinating these inputs the helicopter can make various flights: forward flight, backward flight, sideward flight, hovering, hovering turn, vertical climb, vertical descent, etc. Helicopter flight characteristics are far superior to those of a fixed-wing aircraft. The flight dynamics change according to its flight mode and so a conventional control technique is rather difficult to apply. Several unmanned helicopters have been developed, or are under development in the world. However, to-date, a complete automatic control system has not been realized.

1.1. *Objective of the project*

The project of Sugeno Laboratory aims to develop an intelligent unmanned helicopter that can be remotely controlled by linguistic commands from a ground station with a monitoring system. A necessary and critical aspect of an unmanned helicopter is its tele-control and mission requirements. For this project, the tele-control is to be achieved using fuzzy control theory: Sugeno Laboratory has more than 15 years of experience in this field. Ultimately, our helicopter will incorporate voice-activated commands using natural language such as "fly forward a little bit". The idea is that a relatively inexperienced remote operator can use natural language, voice commands, rather than a couple of joysticks that may require months of training. These commands are naturally 'fuzzy' and hence fit into the fuzzy logic framework nicely.

The voice-activated commands, however, are only part of our objectives. A more important issue is the mission requirements. The unmanned helicopter must have a degree of autonomy for dealing with unforeseen situations such as collision avoidance. To this aim, we are developing a GPS-guided navigation system and an image-based control system.

Flight dynamics and control of a helicopter

The problem with the helicopter flight dynamics is that a helicopter is inherently unstable or a poorly damped system. The flight dynamics are easily affected by the natural environment such as temperature, wind, etc. Its basic aircraft modes are cross-coupled with no clear division into longitudinal and lateral direction modes. In addition to the basic airframe rigid body dynamics, the rotor adds flapping dynamics and the engine adds rotor speed dynamics. Furthermore, there are nonlinear variations in the dynamics with air speed. Hence control of a helicopter, though not impossible, is a difficult one indeed, and careless mistakes and excessive workload of a pilot often cause accidents.

To achieve the linguistic commands implies that we can reduce the necessary number of control inputs from four or five to only one. For example, we just send a single command 'take-off' instead of manipulating a couple of joysticks. To see why this is remarkable, we need a brief review of pilot control actions.

The motion of a helicopter is achieved by (1) cyclically changing the angle of the individual rotor blades of the main-rotor as the rotor rotates (cyclic pitch), (2) tilting the main-rotor path plane, and (3) changing the tail rotor pitch (moves the tail left or right). The helicopter pilot typically has at his/her disposal a cyclic stick to control both fore/aft motions (pitch control) and left/right motions (roll control), a collective pitch lever to control up/down motions (vertical control), and pedals to control right/left yawing motions (yaw control). The helicopter will fall down in a few seconds rather than a few minutes if its pilot takes no action.

A brief history of development

1988-90: A small RC-model-helicopter was used. Two sensors for angular velocities and accelerations were used. Only four variables were measured: pitch rate, roll rate, longitudinal acceleration, and lateral acceleration. A fuzzy controller was installed in a personal computer and connected by a wire to the helicopter. Stable hovering control was achieved.

1990: A software-based-simulator was introduced, which models a real helicopter BK117 and is installed in a supergraphic workstation. Using the simulator, the fuzzy control of forward (backward) flight, leftward (rightward) flight and turning flight were performed.

1991: A real unmanned helicopter *YAMAHA R-50* was introduced. An integrated sensor for attitude angles, attitude angular velocities, and accelerations was introduced. Also velocity sensors of three axes and a height meter were used. Thus among 15 variables necessary for control, 13 variables except a horizontal position were measured. An idea of a hierarchical and modular fuzzy controller was introduced, which was installed in a personal computer still on the ground and wired to the helicopter. The core parts were a flight mode management module in the upper layer and fuzzy control modules in the lower layer. Using R-50, hovering, forward (backward) flight and leftward (rightward) flight were achieved, while the collective pitch control concerning a vertical control was manipulated by a human pilot. The command-based-control of a sequential square flight was first demonstrated.

1992: An on-board fuzzy controller with a fuzzy inference engine was introduced, which was installed in the helicopter. Thus a wireless command-based-control was first achieved. The collective pitch control was then made by a fuzzy controller: take-off and landing were successfully performed. Also two flight modes, hovering turn and rudder turn, were newly added.

1993: The hierarchical fuzzy controller was improved: the total configuration of two layers was almost fixed. Especially a fuzzy controller management module consisting of parameter setting, trim adjustment and coupling compensation were introduced. As for flight modes, a coordinated turn mode was added. Blended flight modes such as a diagonal flight and a turning-up flight were first introduced. A CCD camera was installed in the helicopter. Using image information by a CCD camera, automatic image-based landing was first achieved where the image processing was made on the ground: both image signals and the camera control signals were wirelessly transmitted. An idea of a fly-by-wire controller was introduced.

1994: A differential GPS system to measure a horizontal position was introduced: all 15 variables were measured. Using the GPS, the navigation of the helicopter was successfully achieved. Two modules, namely, navigation and a command interpretation, were added to the upper layer, where a navigation module consists of GPS-guided flights, image-guided flights and programmed flights.

1995: A new on-board fuzzy controller was installed: 256 commands were acceptable. The function of a GPS-guided navigation was improved.

Image-guided tracking of a moving object was successfully performed. An idea of a flight monitoring system was introduced.

1996: The number of fuzzy control rules was increased more than two times to improve the stability under a windy environment. The coupling compensation system was improved. The performance of a GPS-guided navigation was improved. A fly-by-wire controller to assist command-based control was almost completed. A flight monitoring system was successfully tested.

Applications of an unmanned helicopter

There is a multitude of possible applications for an unmanned helicopter beyond the obvious military requirements. Its merit would be inestimable, if the following list of monotonous or dangerous missions could be performed by an autonomous unmanned helicopter.

Concerning monotonous missions, we shall have:
- Taking aerial photograph
- Measuring air pollution
- Agricultural spraying
- Fish finding
- Oil pipeline inspection
- Monitoring of freeway traffic conditions
- Monitoring of railways
- Monitoring of the seacoast

Concerning dangerous missions, we shall have:
- Sea rescue
- Mountain rescue
- Accident monitoring oil tankers, nuclear plants, etc.
- Fire fighting in remote and/or dangerous areas: mountains, high buildings, etc.
- Watching natural phenomena: volcanic eruptions, earthquakes, etc.
- Searching for a crashed aircraft
- Eliminating land mines after a war
- Experimental hazard analysis of a piloted helicopter.

2. Characteristics of a helicopter

A helicopter has several characteristics, which make it one of the most difficult vehicles to control. These characteristics are as follows:

i. It is an unstable and nonlinear system. It is difficult to automate its operation by conventional control.

ii. It has multiple inputs and state variables. There are several cross couplings. The helicopter has six degrees of freedom in its motions: up/down, fore/aft, right/left, pitching, rolling, and yawing as illustrated in Fig. 1. The number of variables necessary for its control is 15, namely, accelerations $(\ddot{x}, \ddot{y}, \ddot{z})$, velocities $(\dot{x}, \dot{y}, \dot{z})$, and positions (x, y, z) in the 3-dimensional space, pitch angle (θ), roll angle (ϕ), yaw angle (ψ), and their angular velocities $(\dot{\theta}, \dot{\phi}, \dot{\varphi})$. It has five control inputs (in the case of an unmanned helicopter, the number may be four).

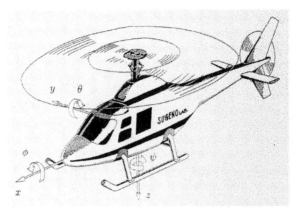

Fig. 1. Degrees of freedom of helicopter.

Throttle: control of rotors
   This is an input for the engine control to drive the rotors. Usually there is a feedback loop for the throttle control to maintain a rotation speed of the main rotor constant.
Collective pitch: control of $\dot{z}$
   This is an input for the climb or descent control by changing the main rotor's lift through the change of the main rotor blade angle.
Longitudinal cyclic pitch: control of θ and $\dot{x}$
   This is an input for the forward or backward flight control by tilting the main rotor path plane forward or backward (pitch angle).
Lateral cyclic pitch: control of ϕ and $\dot{y}$
   This is an input for the rightward or leftward flight control by tilting the main rotor path plane right or left (roll angle).
Antitorque pedals: control of φ
   This is an input for the heading direction control by changing a lift of the tail rotor through the change of the tail rotor blade angle. As for cross couplings, we have, for example, couplings from the longitudinal cyclic pitch to altitude and sideslip: increasing forward speed by changing the longitudinal cyclic pitch may cause a loss of height and a sideslip. One control input to an objective motion causes other different kinds of motions. So we need to control other inputs at the same time to

compensate couplings.

iii. It is easily affected by disturbances. In particular, it is affected by wind and becomes easily unstable. Also in take off and landing, it is affected by ground effect. So a pilot always changes control inputs by finding the trim points of the helicopter (stable attitude) depending on the flight environment such as wind direction, altitude, etc.

iv. Its characteristics change according to environmental conditions. The lift changes with wind, altitude, temperature and humidity.

v. It has many flight modes like hovering, hovering turn, sideward flight, backward flight, vertical climb and descent different from a winged aircraft as shown in Fig. 2. Couplings change according to flight modes and control methods also. Thus, controllers must be designed according to flight modes.

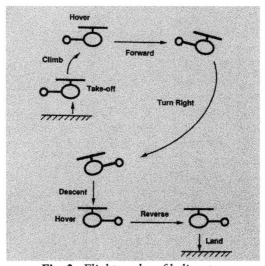

Fig. 2. Flight modes of helicopter.

vi. It is often forced to fly under a severe environment. Different from a winged aircraft, a helicopter flies at very low altitudes, among mountains or buildings, and generally near obstacles. So a pilot has an excessive workload and may cause a fatal failure such as crash by touching a power line.

3. Hardware system

The hardware system for helicopter experiments consists of a helicopter, a command transmitter, a command receiver, a fuzzy controller, various sensors to measure state variables, a differential GPS, an active image sensor with a CCD camera, a telemeter, a FBW controller, and a flight monitoring system.

Helicopter

The helicopter YAMAHA R-50 used in this project is a production model manufactured by Yamaha Motors mainly for agricultural display. The dimensions of the YAMAHA R-50 are: main rotor diameter 3.07 m, overall body length 3.57 m, height 1.1 m and empty weight 44 kg. It is equipped with a water-cooled 2-cycle engine of 12hp, and the payload is 20 kg. Fig. 3 shows an overall view of YAMAHA R-50. We shall replace it with a new model YAMAHA R_{max} very soon.

Fig. 3. YAMAHA R-50.

Fuzzy controller

Fig. 4 shows an on-board-fuzzy controller designed and built at the Sugeno Laboratory.

Fig. 4. On-board-fuzzy controller.

It is equipped with fuzzy inference chips that can store 1500 fuzzy control rules. The inference speed is 1 msec in the case of 60 rules. Besides CPU, ROM and RAM, it has 24 channels-A/D, 8 channels-DA and 8 channels-PWM inputs: it can accept 256 commands.

Integrated inertia sensor
It can measure accelerations, attitude angles and angular velocities using accelerometers and an optical gyro.

Radio wave speed meters
These measure two-dimensional velocities: $\dot{x}$ and $\dot{y}$

Laser height meter
It measures height by using laser: the resolution is 5 cm.

Magnetic azimuth sensor
It measures the heading direction.

Differential GPS8
A Global Positioning System (GPS) measures three-dimensional position and velocities. A differential GPS compensated by another GPS fixed on the ground achieves a high accuracy, for example, with about an error of 25 cm concerning two-dimensional position.

Active image sensor
It measures a distance and a relative velocity of a moving object.

Flight monitoring system
It visually displays the attitudes of the helicopter, its trajectory on a ground-coordinate-map, the reference points such as waypoints and a destination on a helicopter-coordinate-map, and also shows a flight mode, altitude, speed, etc.

Remote control devices
The unmanned helicopter can be remotely controlled from the ground in three ways, which are classified into three levels. At the level 0, we can use a conventional manual radio controller. At the level 1, we can use a Fly-By-Wire controller with two modes: a velocity mode (a) and an attitude mode (b). In the model, we can set desired velocities with respect to x and y, and a desired heading direction (a yaw angle). In the mode 1b, we can set desired attitudes; pitch, roll and yaw angles. In both modes, an up/down velocity is set constant. The controller has a take-off/landing switch, an emergency button to switch to a hovering mode from any flight mode, a stick for velocity or attitude control, and a slide for $\dot{Z}$ control.

Fig. 5 shows FBW controller hardware. At the level 2, we can send flight commands either through a computer keyboard (mode 2a) or a microphone with a voice recognition device (mode 2b).

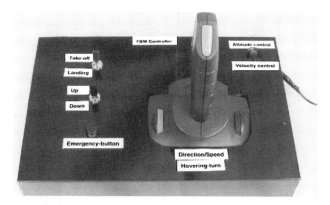

Fig. 5. Fly-By-Wire controller.

4. Software system for helicopter control

Several years ago, we designed a two-layered hierarchical control system that has been and is still being improved every year. The overall architecture was almost completed. The basic idea on the developed software system for helicopter control, i.e., control of a multi-input/multi-output coupled system, is two-way decompositions: vertical and horizontal. Fig. 6 shows the overall architecture. The vertical (hierarchical) decomposition is concerned with multi-tasks control and the horizontal decomposition concerned with multi-variables control.

4.1. *Lower layer*

The lower layer contains a module for fuzzy controller management and a number of fuzzy control modules. Fuzzy control modules are divided into those for longitudinal, lateral, collective, and pedals (rudder) controls: one module corresponds to one control input. Each module consists of "If-Then" fuzzy control rules. The idea taken here is to decompose a multi-variables control into a set of single-variable controls. Then we meet a coupling problem, which is solved in the coupling compensation module. As is seen in Fig. 6, there are some control modules for each single-variable control.

Longitudinal module

This module is used for pitch angle control concerning a forward-rearward movement of the helicopter. There are three modules: x-control, (velocity) slowdown control, $\dot{x}$-control. Fig. 7 shows the fuzzy control structure of the $\dot{x}$-control module. As we see, given a reference forward velocity $R_{\dot{x}}$,

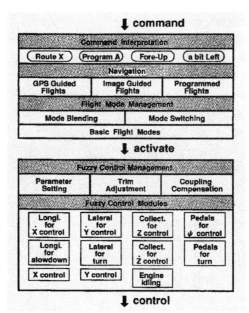

Fig. 6. Hierarchical control system.

the first fuzzy control block, X-velocity, infers a desired pitch angle R_θ using a velocity error $E_{\dot{x}}$ and its derivative $\ddot{x}$ (acceleration). Then the second block, pitch, infers a desired longitudinal input using a pitch angle error E_θ and a pitch rate $\dot\theta$ (pitch angular velocity). We attach gain factors to two inputs and a single output of each block. In the module, the variables T_{pit} and T_{lon} are called trims, which are usually constants. We shall discuss trims later in the section on fuzzy controller management.

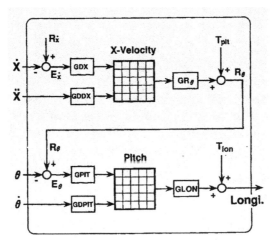

Fig. 7. Longitudinal for x control module.

Basically each fuzzy control block contains 25 fuzzy rules. The number of control rules of this module is 50. A part of those rules is as follows:

$\ddot{x}$ is PB and $E_{\dot{x}}$ is PS → R_θ is PS
$\ddot{x}$ is PB and $E_{\dot{x}}$ is ZO → R_θ is PM
$\ddot{x}$ is PS and $E_{\dot{x}}$ is NS → R_θ is PM
$\ddot{x}$ is PS and $E_{\dot{x}}$ is NB → R_θ is PB
$\ddot{x}$ is ZO and $E_{\dot{x}}$ is PB → R_θ is NM
$\ddot{x}$ is ZO and $E_{\dot{x}}$ is PS → R_θ is NS

The linguistic variables such as PB (*positive big*), PS (*positive small*), ZO (*zero*) in the premise parts are fuzzy variables, while those in the consequent parts are singletons (numerical values). It is well known that generally we do not need fuzzy variables in the consequent parts. The fuzzy variables are associated with normalized triangular or trapezoidal membership functions. We omit actual parameters of fuzzy variables and singletons.

For slowdown control, we just add another fuzzy control block in front of the X-velocity block as is shown in Fig. 8. Given R_x, a destination of x, the slowdown block gives a desired velocity for a slowdown of the helicopter speed. The number of control rules is 55.

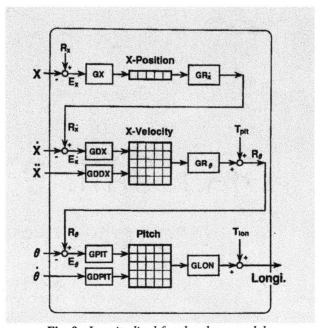

***Fig. 8**. Longitudinal for slowdown module.*

Now for x-control, we replace the X-velocity block with the X-position block as shown in Fig. 9. Given a reference x-position R_x, this block infers a desired pitch angle using an x-position error E_x and a velocity $\dot{x}$. The number of control rules is 50.

Lateral module

This module is used for roll angle control concerning a leftward-rightward movement of the helicopter. There are also three modules: $\dot{y}$-control, coordinated turn, and y-control. We do not use a slowdown control for y-velocity since the helicopter does not fly sideward at a high speed. Instead we need a special control, lateral for turn, for a coordinated turn. We omit here the descriptions of the $\dot{y}$-control and the y-control, which is the same for the longitudinal modules. The number of control rules for the $\dot{y}$-control or the y-control is also 50. The lateral for turn control has a different structure from that of the $\dot{y}$-control or the y-control. The number of control rules is 25. The details will be discussed and shown later in the section on fuzzy controller design.

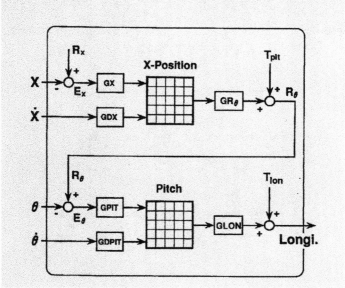

Fig. 9. Longitudinal for x control module.

Collective module

This module is used for vertical control concerning an upward-downward movement. There are three modules: z-control, $\dot{z}$-control and engine idling. Fig. 10 shows the structure of the z-control, which consists of two fuzzy control blocks: z-velocity and altitude.

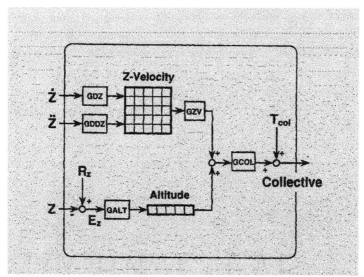

Fig. 10. Collective for z control module.

We first set a reference z-velocity to zero and, given a reference z-position R_z, this module infers a desired collective input using a velocity error $E_{\dot{z}}$, an acceleration $\ddot{z}$, and a position error E_z. The number of control rules is 30. The structure of the $\dot{z}$-control module with 25 control rules is shown in Fig. 11. We shall not need an explanation for this module.

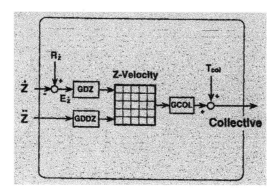

Fig. 11. Collective for z control module.

Pedals (rudder) module

This module is used for yaw angle (φ) control, i.e., heading direction control. There are two modules: pedals for φ-control and pedals for turn. Fig. 12 shows the structure of the pedals for the φ-control module, which infers pedals, input using a yaw angle error E_φ and a yaw rate $\dot{\varphi}$. The number of rules is 25. The pedals for turn module with 30 rules is used for a coordinated turn together with the lateral for turn module which will be shown later in the section on fuzzy controller design.

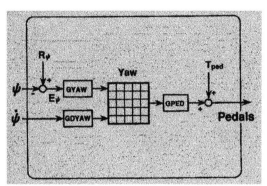

Fig. 12. Pedals for ψ control module.

Fuzzy controller management

This module consists of three submodules: parameter setting, trims adjustment, and coupling compensation. The task of the parameter setting submodule is to store and provide the necessary parameters for the fuzzy control modules. The parameters include the reference inputs, the input and output gains of fuzzy controllers, the parameters of membership functions, and the trims.

There are two kinds of trims: control trims and attitude trims. A control trim implies a desired control input to keep the helicopter at a steady state under a certain environment. An attitude trim also implies a desired attitude for a steady state under a certain environment. For example in Fig. 7, T_{pit} is an attitude trim for a pitch angle and T_{lon} is a control trim for longitudinal control input. In the case of the collective and pedals modules, there is no attitude trim.

The task of trim adjustment submodule is to give the standard trims depending on flight modes such as hovering, forward flight and to adjust these during a flight according to flight conditions. This task is quite important to achieve a stable flight. As a matter of fact, if optimal trims are given, we do not need controllers at a steady state. We shall not discuss this issue in detail in this paper.

The task of coupling compensation submodule is to adjust the control inputs to the helicopter for compensating cross-couplings among state variables. In general, the change of one control input for its corresponding movement causes side effects to the other movements.

In our study we compensate only major couplings caused by a longitudinal control and also a collective control. Because of feedback control, the other minor couplings are to some extent compensated. We shall discuss this further in Section 6.

4.2. Upper layer

The upper layer consists of three modules concerning (i) command interpretation, (ii) navigation, and (iii) flight mode management. Among these, the most important module is flight mode management. In this section, we follow the upper layer from the bottom (iii) to the top (i) in Fig. 6.

Flight mode management module

This module contains three submodules: basic mode, mode blending, and mode switching. The submodule of the basic mode manages the 14 basic flight modes: Idle, Take-off, Land, Hover, Stop, Fore/Aft, Left/Right, Up/Down, L-H-Turn/LR-H-Turn, and L-C-Turn/R-C-Turn. Idle means engine idling which is not a flight mode but taken care of here. L-H-Turn and L-C-Turn mean left-hovering-turn and left-circling-turn, respectively. This module assigns a given flight mode from the command interpretation module to fuzzy control modules through the fuzzy controller management module of the lower layer corresponding four control inputs: longitudinal, lateral, collective, and pedals.

For instance, in the case of hovering, a Hover mode is equally assigned to four fuzzy control modules: the longitudinal for $\dot{x}$-control, the lateral for $\dot{y}$-control, the collective for z-control and the pedals for φ-control. In the case of a forward flight, a Fore mode is, however, assigned only to a longitudinal control module while a Hover mode is assigned to other three control modules: lateral, collective and pedals. This means that the helicopter moves neither left/right nor up/down, and so on. That is, for instance, the lateral control is kept at a Hover mode. Also in the case of an upward flight, an Up mode is assigned to a collective control module and a Hover mode is assigned to other control modules. Further in the case of a left circling turn, first of all a Fore mode is assigned to a longitudinal control module, then an L-C-Turn mode is assigned to the lateral and pedals control modules while a collective control module is maintained at a Hover mode. That is, the helicopter flies forward by a Fore mode without changing its height by a Hover mode and makes a left-circling-turn by an L-C-Turn mode. Among these flight modes, we need special care for Take-off and Landing. We omit its details here.

In this way, many flight modes can be realized by combining a rather small number of lower-level control modules. This is an advantage of a horizontal decomposition.

A complex flight which includes a combination of some basic flight modes is called a blended flight. In a submodule of the mode blending, the flight mode manager blends basic flight modes to achieve a complex flight. For example, in a 3D-Straight mode it blends together Fore, Up, and Right basic flight modes. In this case, (i) the longitudinal for $\dot{x}$-control module is assigned a Fore mode, (ii) the lateral for $\dot{y}$-control module is assigned a Right mode, (iii) the collective for $\dot{z}\,\dot{z}$-control module is assigned an Up mode, and (iv) the pedals for ϕ-control module is assigned a Hover mode.

We can blend physically compatible flight modes (e.g., Fore and Left, Up and Left-Circling -Turn, etc.). However, Right and Left, for example, cannot be blended.

In a submodule of the mode switching we smoothly connect two flight modes (not necessarily different) by inserting another flight mode between them. For example, in the case of Fore-Stop, we insert a Slowdown mode (by activating the longitudinal for slowdown control module) between the Fore and Stop basic flight modes as Fore~Slowdown~Stop. Moreover, in order to change a flight direction, we execute mode switching of the form: Fore→ Turn→ Fore.

Navigation module

This module also contains three submodules: GPS-guided flights, image-guided flights and programmed flights. In the case of the programmed flights, the helicopter flies autonomously according to a prescribed flight program. For example, we can write a program as "take-off, fly up 15 meters, hover for 3 minutes, and then fly 50 meters to the east and stop there".

As for the image-guided flights, at present we can achieve automatic landing and also image-based tracking of a moving object. In the case of the GPS-guided flights, the helicopter flies via some waypoints to a destination using GPS information. The latter two flights will be discussed later.

Command interpretation module

Our system can accept 256 different linguistic commands. These commands can be given either by a microphone or by a computer keyboard as explained before in the section on remote control devices. For example we can speak to the helicopter such as 'Hover', 'Forward', 'Stop', and 'Land'. These commands are classified into seven levels.

The level 7 commands are concerned with GPS-guided flights, the level 6 commands concerned with image-guided flights and the level 5 commands concerned with programmed flights. Then the level 4 commands are fuzzy commands such *as "move left a bit", "make a big left-circling-turn"*. The level 3 commands are concerned with blended flight modes, the level 2 commands concerned with flight mode switching and the level 1 commands concerned with basic flight modes. This module has not been systematized yet. We need a further study to achieve an intelligent remote-flight-control based on linguistic commands.

5. Design of fuzzy controllers

The design of a fuzzy controller is classified into two different tasks: structure design and parameter design. The structure design means to determine the architecture of a controller, the input/output variables of a controller, the format of fuzzy control rules, and the number of rules. The parameter design literally means to determine the optimal parameters for a fuzzy controller. The kinds of parameters have been explained in the previous section.

In our study we take three steps for the design.

(i) We first refer to the knowledge of a pilot, the helicopter dynamics and control engineering. In particular, the structure of a fuzzy controller can be determined by using the above knowledge sources. For example, the longitudinal for $\dot{x}$-control module shown in Fig. 7 has a serial structure of the x-velocity block and the pitch block. This structure can be naturally derived from a physical relation between two state variables $\dot{x}$ and θ. That is, the change of a pitch angle θ of the main rotor path surface causes that of a longitudinal velocity $\dot{x}$ of the helicopter.

(ii) Next we use a software-based helicopter simulator to find the parameters. It models a real helicopter BK117 of Kawasaki Heavy Industries. A simulation software is installed in a super-graphic workstation CRIMSON of Silicon Graphics, which can simulate even wind effects. As an I/O device, the workstation is equipped with multi-channels A/D and D/A converters. So we can directly connect a hardware controller to the simulator and make real-time experiments.

Since the simulator uses a model of a real helicopter, we cannot directly apply the identified parameters to our unmanned helicopter. However, as far as the structure is concerned, we can very well verify its validity by simulation.

(iii) Finally we perform real experiments to adjust the parameters to YAMAHA R-50. As an example, let us consider designing a fuzzy controller of a coordinated turn. There are two kinds of turns: a rudder turn and a bank turn. A typical rudder turn is a hovering turn where a helicopter changes its nose direction in hovering by controlling just a lift of the tail rotor with a pedals-input. On the other hand, in the case of a coordinated turn, a pilot does not use a tail rotor (rudder) for nose control but set a bank angle (roll angle) of the helicopter toward the center of a turn with a lateral input.

We first consider the dynamics of a helicopter. Fig. 13 shows a dynamic balance during a bank turn. From this we find the following three relations

with respect to a dynamic balance:
$$L\cos\phi = wg,$$
$$L\sin\phi = wR\dot\varphi^2,$$
$$\dot{x} = R\dot\varphi,$$
where w is the mass, g is the gravity constant, ϕ the roll angle, $\dot\varphi$ is the yaw rate, R is the radius of a turn, and $\dot{x}$ is the forward velocity. From these it follows that
$$\dot\varphi = g\tan\phi/\dot{x}$$
which implies that a yaw rate $\dot\varphi$ must obey this functional relation with a roll (bank) angle ϕ and a forward velocity $\dot{x}$ in order to make a coordinated turn. Also we know that a side velocity $\dot{y}$ must be zero during an ideal circling turn.

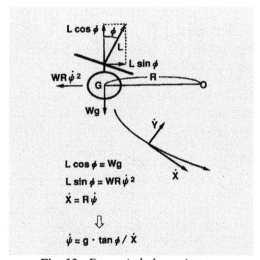

Fig. 13. Dynamic balance in turn.

(1) Now we have found from the above analysis that we should control a yaw rate for a coordinated turn by setting a reference input:
$$R_{\dot\varphi} = g\tan\phi/\dot{x}$$

(2) Next referring to a helicopter handbook, we know that "yaw rate is controlled by pedals".

(3) Also from a helicopter pilot, we can derive the following knowledge:
(3a) "Keep the bank angle constant during a turn."
(3b) "Use pedals to get rid of a slip in the radial direction."

From the above knowledge sources we can find the input-output structures of two control modules.

(i) $\phi \rightarrow$ Lateral from (3a).
(ii) $\dot\varphi, \dot{y} \rightarrow$ Pedals from (2) and (3b).

Fig. 14 shows the lateral control module and the pedals control module for a coordinated turn where a reference input $R_{\dot\phi}$ of the pedals control module is given from (1).

There is a noticeable difference between an ordinary pedals control module shown in Fig. 12 and this pedals module for turn. In the former, the pedals input is used to control a yaw angle and, in the latter, it is used to control a yaw rate and a side velocity. In the lateral control module for turn, the roll angle is controlled by a lateral input and so it gives a side drift $\dot y$ in the radial direction. However, the pedals control module cancels this drift.

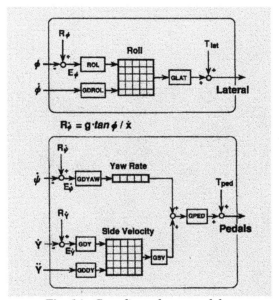

Fig. 14. Coordinated turn modules.

6. Performance of helicopter control system

In this section, we show some performances of the developed control system: basic issues and advanced issues. Concerning advanced issues, we shall discuss robust control, coupling compensation, GPS navigation, and image-guided flights.

6.1. Basic issues of fuzzy control

Fig. 15 shows some trajectories on the phase plane of the roll angular E_ϕ and the roll angular velocity $\Delta E_\phi(\dot\phi)$ taken from the longitudinal for $\dot y$-control module. As we see, the trajectories are drawn on the phase plane of the input-output relation of the controller. In the case of a fuzzy controller the I/O relation is apparently nonlinear compared to a linear case. By watching these trajectories, we can adjust the input/output gains of a controller.

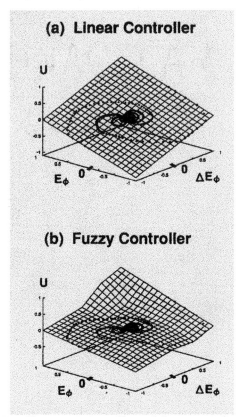

Fig. 15. Characteristics of lateral for y control module.

Taking advantage of nonlinearity of a fuzzy controller, we can easily compensate the non-symmetry of the helicopter dynamics. A single rotor helicopter shows non-symmetry with respect to a left/right movement because of a tail rotor. Fig. 16 shows the integrated drifts (with respect to time) in the y-direction caused by random winds. As is seen in a linear control case, a helicopter moves to the left more than to the right because of non-symmetry. As is seen in a fuzzy control case, adjusting fuzzy control rules and gains of a lateral control module easily compensates this non-symmetry.

Fig. 17 shows the result of a coordinated turn given by a simulation study. As we find, the forward velocity, the roll angle and the yaw rate are well regulated, and the flight trajectory is almost a circle. Fig. 18 shows the result of a 3D straight flight also by a simulation study. Given a 3D-reference velocity v (Fig. 18 (a)), we obtain three reference velocities corresponding to three axes and put these to three control modules: longitudinal, lateral and collective. As shown in Fig. 18 (b) and (c), the helicopter can well realize a 3D straight flight.

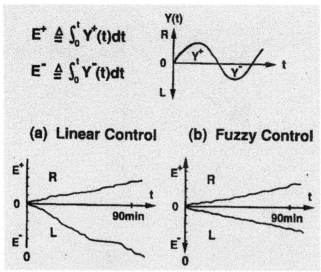

Fig. 16. Integrated drift in y-direction by random wind (around 15m/s).

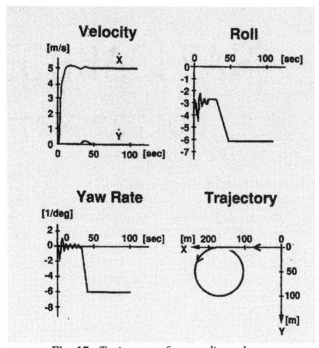

Fig. 17. Trajectory of a coordinated turn.

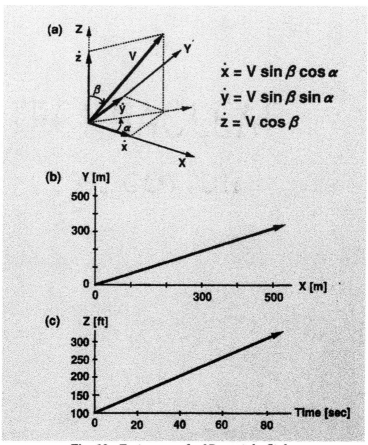

Fig. 18. Trajectory of a 3D straight flight.

6.2. Robust control under a windy environment

It is very important to design a robust controller against a wind disturbance since the helicopter is strongly affected by the wind as discussed in Section 2.

Fig. 19 shows the performance of the collective for z-control module under a windy environment, where a right crosswind of 5 m/sec is put to the helicopter. The real lines show the changes of $\dot{z}$ and z with fuzzy control while the dotted lines show these without fuzzy control.

Next Fig. 20 shows a comparison of linear control and fuzzy control under a windy environment. This figure plots the drifts on the x-y plane caused by random winds: a dot shows a drift in two seconds. Fuzzy control shows much more robustness against winds than linear control. The experiments for this have been performed by the simulator.

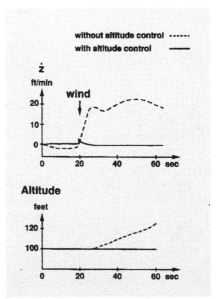

Fig. 19. Hovering with right cross-wind (5 m/sec).

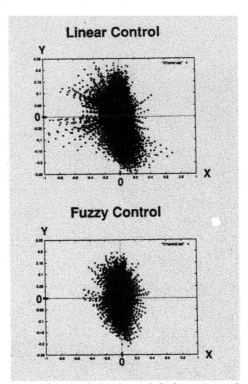

Fig. 20. Robustness of control: horizontal drifts in 2 sec. by random winds (around 15 m/s).

6.3. Coupling compensation

We consider two types of cross-couplings. The first one is concerned with the collective control, which is coupled to the lateral control for y-velocity and the pedal control for yaw angle. For example when a single-rotor helicopter flies up, it moves to the right and the nose turns also to the right. We apply fuzzy control for the modification of a control input against this coupling effect. The I/O structure of a compensator is of a form $(z, \dot{z} \rightarrow \Delta u)$ where Δu is the modification of a lateral or pedals control u.

Fig. 21 shows the lateral control module with a compensator. Fig. 22 shows the effect of fuzzy coupling compensation from the collective to the lateral when the helicopter flies up from a hovering position and stops after a while. The left figure in Fig. 22 shows y-velocity without compensation and the right figure shows it with compensation. We can see a fuzzy compensator works very well. In the figure, we see that starting and stopping the upward movement causes two impulses.

We omit to show the result of a coupling compensation to the pedals control where we use a compensator of the same structure. The effect is almost the same as the lateral control.

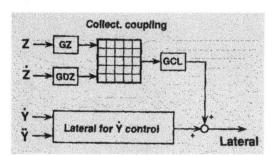

Fig. 21. Lateral control with collective coupling compensator.

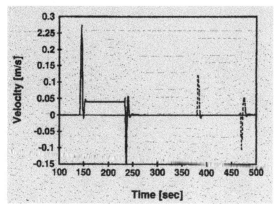

Fig. 22. Effect of coupling compensation from collective to lateral left - without compensation and right - with compensation.

The next coupling to be compensated is concerned with the longitudinal control, which is coupled to the collective, the lateral and the pedals controls as discussed in Section 2. As for compensation to the collective control, we use a structure $(\theta \to \Delta u)$ shown in Fig. 23.

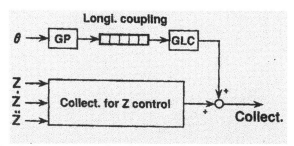

Fig. 23. Collective control with longitudinal coupling compensation.

The effect of longitudinal coupling compensation is shown in Fig. 24. In this figure, the solid line is for the case without compensation and the dashed line is for the case with compensation when the helicopter starts flying forward at 100 km/h. The compensation in this case is not satisfactory. The transient response is not well compensated and there is an offset of 50 cm.

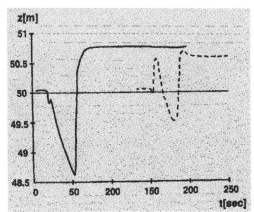

Fig. 24. Effect of coupling compensation from longitudinal to collective: left - without compensation and right - with compensation.

As for the lateral and the pedals, we use a structure of $(z, \dot{z}, \dot{x}^2 \to \Delta u)$ as shown in Fig. 25 where we have two fuzzy control blocks $(\dot{x}^2 \to \Delta u)$ and $(z, \dot{z}^2 \to \Delta u)$. The reason why we put a block of $(z, \dot{z} \to \Delta u)$ is that we cannot well compensate a coupling from the longitudinal control to the collective control as shown above and so we have to consider an indirect coupling such as longitudinal → (collective) → lateral/pedals.

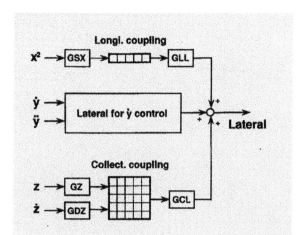

Fig. 25. Lateral control with longitudinal coupling compensation.

Fig. 26 shows the effect of a coupling compensation to the lateral control for y-velocity when the helicopter starts flying forward with 100 km/h from a hovering position: the left figure is without compensation and the right with compensation. We can see that the coupling is very well compensated and the drift of the y-velocity is finally suppressed to zero.

We omit to show the result for compensation to the pedals.

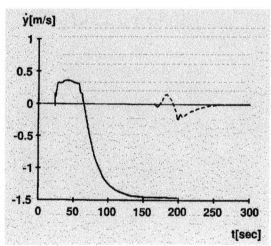

Fig. 26. Effect of coupling compensation from longitudinal to lateral: left - without compensation and right - with compensation.

6.4. GPS-guided navigation

Our unmanned helicopter can autonomously fly from a starting point to a destination via some waypoints using GPS signals. The system has at present the following functions.

Autonomous flight with switching flight modes

Given a waypoint and a destination, the helicopter takes some flight modes from one to another: for example, a forward flight toward a waypoint, a circling turn at an appropriate point so that it can fly over a waypoint, then a forward flight in a different direction toward a destination, a slowdown flight to be able to smoothly stop at a destination and finally a stop mode for hovering at a destination. According to a speed, we can find an optimal point for switching a flight mode. Fig. 27 shows a flight trajectory by a simulation study. Fig. 28 shows an experimental study where the x-y position data with measurement errors given by GPS are plotted.

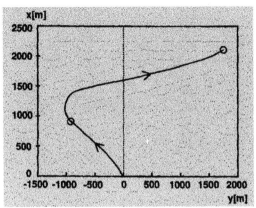

Fig. 27. GPS-guided flight simulation.

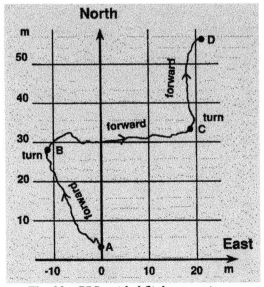

Fig. 28. GPS-guided flight: experiment.

Collision avoidance capability

When the position of an obstacle is given, the helicopter can automatically avoid a collision by changing its flight path. Fig. 29 shows two trajectories with and without an obstacle. Further, it is not difficult to avoid a collision against a mountain or a tall building in three-dimensional space using GPS signals.

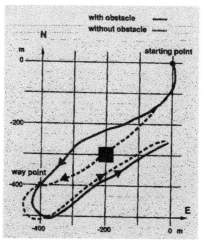

Fig. 29. Collision avoidance by GPS signals.

Automatic heading control

To take an appropriate heading direction is not a trivial task in the case of the helicopter since it can fly to any direction without changing its heading direction. It is better to turn the nose for a destination. There are three types of turns: hovering turn, rudder turn and coordinated turn. According to a present speed, our helicopter automatically selects its turn mode. Among three turns, it is difficult to make a coordinated turn for heading control. We have to set an appropriate bank angle for a coordinated turn according to a speed and then determine a point to stop the turn so that the helicopter finally faces a destination.

Command-based interference during navigation

The helicopter can accept commands during navigation. The commands are those concerning its flight conditions and those concerning its mission. For example we can send commands such as *"fly faster"*, *"fly higher"* and also *"stop the mission and return"*, *"fly to a new destination"* to change its mission.

Adjustment of hovering position after arrival

It is rather difficult to precisely stop at a destination after a high-speed flight. So based on GPS signals, the helicopter automatically adjusts its hovering position and takes a specified heading direction after hovering.

6.5. Image-guided flights

We have so far developed two kinds of guided flights with images by a CCD camera.

Automatic landing

Using image information, the helicopter automatically searches a landmark, approaches it and can land on a landing spot. In the experimental study, a landmark is a white rectangle of dimensions 5 m x 10 m and the landing spot is a red circle with a radius of 1 m. We have obtained satisfactory results by experiments.

Tracking a moving object

The helicopter can track a moving object with a low speed, about 1 m/sec, on the ground. We use an active image sensor to detect a distance between an object and the helicopter, and also a relative velocity of an object. The most important point here is to search an object when it is missed. Fig. 30 shows a diagram of an image-based tracking system. In the experiments, the helicopter could track successfully a remotely controlled vehicle.

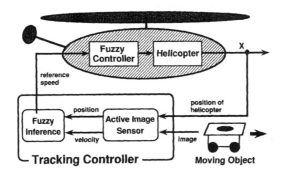

Fig. 30. Image-based tracking system.

Active image sensor

A CCD camera is used for an active image sensor to measure both distance and velocity. Two servo controllers can control the tilt and pan angles of the camera. The direction of the camera is controlled so that it always catches an object at the center of the image. So by measuring a displacement of a detected object from the center of an image and a height of the camera from the ground, we calculate the distance between an object and the helicopter. Also by measuring the pitch angular velocity of the camera when it is tracking an object, we can calculate a relative velocity of an object. Though an image processor can deal with 30 frames per second, this active image sensor needs 0.5 sec to measure distance and velocity because of noise. We are going to improve the performance of measurements from 0.5 sec to 0.1 sec.

7. Conclusions

We have successfully developed an intelligent unmanned helicopter by applying a fuzzy control technique. The helicopter can be remotely controlled by giving linguistic commands from a ground station.

To fly the helicopter in the invisible area, GPS-guided flights and image-guided flights have been studied with a flight monitoring system. So far command-based flights have been fully achieved and important control problems such as coupling compensation and robustness under a windy environment have been solved. As for GPS-guided flights, we need more experimental studies in the invisible area. Also we have to improve the performance of image-guided flights to increase the autonomy of the helicopter. We need a further study concerned with the reliability of the total system to apply this sort of an unmanned helicopter to a real use.

Acknowledgments

The author thanks his students J. Nishino, T. Hyakutake, Y. Saitoh, H. Miwa, H. Akagi, I. Hirano, S. Tani, S. Nakamura, S. Kotsu, M. Uechi, M. Nakamura and A. Kageyama who worked very hard for this project. This project was partly supported by STA and MITI, Japan.

References

[1] J. Nishino, T. Murofushi and M. Sugeno, "A study on hovering of a model helicopter using fuzzy control," *Proc. of 6th Fuzzy System Symposium*, 559/560, 1990 (in Japanese).
[2] T. Hyakutake, T. Murofushi and M. Sugeno, "Fuzzy control of a simulated helicopter," *Proc. of 7th Fuzzy System Symposium*, 35/38, 1991 (in Japanese).
[3] M. Sugeno et al., "Helicopter flight control based on fuzzy logic," *Proc. of First International Fuzzy Engineering Symposium*, 1120/1121, Yokohama, 1991.
[4] M. Sugeno, J. Nishino and H. Miwa, "A hierarchical structured fuzzy controller for an unmanned helicopter," *Proc. of 8th Fuzzy System Symposium*, 53/56, 1992 (in Japanese).
[5] Y. Saitoh, I. Hirano and M. Sugeno, "Flight control of an unmanned helicopter based on fuzzy control," *Proc. of 9th Fuzzy System Symposium*, 37/40,1993 (in Japanese).
[6] M. Sugeno, M. F. Griffin, A. Bastian, "Fuzzy hierarchical control of an unmanned helicopter," *Proc. of 5th IFSA World Congress*, 179/182, Seoul, 1993.
[7] M. Sugeno et al., "Issues for blending fuzzy controllers in hierarchical systems," *Proc. of First Asian Symposium*, Singapore, 1993.
[8] H. Akagi, S. Tani and M. Sugeno, "Intelligent flight control of an unmanned helicopter," *Proc. of 10th Fuzzy System Symposium*, 745/748,1994 (in Japanese).
[9] M. Sugeno et al., "Development of an intelligent unmanned helicopter," *Proc. of 4th FUZZ-IEEE and 2nd IFES*, Vol.5, 33/34, Yokohama, 1995.
[10] M. Sugeno et al., "Intelligent control of an unmanned helicopter based on fuzzy logic," *Proc. of American Helicopter Society 51st Annual Forum*, Texas, 1995.
[11] I. Hirano, S. Kotsu and M. Sugeno, "GPS and image guided landing of an unmanned helicopter," *Proc. of 11th Fuzzy System Symposium*, 389/392, 1995 (in Japanese).
[12] M. Sugeno, "Navigation of an unmanned helicopter," *Proc. of Canada-Japan Bilateral Workshop*, 8/11, Toronto, 1996.
[13] S. Nakamura and M. Sugeno, "Tracking control of an unmanned helicopter using visual image information," *Proc. of 12th Fuzzy System Symposium*, 585/588, 1996 (in Japanese).
[14] S. Tani, M. Nakamura and M. Sugeno, "Intelligent navigation of an unmanned helicopter," *ibid.*,

589/592, 1996 (in Japanese).
[15] M. Sugeno, "Fuzzy control of an unmanned helicopter toward real applications," *Proc. of ISAI/IFIS*, 22/23, Cancun, 1996.
[16] M. Sugeno, "Recent advances in fuzzy control of an unmanned helicopter," *Proc. of 2nd Asian Fuzzy Systems Symposium*, Taiwan, 1996.
[17] M. Sugeno, "Development of an unmanned helicopter: the present status and some research issues," *Proc. of International Workshop on Breakthrough Opportunities for Fuzzy Logic*, 30/33, Yokohama, 1996.
[18] A. Kageyama and M. Sugeno, "FBW control of an unmanned helicopter using a flight monitor," *Proc. of 13th Fuzzy System Symposium*, 271/274, 1997 (in Japanese)
[19] S. Nakamura, M. Uechi and M. Sugeno, "Intelligent navigation of an unmanned helicopter using GPS and image information," *ibid.*, 275/278, 1997 (in Japanese).

Multi-Dimensional Fuzzy Reasoning[*]

Michio Sugeno and Tomohiro Takagi
Department of Systems Science,
Graduate School of Science and Engineering,
Tokyo Institute of Technology,
4259 Nagatsuta, Midori-ku, Yokohama 227, Japan

Received May 1981
Revised December 1981

Abstract

This paper suggests a method of multi-dimensional fuzzy reasoning concerned with both modus ponens and modus tollens. It also discusses an example to show how the method works.

Keywords: Lukasiewicz's infinite valued logic, fuzzy logic, fuzzy implication, multi-dimensional fuzzy reasoning.

1. Introduction

A considerable number of studies on fuzzy reasoning, have been reported, e.g., [1, 2, 3, 5], since Zadeh presented the compositional rule of inference. Fuzzy reasoning has been especially applied for fuzzy control and diagnosis, and lately for fuzzy modeling. However, there have been few discussions on it in the multi-dimensional case. This paper suggests a method of multi-dimensional fuzzy reasoning which is of great importance particularly when we deal with fuzzy modeling of systems. It also shows numerical examples which claim the validity of the proposed method.

In general there are two methods, as is well known, on fuzzy reasoning: one is based on the compositional rule of inference and the other on fuzzy logic with such

---

[*]©1983 *North-Holland. Retyped with written permission from Fuzzy Sets and Systems,* **9** *(1983) 313-325.*

and such a base logic, e.g., Tsukamoto's method [4].

Since a multi-dimensional implication such as "if (x is A, y is B) then z is C" where A, B and C are fuzzy sets is not merely a collection of one-dimensional implications, a conventional interpretation is usually taken in the multidimensional case. For example in the compositional rule of inference, the above two-dimensional implication is translated into:

(1) if x is A and y is B then z is C,

or (2) if x is A then if y is B then z is C.

In Tsukamoto's method, it is decomposed into:

(1) if x is A then z is C,

and (2) if y is B then z is C,

and the intersection $C' \cap C''$, where C' is the inferred value from the first implication and C'' from the second implication, is taken for the consequence of reasoning. It is clear that those methods do not always express real situations, since a pair of propositions (x is A, y is B) could be any fuzzy relation "(x, y) is R" with respect to x and y. For example there might be a functional relation such that $z = x^2 y$ behind a fuzzy implication.

Further, in the multi-dimensional case we have some difficulties in applying the compositional rule of inference, because we have to deal with multi-dimensional fuzzy relations.

So we use in this paper Tsukamoto's method by which we can do without the fuzzy relational matrix and present an algorithm of fuzzy reasoning in the general case. The algorithm is very similar to that we use to approximate a function f(x, y) by a linear function.

2. Fuzzified Lukasiewicz's logic and its reasoning

In this section we shall briefly introduce Tsukamoto's method of fuzzy reasoning.

The truth value of implication in Lukasiewicz's infinite-valued logic is expressed as

$$/A \to B/ = (1 - /A/ + /B/) \wedge 1, \qquad (1)$$

where $/A/ \in [0, 1]$ is the truth value of a proposition A. Let P, Q be fuzzy propositions and $\underline{P}$, $\underline{Q}$ be linguistic truth values that are fuzzy subsets of $[0, 1]$ as usual. Then $\overline{P \to Q}$ is derived by the extension principle from (1),

$$\overline{P \to Q} = (1 - \underline{P} + \underline{Q}) \wedge 1. \qquad (2)$$

In Eq. (2) the symbols -, + and $\wedge$ are the extended ones. In the sequel we shall use the extended operations for fuzzy sets without a comment. For calculation it is better to take a strong α- cut of Eq. (2). We have

$$\overline{P \to Q}_\alpha = (1 - \underline{P}_\alpha + \underline{Q}_\alpha) \wedge 1. \qquad (3)$$

where $P_\alpha = \{u \mid h_P(u) > \alpha\}$ and $h_P(u)$, $u \in [0, 1]$, is the membership function of $\underline{P}$. When $\overline{P \to Q}$ and $\underline{P}$ are given, $\underline{Q}$ is easily obtained by solving Eq. (3).

In many cases we can assume that $\overline{P \to Q}$ is normal, its membership function is non-decreasing in its domain [0, 1] and also $\underline{P}$ is normal and convex. Denote $R = P \rightarrow Q$. Then it follows from the above assumptions that

$$R_\alpha = (r(\alpha), 1], \tag{4}$$
$$P_\alpha = (p_1(\alpha), p_2(\alpha)) \tag{5}$$

where $r(\alpha) \in [0,1]$ is determined from h_R and $p_1(\alpha), p_2(\alpha)$ from h_P. Solving Eq. (3), $\underline{Q}_\alpha$ is obtained as:

$$\underline{Q}_\alpha = ((p_1(\alpha) + r(\alpha) - 1) \vee 0, 1]. \tag{6}$$

Here we should take a caution of the term $-P_\alpha$ in Eq. (3). For simplicity suppose an equation $\underline{R}_\alpha = -\underline{P}_\alpha + \underline{Q}_\alpha$. If we directly solve $\underline{Q}_\alpha$ according to extended operations, we do not have $\underline{Q}_\alpha = \underline{R}_\alpha + \underline{P}_\alpha$ More simply it is not even true that $(\underline{Q}_\alpha - \underline{P}_\alpha) + \underline{P}_\alpha = \underline{Q}_\alpha$. So we should transfer P_α from the right-hand side to the left in Eq. (3). This sort of problem will be also discussed in the Appendix. The solution in Eq. (6) is the same as that obtained by solving Eq. (1) first and then applying the extension principle. Now h_Q is drawn since Eq. (6) implies that h_Q is a non-decreasing function.

Fuzzy reasoning based on modus ponens is carried out as follows, where modus ponens is written as: $\dfrac{P', P \rightarrow Q}{Q'}$

(1) Given a premise P', set $P' \cong P$ is $\underline{P}$. By converse of truth qualification, $\underline{P}$ is found as

$$\underline{P} = h_P(P') \tag{7}$$

where h_P is the extension of h_P for a fuzzy set P'.

(2) Calculate $\underline{Q}$ from $\underline{P}$ and $P \rightarrow Q$ which is assumed to be given.

(3) Set $Q' = \underline{Q}$ is Q. Then Q' is the consequence of fuzzy reasoning. By truth qualification, Q' is obtained as

$$Q' = h_Q^{-1}(\underline{Q}), \tag{8}$$

where h_Q^{-1} is also the extended one.

Let us consider a special case where $\underline{R}$ is u-true and P' is a singleton x_0. Then what we have in Eq. (4) is:

$$r(\alpha) = \alpha \tag{9}$$

and from Eqs.(5) and (7)

$$p_1(\alpha) = h_P(x_0) \text{ for all } \alpha \in [0, 1). \tag{10}$$

Substituting these into Eq. (6), it follows from Eq. (8) that

$$h_{Q'}(y) = (h_Q(y) - h_P(x_0) + 1) \wedge 1 \tag{11}$$

Next let us consider the inverse problem, i.e., fuzzy reasoning based on modus tollens which is written as: $\dfrac{Q', P \rightarrow Q}{P'}$

when $P \to Q$ and Q are given, P is bound under similar assumptions.
$$\underline{P_\alpha} = (0, (q_2(\alpha) - r(\alpha) + 1) \wedge 1] \tag{12}$$
where
$$\underline{Q_\alpha} = (q_1(\alpha), q_2(\alpha)). \tag{13}$$
The algorithm to find P' when Q' is given, is the same as in modus ponens. In particular when $P \to Q$ is u-true and Q' is a singleton y_0, we have,
$$h_{P'}(x) = -h_P(x) + h_Q(y_0) + 1) \wedge 1 \tag{14}$$

3. Multi-dimensional reasoning

As stated in the Introduction, we suggest in this section a method of multi-dimensional fuzzy reasoning. Unfortunately we have not had a general method in the multi-dimensional case so far. Of course we can use the compositional rule of inference. However there are two difficulties: one is to identify or construct a fuzzy relation not merely as, for example, a Cartesian product of some number of fuzzy sets, and the other is to calculate a multi-dimensional fuzzy relational composition, for example, in order to compute a fuzzy control algorithm in real time.

The idea suggested here is quite simple and reasonable, which is very similar to linear interpolation.

We should first consider that we cannot logically infer anything from a single implication, say, $A \to B$ unless the same premise to A is given. This is also somewhat true in fuzzy reasoning. This is the reason why there are some persons who translate $A \to B$ into a fuzzy relation $R = "A \to B$ and not $A \to V"$ where V is the universe of discourse of B. Suppose "x is big y is small". Then no one can infer the value of y from this single implication when, for example, x is medium. We should have at least two different implications, $A_1 \to B_1$, and $A_2 \to B_2$ for one-dimensional reasoning. What we can do then is to infer something when we are given a premise lying between A_1 and A_2. This reasoning must be carried out using some method like interpolation.

Apart from computational technique, it is sufficient for the purpose to discuss a two-dimensional case. Thus we can start with 4 implications such that

$$(x \text{ is } A_1, y \text{ is } B_1) \to z \text{ is } C_{11,}$$
$$(x \text{ is } A_1, y \text{ is } B_2) \to z \text{ is } C_{12,}$$
$$(x \text{ is } A_2, y \text{ is } B_1) \to z \text{ is } C_{21,}$$
$$(x \text{ is } A_2, y \text{ is } B_2) \to z \text{ is } C_{22,}$$

The important point we want to stress is that we cannot translate $(x$ is A, y is $B)$ into, for example, "(x, y) is $A \times B$" where $A \times B$ is Cartesian product, since generally a pair of two propositions $(x$ is A, y is $B)$ could be any relation as mentioned in the Introduction. This means that we should deal with $(x$ is A, y is $B)$ directly without modification.

The situation of four implications is shown in Fig. 1.

Multi-Dimensional Fuzzy Reasoning 49

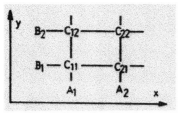

Fig. 1. Four implications.

Our problem is to infer "z is C" from a given premise (x is A, y is B) where A is assumed to be between A_1 and A_2 and also B between B_1 and B_2. When we apply fuzzy reasoning introduced in the previous section to this problem, we face a difficulty to find τ such that (x is A, y is B) ≅ (x is A_1, y is B_1) is τ, where τ is a linguistic truth value. It is impossible to solve the above semantical equation unless the relation between x and y is given. That is, generally the truth value of a pair of propositions is expressed using those of individual propositions as

$$(A_1, B_1) = f(\underline{A_1}, \underline{B_1})$$

where a function f of arbitrary and $\underline{A_1}$, $\underline{B_1}$ is unknown. Our method can do without f.

The outline of the algorithm is shown in Fig. 2. First let us infer the value of z at the point (A_1, B) from (A_1, B_1) as indicated by an arrow in Fig. 2. Modus ponens is written

$$\frac{(x \text{ is } A_1, y \text{ is } B), \ (x \text{ is } A_1, y \text{ is } B_1) \to z \text{ is } C_{11}}{z \text{ is } C'_{11}}$$

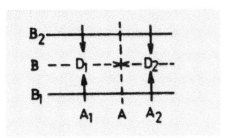

Fig 2. Outline of algorithm.

Here C'_{11} is easily obtained by one-dimensional reasoning shown in the previous section since A_1 is fixed along the arrow. That is, set

$$(x \text{ is } A_1, y \text{ is } B) \cong (x \text{ is } A_1, y \text{ is } B_1) \text{ is } \tau \qquad (15)$$

then clearly

$$y \text{ is } B \cong y \text{ is } B_1 \text{ is } \tau \qquad (16)$$

for the same τ.

Let us next infer z at the same point (A_1, B) this time from (A_1, B_2). That is

$$\frac{(x \text{ is } A_1, B), \ (x \text{ is } A_1, B_2) \to z \text{ is } C_{12}}{z \text{ is } C'_{12}}$$

Now let

$$D_1 = \frac{1}{2}(C'_{11} + C'_{12}). \tag{17}$$

Then we obtain a new implication with respect to the point (A_1, B) such that
$$(x \text{ is } A_1, y \text{ is } B) \to z \text{ is } D_1.$$
Here D_1 may be obtained in another way: $D_1 = .C'_{11} \cap C'_{12}$. For the point (A_2, B) we can also obtain an implication such that
$$(x \text{ is } A_2, y \text{ is } B) \to z \text{ is } D_2.$$

Finally if we follow the same procedure along the dotted arrows in Fig. 2, we can infer the value of z at (A, B). So the method is just like linear interpolation. In the following section we shall show how well the method works.

Let us consider the inverse problem such as finding x when "y is B" and "z is C" are given. The first part of the algorithm is the same as in the previous case. We start from the part where D_1 and D_2 in Fig. 2 have been obtained. That is, we have two implications
$$(x \text{ is } A_1, y \text{ is } B) \to z \text{ is } D_1,$$
$$(x \text{ is } A_2, y \text{ is } B) \to z \text{ is } D_2,$$
The second half of the algorithm is illustrated this time in the (y, z) plane as in Fig. 3. Now consider the following modus tollens:

$$\frac{z \text{ is } C, \ (x \text{ is } A_1, y \text{ is } B) \to z \text{ is } D_1}{(x \text{ is } A'_1, y \text{ is } B)}$$

Since "y is B" is fixed, the problem is also reduced to one-dimensional modus tollens. So "x is A_1" can be inferred, which is shown by the lower arrow in Fig. 3. Along the upper arrow, "x is A_2" is also obtained. By setting $A = (A'_1 + A'_2)/2$, "x is A" is finally inferred from $(y \text{ is } B, z \text{ is } C)$.

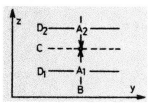

Fig. 3. *Outline of algorithm.*

4. Illustrative example

4.1. *Reasoning based on modus ponens*

Let us consider the following linear functional relation
$$z = x + ay \tag{18}$$
to examine the validity of the method. This relation is understood as an underlying relation which we want to express by a set of fuzzy implications. For simplicity fuzzy

variables used in implications are chosen as

$$P_1(x) = \frac{1}{2\bar{x}} x + \frac{1}{2}, \quad x \in [-\bar{x}, \bar{x}], \tag{19}$$

$$N_1(x) = P_1(-x), \tag{20}$$

$$P_2(y) = \frac{1}{2\bar{y}} y + \frac{1}{2}, \quad y \in [-\bar{y}, \bar{y}], \tag{21}$$

$$N_2(y) = P_2(-y), \tag{22}$$

where P and N imply 'positive' and 'negative', respectively, and $P_i(x)$ is the membership function of P_i. Those are shown in Figs. 4 and 5. In the sequel the membership function of a fuzzy set A will be written $A(x)$.

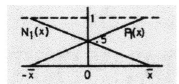 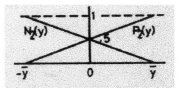

Fig. 4. *Membership functions of P_1, N_1.* **Fig.5.** *Membership functions of P_2, N_2.*

Let us further assume that the value of z, for example, at the point (x is P_1, y is P_2) is precisely given according to the underlying relation shown in Eq. (18). Then we have an implication

$$(x \text{ is } P_1, y \text{ is } P_2) \rightarrow z \text{ is } P_1 + aP_2.$$

If we denote $P_3 = P_1 + aP_2$, then by a simple calculation

$$P_3(z) = (P_1 + aP_2)(z) = \frac{1}{2(\bar{x} + a\bar{y})} z + \frac{1}{2}. \tag{23}$$

The method of calculation such as $P_1 + aP_2$ is discussed in the Appendix.

In a similar manner, we have implications for the other three points as is shown in Fig. 6. They are

$(x \text{ is } N_1, y \text{ is } N_2) \rightarrow z \text{ is } N_3,$
$(x \text{ is } N_1, y \text{ is } P_2) \rightarrow z \text{ is } P_4,$
$(x \text{ is } P_1, y \text{ is } N_2) \rightarrow z \text{ is } N_4,$
$(x \text{ is } P_1, y \text{ is } P_2) \rightarrow z \text{ is } P_3,$

where

$$P_3(z) = \frac{1}{2(\bar{x} + a\bar{y})} z + \frac{1}{2}, \tag{24}$$

$$N_3(z) = P_3(-z), \tag{25}$$

$$P_4(z) = \frac{1}{2(a\bar{y} - \bar{x})} z + \frac{1}{2}, \tag{26}$$

$$N_4(z) = P_4(-z), \tag{27}$$

and $a\bar{y} + \bar{x} \neq 0$, $a\bar{y} - \bar{x} \neq 0$ are assumed of course.

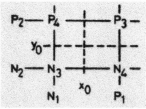

Fig. 6. Situation of implications.

Now let us consider a special premise such as (x is x_0, y is y_0) where x_0 and y_0 are singletons. What we expect by this is to obtain $z_0 = x_0 + ay_0$ from our fuzzy reasoning. According to the algorithm described in the previous section, we have two modus ponens along the lines reaching (x is P_1, y is y_0) in Fig. 6 such that

$$\frac{(P_1, y_0), (P_1, P_2) \to P_3}{P_3'}, \quad \frac{(P_1, y_0), (P_1, N_2) \to N_4}{N_4'}.$$

Assuming that the truth values of all the implications are u - true, that is, $r(\alpha) = \alpha$ in (4), we obtain from the procedure of fuzzy reasoning shown in Section 2

$$P_3'(z) = P_3(z) + 1 - P_2(y_0)$$
$$= \frac{1}{2(a\overline{y} + \overline{x})} z - \frac{1}{2y} y_0 + 1, \quad (28)$$

$$N_4'(z) = N_4(z) + 1 - N_2(y_0)$$
$$= \frac{1}{2(a\overline{y} - \overline{x})} z + \frac{1}{2y} y_0 + 1. \quad (29)$$

They are shown in Fig. 7.

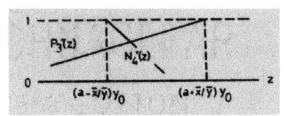

Fig. 7. Membership functions of P'_3 and N'_4.

Let $K = \frac{1}{2}(P_3' + N_4')$, (30)

then, $K(z) = \frac{1}{2\overline{x}}(z + 2\overline{x} - ay_0)$. (31)

The result is shown in Fig. 8.
We also have two modus ponens for the point (N_1, y_0) such that.

$$\frac{(N_1, y_0), (N_1, P_2) \to P_4}{P_4'}, \quad \frac{(N_1, y_0), (N_1, N_2) \to N_3}{N_3'}.$$

Let,

$$H = \tfrac{1}{2}(P'_4 + N'_3), \tag{32}$$

then we also obtain

$$H(z) = -\frac{1}{2\bar{x}}(z - 2\bar{x} - ay_0) \tag{33}$$

which is shown in Fig. 9.

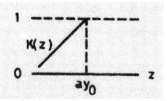

Fig. 8. Membership function of K.   **Fig. 9.** Membership function of H.

Now the new implications with respect to (P_1, y_0) and (N_1, y_0) are
$(x$ is P_1, y is $y_0) \to z$ is K
$(x$ is N_1, y is $y_0) \to z$ is H
Finally we have the following modus ponens along the dotted lines in Fig. 6.

$$\frac{(x_0, y_0), (P_1, y_0) \to K}{K'}, \quad \frac{(x_0, y_0), (N_1, y_0) \to H}{H'}.$$

where H' and K' are calculated as

$$K'(z) = K(z) - P_1(x_0) + 1, \tag{34}$$
$$H'(z) = H(z) - N_1(x_0) + 1, \tag{35}$$

They are shown in Fig. 10.

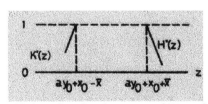

Fig. 10. Membership functions of K' and H'.

Let

$$L = \tfrac{1}{2}(K' + H'), \tag{36}$$

then L is the final consequence of fuzzy reasoning, which is just found to be a singleton as is shown in Fig. 11. That is

$$L = x_0 + ay_0, \tag{37}$$

which gives the precise value of z at (x_0, y_0).

Thus we can state that our method of multi-dimensional fuzzy reasoning represents precisely a linear functional relation. We should recall that we have used only four implications at all.

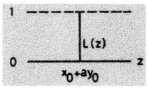

Fig. 11. *Membership functions of L.*

4.2. Reasoning based on modus tollens

Now let us consider the inverse problem of that discussed in the previous section. Our problem is to find x when (y is y_0, z is z_0) is given. According to the algorithm, we can start from Fig. 12.

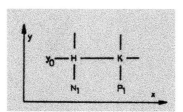

Fig. 12. *Situation of two implications.*

We have already obtained two implications on the line $y = y_0$ in Fig. 12 such as

(x is N_1, y is y_0) $\to z$ is H,
(x is P_1, y is y_0) $\to z$ is K

where H and K are given in Eqs. (32) and (30).

When (y_0, z_0) is given, we can consider the following two modus tollens.

$$\frac{z_0,\ (P_1, y_0) \to H}{(P_1'', y_0)}, \quad \frac{z_0,\ (N_1, y_0) \to K}{(N_1'', y_0)}$$

They are illustrated on the (y, z) plane in Fig. 13.

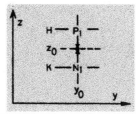

Fig. 13. *Two modus tollens.*

We have from (14)

$$P_1''(x) = -P_1(x) + K(z_0) + 1, \tag{38}$$

$$N_1''(x) = -N_1(x) + H(z_0) + 1. \tag{39}$$

Let

$$S = \tfrac{1}{2}(P_1'' + N_1''), \tag{40}$$

then it follows that

$$S = z_0 - ay_0 \tag{41}$$

which also gives the precise value of x, i.e., $x_0 = z_0 - a\,y_0$.

Conclusions

We have discussed some problems concerned with multi-dimensional fuzzy reasoning and proposed a method. The method enables us to precisely represent any linear functional relation. On the other hand it is very difficult to find a fuzzy relation in a compositional rule of inference which can represent a linear functional relation under the condition that four implications, for example, are given as discussed in the paper.

The purpose of our research is to give a new mathematical tool to describe systems, or in other words to give a way to fuzzy modeling of systems. It is of crucial importance for general purpose to examine the ability of the description of such and such a mathematical tool. So we have discussed our method in detail.

References

[1]. J.F. Baldwin, A model of fuzzy reasoning through multi-valued logic and set theory, *Int. J. Man-Machine Studies* **11** (1979) 351-380.
[2]. J.F. Baldwin, Fuzzy logic and approximate reasoning for mixed input arguments, *Int. J. Man-Machine Studies* **11** (1979) 381-396.
[3]. S. Fukami, M. Mizumoto and K. Tanaka, Some considerations on fuzzy conditional inference, *Fuzzy Sets and Systems* **4** (1980) 243-273.
[4]. Y. Tsukamoto, An approach to fuzzy reasoning method, in: M.M. Gupta, Ed. *Advances in Fuzzy Set Theory and Applications* (North-Holland), Amsterdam, 1979) 137-149.
[5]. L.A. Zadeh, Fuzzy logic and approximate reasoning, Synthese **30** (1975) 407-428.

Appendix

Addition and subtraction in a class of fuzzy numbers with linear monotone membership functions

Here we deal with extended operations in a special class of fuzzy numbers. Suppose a class of fuzzy sets as are shown in Fig. 14 where fuzzy sets P and N have linear monotone membership functions $P(x)$ and $N(x)$, respectively.

$$P(x) = \frac{1}{a_2 - a_1}(x - a_1), \quad x \in [a_1, a_2],$$

$$N(y) = -\frac{1}{b_2 - b_1}(y - b_2), \quad y \in [b_1, b_2].$$

Their inverse functions are denoted by P^{-1} and N^{-1}.

$$\begin{aligned}P^{-1}(\alpha) &= (a_2 - a_1)\alpha + a_1,\\ N^{-1}(\alpha) &= -(b_2 - b_1)\alpha + b_2\end{aligned} \quad \alpha \in [0,1).$$

There are two types of fuzzy numbers, P and N, in the class. A singleton is considered as a special case of P or N.

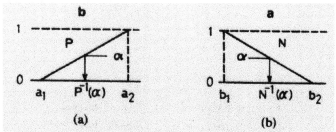

Fig 14. $P(x)$ and $N(x)$, and their inverse functions P^{-1} and N^{-1}.

It is well known that a set of fuzzy numbers with an operation 'addition' does not form a group. For example, let
$$X = P + N.$$
According to the extension principle, taking α-cut, we have
$$\begin{aligned} X_\alpha &= P_\alpha + N_\alpha \\ &= (P^{-1}(\alpha), a_2] + [b_1, N^{-1}(\alpha)) \\ &= (P^{-1}(\alpha) + b_1, N^{-1}(\alpha) + a_2). \end{aligned}$$
If we subtract P from X, then
$$\begin{aligned} X_\alpha - P_\alpha &= (P^{-1}(\alpha) + b_1, N^{-1}(\alpha) + a_2) - (P^{-1}(\alpha), a_2] \\ &= (P^{-1}(\alpha) + b_1 - a_2, N^{-1}(\alpha) - P^{-1}(\alpha) + a_2) \\ &\neq N_\alpha. \end{aligned}$$
That is, $(P + N) - P \neq N$.

Next let us take another more simple example. Let $Y = 2P - P$, then it is natural to expect $Y = P$. However, we have
$$\begin{aligned} Y_\alpha &= 2P_\alpha - P_\alpha \\ &= (2P^{-1}(\alpha), 2a_2] - (P^{-1}(\alpha), a_2] \\ &= (2P^{-1}(\alpha) - a_2, 2a_2 - P^{-1}(\alpha)] \\ &\neq P_\alpha \end{aligned}$$
That is, $Y \neq P$.

As for addition, it is true that $3P = 2P + P$, for example. Notice that $P + N$ is an operation of the kind $2P-P$ since N is a fuzzy number of the type '$-P$'. In general we have to take caution of subtraction defined by the extension principle. We should indirectly calculate $A - B$ by finding X such that $A = X + B$.

So let us try to find Y such that
$$2P = Y + P$$
Setting, $Y_\alpha = (y_1(\alpha), y_2(\alpha)]$ we have
$$\begin{aligned} (2P^{-1}(\alpha), 2a_2] &= (y_1(\alpha), y_2(\alpha)] + (P^{-1}(\alpha), a_2] \\ &= (y_1(\alpha) + P^{-1}(\alpha), y_2(\alpha) + a_2]. \end{aligned}$$

This implies that $Y = P$ since $y_1(\alpha) = P^{-1}(\alpha)$ and $y_2(\alpha) = a_2$ are obtained.

Now we present new extended operations in a class of fuzzy numbers with linear monotone membership functions which give a group structure.

Addition:
$$X = P + N \leftrightarrow X^{-1}(\alpha) = P^{-1}(\alpha) + N^{-1}(\alpha)$$

Inverse:
$$X = -P \leftrightarrow X^{-1}(\alpha) = -P^{-1}(\alpha).$$

Scalar multiplication:
$$X = aP \leftrightarrow X^{-1}(\alpha) = aP^{-1}(\alpha)$$

The results of these operations belong to the same class; the operations are closed. It is valid that $(P + N) - P = N$, $2P - P = P$ and $3P = 2P + P$, etc. It is easily seen that this class of fuzzy numbers with addition yields a commutative group.

A New Approach to Design of Fuzzy Controller[*]

M. Sugeno and T. Takagi
*Department of Systems Science
Tokyo Institute of Technology
4259 Nagatsuta, Midori-ku
Yokohama 227, Japan*

1. Introduction

We have heuristically designed fuzzy controllers thus far since we have lacked a fuzzy model of a system. The authors have recently developed a method of multi-dimensional fuzzy reasoning that enables one to build a dynamic model of a system just as we do in terms of differential equations. This paper presents a new idea of designing a fuzzy controller based on a fuzzy model of a system.

It is necessary for understanding the idea to have such concepts as fuzzy reasoning based on Lukasiewicz's infinite valued logic, truth qualification in fuzzy logic and multi-dimensional fuzzy reasoning, etc. These are briefly described in the Appendix.

2. Derivation of Fuzzy Control Rules

Though it is most important to identify a system structure and build its model, let us begin with a given fuzzy model of a system, i.e., a set of fuzzy implications such that

$$(u_n \text{ is } P, y_{n-1} \text{ is } P) \to y_n \text{ is } P^2 \tag{1}$$

$$(u_n \text{ is } N, y_{n-1} \text{ is } P) \to y_n \text{ is } P^{1/2} \tag{2}$$

$$(u_n \text{ is } P, y_{n-1} \text{ is } N) \to y_n \text{ is } N^{1/2} \tag{3}$$

$$(u_n \text{ is } N, y_{n-1} \text{ is } N) \to y_n \text{ is } N^2 \tag{4}$$

where the linguistic truth values of all implications are assumed to be u-true.

---

[*] © 1983 Plenum Press. Retyped with written permission from *Advances in Fuzzy Sets, Possibility Theory and Applications* (P.P. Wang, Ed.), 325-334.

Here P and N are fuzzy variables implying "positive" and "negative" respectively. P^2 and $P^{1/2}$ imply "very positive" and "rather positive" or something like that. Those membership functions are:

$$P(x) = \frac{1}{2a}x + \frac{1}{2} \tag{5}$$

$$N(x) = P(-x) \tag{6}$$

$$P^2(x) = P(x)^2 \tag{7}$$

$$P^{1/2}(x) = P(x)^{1/2} \tag{8}$$

where x stands for u_n, y_{n-1}, and $y_n \in [-a, a]$.

The above fuzzy model represents a system with first order delay in discrete time. When (u_n, y_{n-1}) is given, the next y_n is easily calculated by multi-dimensional fuzzy reasoning (see Appendix). Fuzzy implications in a fuzzy model may be called fuzzy system behaviors, where each implication corresponds to a local behavior of a system.

Now we deal with the case that control objective is to set the output at zero. Let us design fuzzy controllers with respect to e_{n-1}, and e_n where $e_{n-1} = -y_{n-1}$ and $e_n = -y_n$. For simplicity we use y_{n-1}, and y_n instead of e_{n-1}, and e_n.

Controller 1

Assume that (y_{n-1}, y_n) is observed ("positive", "very positive") as is shown in Fig. 1. Then we have to decrease the output at $n+1$ period. For example y_{n+1} would be desirable for the control objective to become "positive". Then the problem is what u should be like.

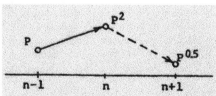

Fig. 1. Control 1 when (y_{n-1}, y_n) is (P, P^2).

Let

$$(u_{n+1} \text{ is } X, y_n \text{ is } P^2) \rightarrow y_{n+1} \text{ is } P \tag{9}$$

The clue to finding X which decreases y from P^2 to P lies at the second behavior, i.e., Eq. (2) in the fuzzy model. X is easily found by comparing the variables in Eq. (2) with those in Eq. (9).

Set the semantical equations such that

$$P \simeq P^{1/2} \text{ is } \tau_1 \tag{10}$$

$$P^2 \simeq P \text{ is } \tau_2 \tag{11}$$

$$X \simeq N \text{ is } \tau_3 \tag{12}$$

where τ_i is a linguistic truth value. Then τ_3 would be approximately equal τ_1 and τ_2. In this case we can find that $\tau_1 = \tau_2 =$ "very true". It follows from truth qualification that $X = N^2$ by letting $\tau_3 =$ very true. Also when (y_{n-1}, y_n) is (N, P)

as is shown in Fig. 2, let us decrease y from P to $P^{1/2}$. The value u_{n+1} is found to be N in a similar manner. Finally we obtain the following four control rules:

$$(y_{n-1} \text{ is } P, y_n \text{ is } P^2) \rightarrow u_n \text{ is } N^2 \tag{13}$$
$$(y_{n-1} \text{ is } P, y_n \text{ is } N) \rightarrow u_n \text{ is } P \tag{14}$$
$$(y_{n-1} \text{ is } N, y_n \text{ is } P) \rightarrow u_n \text{ is } N \tag{15}$$
$$(y_{n-1} \text{ is } N, y_n \text{ is } N^2) \rightarrow u_n \text{ is } P^2, \tag{16}$$

where the truth values of the control rules are also assumed to be u-true.

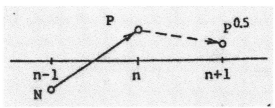

Fig. 2. Control 1 when (y_{n-1}, y_n) is (N, P).

Controller 2

Let us design another controller under the idea shown in Figs. 3 and 4. The procedure is the same as in Controller 1. We obtain

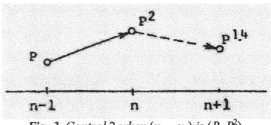

Fig. 3. Control 2 when (y_{n-1}, y_n) is (P, P^2).

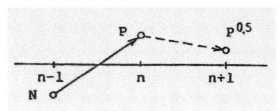

Fig. 4. Control 2 when (y_{n-1}, y_n) is (N, P).

$$(y_{n-1} \text{ is } P, y_n \text{ is } P^2) \rightarrow u_n \text{ is } N^{1/2} \tag{17}$$
$$(y_{n-1} \text{ is } P, y_n \text{ is } N) \rightarrow u_n \text{ is } P \tag{18}$$
$$(y_{n-1} \text{ is } N, y_n \text{ is } P) \rightarrow u_n \text{ is } N \tag{19}$$
$$(y_{n-1} \text{ is } N, y_n \text{ is } N^2) \rightarrow u_n \text{ is } P^{1/2}. \tag{20}$$

Controller 3

The following control rules are also derived by referring to Figs. 5 and 6.

$$(y_{n-1} \text{ is } P, y_n \text{ is } P^2) \rightarrow u_n \text{ is } N^{3/2} \tag{21}$$

$$(y_{n-1} \text{ is } P, y_n \text{ is } N) \rightarrow u_n \text{ is } P^{3/2} \tag{22}$$

$$(y_{n-1} \text{ is } N, y_n \text{ is } P) \rightarrow u_n \text{ is } N^{3/2} \tag{23}$$

$$(y_{n-1} \text{ is } N, y_n \text{ is } N^2) \rightarrow u_n \text{ is } P^{3/2}. \tag{24}$$

As is easily seen, two implications, the first and the fourth, in the fuzzy model are not necessary to derive fuzzy control rules since those describe diverging behavior of the process. In general, finding X in Eq. (9) is not always easy since τ_1 and τ_2 may be different from each other. We omit a general method in this paper.

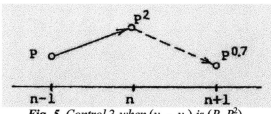

Fig. 5. Control 3 when (y_{n-1}, y_n) is (P, P^2).

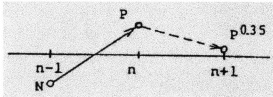

Fig. 6. Control 4 when (y_{n-1}, y_n) is (N, P).

3. Simulation

Let us apply three controllers to the process. The results are illustrated in Fig. 7. As is seen in Fig. 7, the first controller is the best. Next let us examine the ability of each controller by putting pure dead time to the process. Two dead times of $n_d = 1$ and 4 are put into the process. The results are shown in Figs. 8 and 9. The third controller shows the best performance when $n_d = 1$ and the second one the best when $n_d = 4$. The results of the first and the third controller are omitted in Fig. 9 since those are something mad.

The reason why the second controller is the best when $n_d = 4$ is found by comparing Fig. 3 with Figs. 1 and 5. That is, one should not decrease y too much when y is changing from "positive" to "very positive" if the process has dead time: the excessive decrease of y causes oscillation.

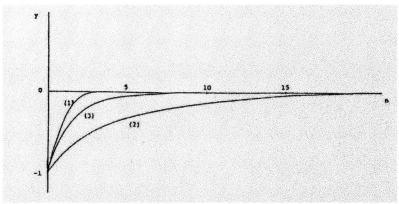

Fig. 7. *Fuzzy controls of process with no dead time.*

4. Conclusions

We have presented a method to derive fuzzy control rules from a fuzzy model of a system. The results of simulation are satisfactory. Together with a method of fuzzy modeling in terms of fuzzy implications, the present method is expected to be very powerful. It gives a way to analyze theoretically a fuzzy control system and make clear such problems as control performance and stability, etc.

The philosophy of derivation of fuzzy control rules is the same even if a system has nonlinearity. A nonlinear system as well as a linear system can be described by a set of fuzzy system behaviors. There is no difference between them except the number of fuzzy behaviors of a system. Fuzzy control rules depending on control object are derived by referring some, not all, of fuzzy system behaviors, where we can easily consider necessary situations for the design purpose as is shown in Figs. 1 and 2.

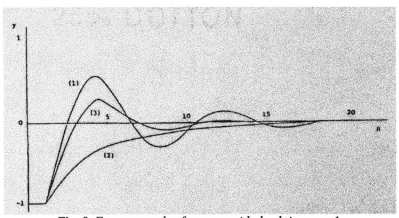

Fig. 8. *Fuzzy controls of process with dead time $n_d = 1$.*

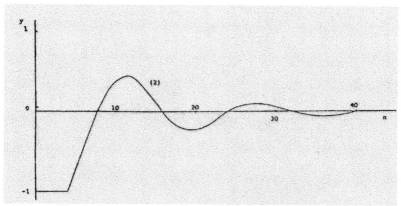

Fig. 9. *Fuzzy controls of process with dead time* $n_d = 4$.

References

[1] M. Sugeno and T. Takagi, Multi-dimensional fuzzy reasoning, *Submitted to International Journal of Fuzzy Sets and Systems*.
[2] L. Zadeh, Fuzzy logic and approximate reasoning, *Synthese*, **30**, 407/428, 1975.
[3] Y. Tsukamoto, An Approach to Fuzzy Reasoning Method, in *Advances in Fuzzy Set Theory and Applications*, M. M. Gupta et al., ed., 1979.

Appendix

Fuzzified Lukasiewicz's logic and its reasoning

The truth value of implication in Lukasiewicz's infinite valued logic is expressed as

$$/A \to B/ = (1 - /A/ + /B/) \wedge 1,$$

where $/A/ \in [0, 1]$ is the truth value of a proposition A.

Let P, Q be fuzzy propositions and $\underline{P}$, $\underline{Q}$ be linguistic truth values that are fuzzy subsets of $[0, 1]$ as usual. Then $\underline{P} \to \underline{Q}$ is derived by extension principle from the above equation.

$$\underline{P} \to \underline{Q} = (1 - \underline{P} + \underline{Q}) \wedge 1$$

where the operations -, + and $\wedge$ are extended ones for fuzzy sets.

For calculation it is better to take strong α - cut. We have

$$\underline{P} \to \underline{Q}_\alpha = (1 - \underline{P}_\alpha + \underline{Q}_\alpha) \wedge 1$$

where $\underline{P}_\alpha = \{u| \, h_{\underline{P}}(u) > \alpha\}$ and $h_{\underline{P}}(u)$, $u \in [0,1]$, is the membership function of $\underline{P}$.

In many cases we can assume that $\underline{P} \to \underline{Q}$ is normal, its membership function is non-decreasing in its domain $[0, 1]$ and also $\underline{P}$ is normal and convex. Denote $R = \underline{P} \to \underline{Q}$. Then it follows from the above assumptions that

$$\underline{R}_\alpha = (r(\alpha), 1]$$
$$\underline{P}_\alpha = (p_1(\alpha), p_2(\alpha)),$$

where $r(\alpha) \in [0, 1]$ is determined from h_R and $p_1(\alpha)$, $p_2(\alpha)$ from $h_{\underline{P}}$. Now $\underline{Q}_\alpha$ is obtained as

$$\underline{Q}_\alpha = ((p_1(\alpha) + r(\alpha) - 1) \vee 0, 1].$$

Then $\underline{h}_Q$ is easily drawn since the above expression implies that $\underline{h}_Q$ is a non-decreasing function.

Fuzzy reasoning based on an implication $P \to Q$ with $\underline{P \to Q}$ is carried out as follows, where fuzzy modus ponens is written

$$\frac{x \text{ is } P', x \text{ is } P \to y \text{ is } Q}{y \text{ is } Q'}$$

1) Given a premise P', set $P' \simeq P$ is $\underline{P}$. By converse of truth qualification, $\underline{P}$ is found as
$$\underline{P} = h_P(P'),$$
where h_p is the extended function of h_p for a fuzzy set P'.

2) Calculate Q from $\underline{P}$ and $\underline{P \to Q}$ as shown above.

3) Set $Q' = Q$ is Q. Then Q' is the consequence of fuzzy reasoning. By truth qualification, Q' is obtained as
$$Q' = h_Q^{-1}(Q),$$
where h_Q^{-1} is also the extended one.

Multi-dimensional fuzzy reasoning

Apart from computational technique, it is sufficient for the purpose to deal with the two dimensional case. Let us start with four implications such that

$(x \text{ is } A_1, y \text{ is } B_1) \to z \text{ is } C_{11}$
$(x \text{ is } A_1, y \text{ is } B_2) \to z \text{ is } C_{12}$
$(x \text{ is } A_2, y \text{ is } B_1) \to z \text{ is } C_{21}$
$(x \text{ is } A_2, y \text{ is } B_2) \to z \text{ is } C_{22},$

The situation of four implications is shown in Fig. A1.

Our problem is to infer "z is C" from a given premise $(x \text{ is } A, y \text{ is } B)$ where A is assumed to be between A_1 and A_2 and also B between B_1 and B_2.

The outline of the algorithm is shown in Fig. A2. First let us infer the value of z at the point (A_1, B) from (A_1, B_1) as indicated by an arrow in Fig. A2. Modus ponens is written as:

$$\frac{(x \text{ is } A_1, y \text{ is } B), (x \text{ is } A_1, y \text{ is } B_1) \to z \text{ is } C_{11}}{z \text{ is } C_{11}'}$$

Here C_{11}' is easily obtained by one-dimensional reasoning since A_1 is fixed along the arrow. That is, set

$$(x \text{ is } A_1, y \text{ is } B) \simeq (x \text{ is } A_1, y \text{ is } B_1) \to \text{ is } \tau$$

then clearly

$$y \text{ is } B \simeq y \text{ is } B_1 \text{ is } \tau$$

for the same τ.

Let us next infer z at the same point (A_1, B) this time from (A_1, B_2). That is,

$$\frac{(x \text{ is } A_1, y \text{ is } B), (x \text{ is } A_1, y \text{ is } B_2) \to z \text{ is } C_{12}}{z \text{ is } C_{12}'}$$

Now let

$$D1 = \frac{1}{2}(C_{11}' + C_{12}'),$$

where scalar multiplication and addition are the extended operations. Then we obtain a new implication with respect to the point (A_1, B) such that

$(x \text{ is } A_1, y \text{ is } B) \to z \text{ is } D_1$.

Here D_1, may be obtained in another way: $D_1 = C_{11}' \cap C_{12}'$. For the point (A_2, B) we can also obtain an implication such that

$(x \text{ is } A_2, y \text{ is } B) \to z \text{ is } D_2$.

Finally if we follow the same procedure along the dotted arrows in Fig. A2, we can infer the value of z at (A, B).

In general there may be a certain number of implications available at a different situation from Fig. A1. The algorithm is easily extended to this case.

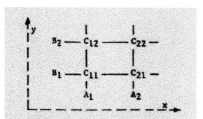   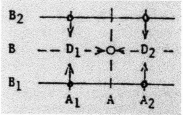

Fig. A1. *Situations of four implications.*   **Fig. A2.** *Outline of algorithm.*

Fuzzy Control of Model Car*

M. Sugeno and M. Nishida
*Department of Systems Science,
Tokyo Institute of Technology,
4259 Nagatsuta, Midori-ku,
Yokohama, 227 Japan*

*Received January 1984
Revised August 1984*

Abstract

This paper discusses fuzzy control of a model car. Fuzzy control rules are derived by modeling an expert's driving actions. Experiments are performed using a model car with a sensing unit and a micro-computer.

Keywords: fuzzy control, fuzzy control rules, human operator's model, micro-computer control, car driving.

1. Introduction

Fuzzy control is one of the most interesting fields where fuzzy theory can be effectively applied. Recently some practical applications of fuzzy control [1-6] have been reported.

As far as fuzzy control is concerned, we can say that our main interest is now towards its applications. When we intend to apply fuzzy control to an industrial process, one of the key problems to be solved is to find fuzzy control rules. There are a number of ways to find fuzzy control rules based on:
    (1) the operator's experience,
    (2) the control engineer's knowledge,
    (3) fuzzy modeling of the operator's control actions,
    (4) fuzzy modeling of the process.

This paper discusses the fuzzy control of a model car where control rules are derived by modeling an expert's driving actions. Most of fuzzy controllers have

---

*© 1985 North-Holland. Retyped with written permission from
Fuzzy Sets and Systems, **16** (1985) 103-113.*

been designed by the first and the second methods so far. Those two methods are heuristic in their natures, which are very difficult to be generalized. We often face the case where an operator cannot tell linguistically what actions he takes in such and such a situation. For instance, consider when we drive a car. We know car driving techniques using our hands and legs rather than by our brain. In fact we can easily suppose that it would be difficult to make skillful control rules based on a driver's knowledge. The first and second methods are not suitable in this control situation. Also, if a process is complex, it is difficult to design a fuzzy controller even from a control engineer's point of view. In such cases it is very useful to develop a method to derive fuzzy control rules by modeling operator's control actions, if there is a good reason to believe that an operator does an excellent control. If it is not the case, we have to take the fourth method.

2. Preliminaries

Fuzzy control rules and reasoning method

In this study we use the following fuzzy control rules for a multi-input and single-output controller. The i-th control rule is of the form

$$R^i : x_1 \text{ is } A_1^i, x_2 \text{ is } A_2^i, \cdots, x_n \text{ is } A_n^i$$
$$\rightarrow y^i = p_0^i + p_1^i x_1 + \cdots + p_n^i x_n \tag{1}$$

where the A^i_j's are fuzzy variables and y^i is the output of the i-th control rule determined by a linear equation with coefficients p^i_j. The membership function of a fuzzy set A is simply written $A(x)$ and it is composed of straight lines.

When the inputs $x^0_1 - x^0_n$ are given, the truth value of the premise of the i-th rule is calculated as

$$w^i = \bigwedge_{i=1}^{n} A_i^j(x_j^0) \tag{2}$$

and the output y^0 is inferred from m rules by taking the weighted average of the y^i's:

$$y_0 = \sum_{i=1}^{m} w^i y^i \Big/ \sum_{i=1}^{m} w^i \tag{3}$$

Identification algorithm of fuzzy control rules

Suppose that we are given input-output data $x_{1k}, x_{2k}, x_{3k}, \ldots, x_{nk}, y_k$ taken from operator's control actions. We then have to identify:
(1) the number of fuzzy partitions of the input space, for example, small, medium and big for x_1, and small and big for x_2, etc.,
(2) the membership functions of those fuzzy variables,
(3) the coefficients in the consequences of the rules the number of which is $m \times (n+1)$.

These are identified so that the output error is minimized where we mean by the output error, the difference between the output of a model and that of a process. The first problem, i.e., fuzzy partition, has no general way to be solved since it is a combinatorial problem. Engineering sense is, however, available and a heuristic search of the optimal partition is also possible. Provided that fuzzy partitions and the membership functions are given, the coefficients $p^i_0 - p^i_n$ are easily identified using the output Eq. 3 so as to minimize the output error. We can use the so-called stable Kalman filter. To identify the membership functions, we can use the complex method to search optimal parameters since the performance index, i.e., the output error, is nonlinear with respect to the parameters of the membership functions. Here we have to iterate the identification process of the coefficients since those depend on the membership functions in the premises. Note that each membership function is determined by two or three parameters which give the maximal grade 1 and the minimal grade 0. See [6] for the details.

3. Derivation of control rules

The aim of this study is to control the steering handle of a model car so that it smoothly runs through a crank-shaped course.

A computer model of a car has been made in a micro-computer to find fuzzy control rules. Man can control its steering handle, and the trajectory of the car is displayed on a monitor TV screen. The speed of the car is kept constant throughout this study. So the control input to the car is only the angle of the steering handle.

Figure 1 shows an example of the trajectories of a car on a crank-shaped course controlled by an experienced person. About 500 input-output data have been taken from 18 trajectories. As the inputs to a human driver, four variables x_1- x_4, have been chosen as are seen in Figure 2 where

x_1 = distance from entrance of corner,
x_2 = distance from inner wall,
x_3 = direction (angle) of car,
x_4 = distance from outer wall.

The output of a driver, y, is the movement of the steering handle, i.e., the moved angle from its home position. Using the input-output data x_{1k}, ...x_{4k} and y_k, expert's driving actions are modelled in the form of 20 control rules. As for fuzzy partitions of the input space, we choose three fuzzy variables for x_1, two for x_2, three for x_3, and one for x_4 by observing the expert's actions.

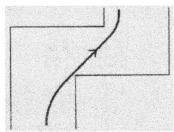

Fig. 1. *Trajectory of an experienced driver's car control.*

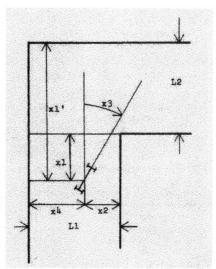

Fig. 2. Input variables to a driver.

Now we can identify the membership functions in the premises and the coefficients in the consequences. Figure 3 shows the identified membership functions associated with linguistic labels.

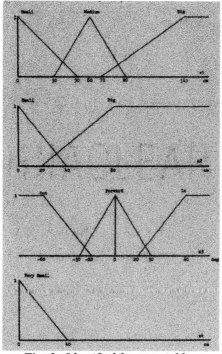

Fig. 3. Identified fuzzy variables.

Table 1. *Fuzzy Variables in Premises of Control Rules*

Rule	x_1	x_2	x_3	x_4
R^1	-	-	Out	Very Small
R^2	-	-	Forward	Very Small
R^3	Small	Small	Out	-
R^4	Small	Small	Forward	-
R^5	Small	Small	In	-
R^6	Small	Big	Out	-
R^7	Small	Big	Forward	-
R^8	Small	Big	In	-
R^9	Medium	Small	Out	-
R^{10}	Medium	Small	Forward	-
R^{11}	Medium	Small	In	-
R^{12}	Medium	Big	Out	-
R^{13}	Medium	Big	Forward	-
R^{14}	Medium	Big	In	-
R^{15}	Big	Small	Out	-
R^{16}	Big	Small	Forward	-
R^{17}	Big	Small	In	-
R^{18}	Big	Big	Out	-
R^{19}	Big	Big	Forward	-
R^{20}	Big	Big	In	-

Tables 1 and 2 show the premises of the control rules and the identified coefficients in the consequences, respectively. The rules R^1 and R^2 are for keeping the car away from the outer wall. The others are for driving it smoothly.

Now computer simulation of car driving has been done using those control rules. Figures 4a, 4b and 4c show the results of the simulation in some different situations. We can see that the derived control rules work very well.

4. Control of model car

Since the computer simulation was successful, the authors have made a model car and tried its fuzzy control using a micro-computer. The configuration of the model car and the micro-computer for control are the following.

Body: length 40 cm, width 21 cm, weight 2 kg.
Drive force: DA motor with gear ratio 1: 230.
Front wheels control: servo-motor.
Speed: 100 cm/min.

Table 2. Coefficients in Consequences of Control Rules

Rule	p_0	p_1	p_2	p_3	p_4
R^1	3.000	0.000	0.000	-0.045	-0.004
R^2	3.000	0.000	0.000	-0.030	-0.090
R^3	3.000	-0.041	0.004	0.000	0.000
R^4	0.303	-0.026	0.061	-0.050	0.000
R^5	0.000	-0.025	0.070	-0.075	0.000
R^6	3.000	-0.066	0.000	-0.034	0.000
R^7	2.990	-0.017	0.000	-0.021	0.000
R^8	1.500	0.025	0.000	-0.050	0.000
R^9	3.000	-0.017	0.005	-0.036	0.000
R^{10}	0.053	-0.038	0.080	-0.034	0.000
R^{11}	-1.220	-0.016	0.047	-0.018	0.000
R^{12}	3.000	-0.027	0.000	-0.044	0.000
R^{13}	7.000	-0.049	0.000	-0.041	0.000
R^{14}	4.000	-0.025	0.000	-0.100	0.000
R^{15}	0.370	0.000	0.000	-0.007	0.000
R^{16}	-0.900	0.000	0.034	-0.030	0.000
R^{17}	-1.500	0.000	0.005	-0.100	0.000
R^{18}	1.000	0.000	0.000	-0.013	0.000
R^{19}	0.000	0.000	0.000	-0.006	0.000
R^{20}	0.000	0.000	0.000	-0.010	0.000

Sensor: supersonic sensor driven by a stepping motor for the measurements of distance and direction.
Micro-computer: CPU 8080 with 4K BASIC.

Figures 5a and 5b show the appearance of the model car. The fuzzy control algorithm is programmed in BASIC language and installed in the microcomputer. The size of the program is about 100 steps in BASIC.

In the experiments the micro-computer is not mounted on the car and is connected by a flat cable because of an electric power supply problem. We can see the cable connected to the back side of the car in Figure 5a. The other interfaces for the inputs and outputs of the micro-computer are mounted on the car as are seen in Figure 5b.

We shall now look at the sensing devices.

Measurement of distance

A sensor is made by using a 40 KHz supersonic transducer. It can measure distance by counting the time difference between the time of emitting supersonic wave and that of receiving its echo reflected by the outer or inner wall. For the purpose of the measurements the wall is placed along the driving course. The sensor can measure at most 2 m distance. Its resolution is about 2 cm.

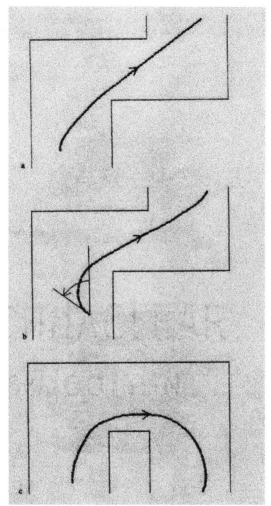

Fig. 4(a,b,c). Results of computer simulation.

Measurement of direction

The supersonic sensor is rotated just like a radar by a stepping motor. We can see the sensing unit on the roof of the car in Figure 5a. When the sensing unit is put on the position in parallel with the wall, it can receive supersonic echo. Since a supersonic wave has a good directivity, the direction of the car can be measured by counting the number of pulses put into a stepping motor between the home position of the sensing unit and the position where it can catch the echo. The stepping motor is rotated by 3.6° with 1 pulse. The resolution of the sensor is 7.2°.

Control of front wheels

A servomotor is used to control the front wheels. The output of the fuzzy controller is transmitted to a servomotor through a D/A converter. The resolution of this control is about 0.86° bit. When the rotation angle of the front wheels is 1°, the change of the direction of the car is about 1/60° after 1 sec run. Therefore the universe of discourse for fuzzy variables has to be adjusted by considering the above mentioned fact and also the real situation of the driving course. However, the problem is quite simple because it is merely the modification of controller's gain factors.

Experiments

Fuzzy control experiments have been performed on a test course with 1-m width. Figures 6a-6d show the trajectories of the model car in some different situations. As we can see, the car can run very smoothly through a crank-shaped course. However, it has occasionally failed because of sensing error.

Fig. 5(a,b). Model car.

5. Conclusions

It has been found that the way to derive fuzzy control rules from an experienced operator's control actions is very useful to design a fuzzy controller. The authors are now applying this sort of fuzzy control to put a car into a garage, where the speed of a car also has to be controlled.

The authors express their acknowledgements to Mr. T. Tanaka and Dr. T. Takagi for their help.

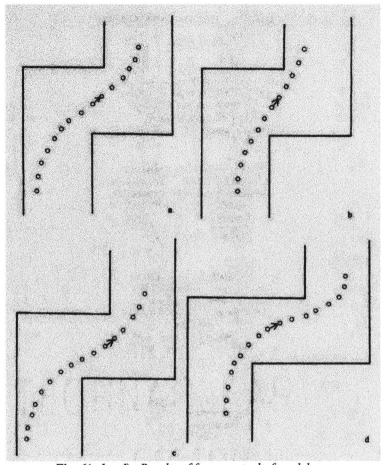

Fig. 6(a,b,c,d). Results of fuzzy control of model car.

References

[1] M.J. Flanagan, On the application of approximate reasoning to control of the activated sludge process, *Proc. Joint Automatic Control Conf.*, San Francisco (1980).
[2] R.M. Larsen, Industrial applications to fuzzy logic control, *Int'l Journal of Man-Machine Studies* 12 (1980) 3-10.
[3] L.P. Holmblad and J.-J. Ostergaard, Control of a cement kiln by fuzzy logic, in: M.M. Gupta and E. Sanchez, Eds., *Fuzzy Information and Decision Processes* (North-Holland, Amsterdam, 1982) 398-399.
[4] G. Bartolini, G. Casalino, F. Davoli, M. Mastretta, R. Minciardi and E. Morten, Development of performance adaptive fuzzy controllers with application to continuous casting plants, in: R. Trappl, Ed., *Cybernetics and Systems Research*

(North-Holland, Amsterdam, 1982) 721-728.
[5] R.E. King, Fuzzy logic control of a cement kiln precalciner flash furnace, *Proc. IEEE conf. on Applications of Adaptive and Multivariable Control*, Hull (1982) 56-59.
[6] T. Takagi and M. Sugeno, Derivation of fuzzy control rules from human operator's control actions, *IFAC Symposium on Fuzzy Information, Knowledge Representation and Decision Analysis*, Marseille (1983).

Chapter 6

Fuzzy Algorithmic Control of a Model Car By Oral Instructions[*]

M.Sugeno, T. Murofushi, T. Mori, T. Tatematsu, and J. Tanaka
Department of Systems Science,
Tokyo Institute of Technology,
4259, Nagatsuta, Midori-ku,
Yokohama 227, Japan

Received October 1987

Abstract

This paper is concerned with an experimental study of fuzzy algorithmic control of a model car. A fuzzy algorithmic controller is designed by referring to an operator's experience and knowledge. The car is equipped with ultrasonic sensing devices and a micro-computer. It is controlled by oral instructions such as "go straight", "turn right", "enter garage", and "go out of garage".

Keywords: control theory, linguistic modeling, fuzzy algorithm, car control.

1. Introduction

So far we have studied fuzzy control of a model car: (i) moving a car through a turn [2] and parking a car [3], and (ii) linguistic control [4]. In the first study, its purposes were to learn fuzzy control rules by observing an expert's way to drive a car and to execute such control rules by an actual car to see how well they work. Only the angle of the front wheels has been controlled in moving a car through a turn, but the speed of the car has also been controlled in parking it. In the second study, its main purpose was to move a car by linguistic instructions such as *"turn left"*, *"change direction"*. We have not applied ordinary rule-based control but used a fuzzified

[*]©1989 North-Holland. Retyped with written permission from *Fuzzy Sets and Systems, 32* (1989) 207-219.

control equation for a steering wheel.

Following those studies, we adopt fuzzy algorithmic control in this study. The idea of fuzzy algorithm was first proposed by Zadeh [6]. "A fuzzy algorithm is an ordered set of fuzzy instructions upon which execution yields an approximate solution to a specified problem" [7]. Much of what we do are fuzzy algorithmic routines: "we employ fuzzy algorithms both consciously and subconsciously when we walk, drive a car, search for an object, tie a knot, park a car, cook a meal, find a number in a telephone directory, etc." [7]. These examples shown by Zadeh are too complex or too ill-defined for precise control.

Throughout past experiences, we have learned that it is better to use a fuzzy algorithmic control method rather than to use ordinary rule-based control methods as far as car control is concerned. When we drive a car, we take *sequential actions* concerned with both speed and steering regulations in a macroscopic movement. For example, in turning at a corner, you drive straight until your car comes close to the corner, and then you slow down; next, you turn the wheel and keep it until the car almost finishes turning; last, you turn back the wheel and accelerate.

2. Model car control system

We design a fuzzy algorithmic controller to control the angle of the front wheels and the speed of a model car, by which the car can execute macro instructions such as *"go straight"*, *"turn right"*, *"enter garage"*, and *"go out of garage"*. We perform experiments using an actual model car (Figure 1) on such a course as is illustrated in Figure 2.

Fig. 1. The model car.

Figure 3 shows the outline of information flow of the model car control system. An operator first gives a linguistic instruction through a speech recognition device to the car. Then, by considering its present states and surroundings, the car determines its move. The car executes macro-instructions referring to information obtained by the sensing devices.

Figure 4 shows the hardware configuration of the system. The operator's instructions are recognized by the speech recognition device and put into a personal computer. In this personal computer, an instruction is interpreted and actions to be taken are sequentially determined according to a fuzzy algorithm. Then the signals concerned with actions are transmitted by a wireless transmitter from the personal computer to the car. The car is equipped with an on-board microcomputer which controls the car speed, the angle of the front wheels, and the sensors. The information obtained by the sensors is also transmitted from the car to the personal computer.

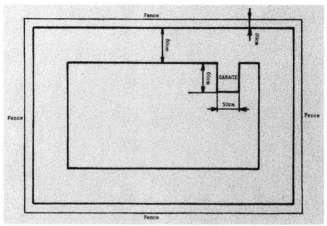

Fig. 2. Experimental course.

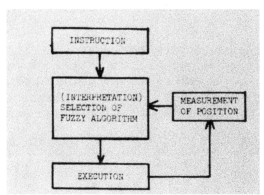

Fig. 3. Information flow of model car control.

2.1. The model car

The model car has the following configuration.

Body: length 56 cm, width 28 cm, height 43 cm, weight 7 kg.
Drive force of rear wheels: DC servomotor.
Steering control of front wheels: DC servomotor.
Speed: forward 20-50 cm/s, backward 20 cm/s.
Sensor: two ultrasonic wave sensors driven by a DC servomotor
for the measurements of distance and direction.
Controller: on-board micro computer (6809) with 8 KB ROM and 2 KB

RAM, personal computer (NEC PC-9801).

Measurement of distance and direction: The following variables are measured with the sensing device shown in Figure 5. x: front fence distance, y: side fence distance, and θ: heading angle.

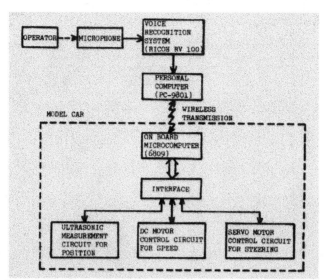

Fig. 4. Hardware configuration of model car control system.

Two sensors are put at right angles to each other: one measures x and the other y as shown in Figure 6. A sensor consists of a 40 KHz ultrasonic transducer. It can measure distance by counting the time difference between the time of emitting an ultrasonic wave and that of receiving its echo reflected by the front or the side fence. The resolution of sensing is about 1 cm. The sensor can measure at most 2 m distance. This is the reason why the garage is very near a corner in Figure 2.

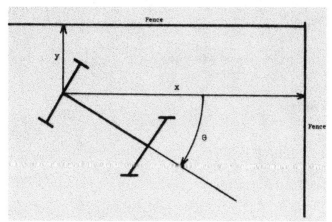

Fig. 5. Measurement variables.

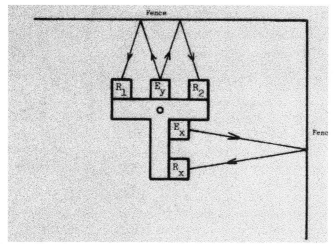

Fig. 6. The ultrasonic sensors; E_x, E_y: Emitter, R_x, R_1, R_2: Receiver.

The sensing unit adjusts itself to the front and side fences by turning clockwise or counterclockwise as is shown in Figure 7. That is, for any direction of the car the two sensors are always kept parallel with the corresponding fences. This is carried out by the mechanism shown in Figure 7 where the position of the sensors is controlled so that the distances measured by the receivers R_1 and R_2 are equal to each other. With the aid of this mechanism, the heading angle θ can be measured by using the potentiometer: the rotation of the sensing unit is measured by the potentiometer. The resolution is about 3^0.

The information which cannot be sensed with the ultrasonic sensors, the width of the road, the location of the garage, and the space between the fence and the road, are given to the controller beforehand.

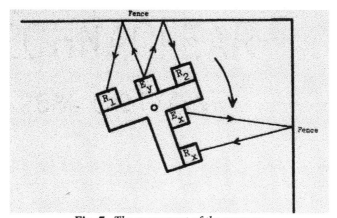

Fig. 7. The movement of the sensors.

Control of front wheels: The angle of the front wheels is controlled by a DC servomotor. The output of the fuzzy controller is transmitted to the motor through a D/A converter. The resolution of this control is about 0.86^0/bit. When the rotation angle of the front wheel is 1, the change of the direction of the car is about $1/60^0$ after 4 cm run.

Control of car speed: Car speed is controlled by changing the rotation speed of the DC servomotor, not by changing gears.

2.2. Instructions and interpretation

We used the following 14 instructions.
  (i) macro-instructions: *go straight, turn right, enter garage, go out of garage.*
  (ii) micro-instructions: *start, stop, go back, speed up, slow down, steer right, steer middle, steer left, fix sensor.*
  (iii) conjunction for macro-actions: *then.*

A macro-instruction makes the car take a series of actions. If the instruction *"turn right"* is given, the car executes sequential actions as mentioned at the end of Section 1. If *"go straight"* is given, the car moves straight along the side fence by rule-based control.

A micro-instruction makes the car take a single action. For example, if *"steer right"* is given, the car turns the wheel right and keeps it. The instruction *"fix sensor"* stops the movement of the ultrasonic sensing devices; this instruction is used in the case that the car is controlled only by micro-instructions. By giving micro-instructions, we can move the car like an ordinary radio-controlled car.

When an instruction is given, an action to be taken is one of the following:
  (1) the instruction is executed immediately,
  (2) the instruction is executed after the execution of the preceding macro instruction is finished,
  (3) the instruction is rejected or disregarded.

For example, let us assume that the instruction *"go straight"* is given while the car is executing the algorithm for *"turn right"*. If the car is still before a corner, the execution of *"turn right"* is interrupted and *"go straight"* is executed. If the car is turning a corner, the car executes *"go straight"* after it finishes turning. Further if *"go out of garage"* is given while the car is executing *"turn right"*, then the car always rejects it. The micro-instructions except *"start"*, *"speed up"* and *"slow down"* always interrupt the execution process of a macro-instruction.

Using the conjunction *"then"*, we can give a number of sequential macro instructions such as *"go straight" "then" "turn right" "then" "enter garage"*.

2.3. Control

The macro-instructions *"turn right"*, *"enter garage"*, and *"go out of garage"*, are executed by a fuzzy algorithmic method. For example, the flowchart of the fuzzy algorithm for *"turn right"* is shown in Figure 8: the car goes straight on until it comes near a corner, and then the car reduces its speed; the car goes forward until the turning point, and then the front wheels are turned to the right; the angle of the front wheels is kept until the car almost finishes turning; and last the car comes into the actions for the macro-instruction "*go straight*". The flowcharts of the other instructions are omitted because of their complexity.

The branching conditions in Figure 8 except "Failed?" are fuzzy. In our study fuzzy branchings are executed by setting thresholds. That is, for example, the instruction "IF x is *small* THEN **stop** ELSE go to 7" is equivalent to "IF $\mu_{small}(x) > \alpha$ THEN **stop** ELSE go to 7", where μ_{small} is a membership function representing *small* and α is a given threshold.

Let us consider the fuzzy condition "Turning Point?" in Figure 8.

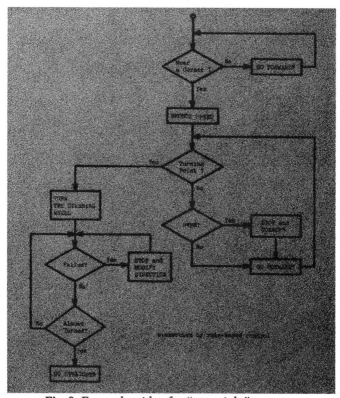

Fig. 8. Fuzzy algorithm for "turn right".

It is expressed in the programming language C as:
   b_OR(l_AND(*near*( ), *right_facing*( )),
         l_AND(NOT(*near*( )), *left-facing*( ))) IS VERY_TRUE.
This statement means that "the car is *near a corner* and the direction of the car is *right*, or the car is not *near a corner* and the direction is *left*". The values of the functions near ( ), right_facing ( ), and left_facing ( ) are the degrees of the truth of the statements "the car is *near a corner*", "the direction of the car is *right*", and "the direction of the car is *left*", respectively. The fuzzy sets *labeled near a corner, right,* and *left* are shown in Figure 9. l_AND, b_OR, and NOT are logical connectives

defined as 1_AND(x, y) = min(x, y), b_OR(x, y) = min(1, $x + y$), and NOT(x) = 1 - x. "IS VERY_TRUE" means "is greater than 0.7"; the threshold is 0.7.

Ordinary rule-based control is applied to execute the macro-instruction "*go straight*" and such instructions in fuzzy algorithms as "*go forward*" and "*go back*" are shown in Figure 8. The angle of the front wheels is controlled so that the car moves on a reference line, which is imaginarily put parallel to the side fence. Figure 10 illustrates the variables used in the rules. d: the distance between the reference line and the car, θ: the angle between the reference line and the car, and ϕ: the angle of the front wheels.

The fuzzy control rules for going forward are the following:

Rule 1: IF d is *left* and θ: is *middle* THEN ϕ is *right*,
Rule 2: IF d is *right* and θ: is *middle* THEN ϕ is *left*,
Rule 3: IF θ: is *left* THEN ϕ is *right*,
Rule 4: IF θ: is *right* THEN ϕ is *left*,

where the fuzzy sets are shown in Figure 11. Rule 1 and Rule 2 move the car close to the reference line, and Rule 3 and Rule 4 correct the direction of the car. The reasoning method is the second one described in [5]: a fuzzy set in a consequence of a rule has a monotone membership function and defuzzification is computed as a weighted mean (see Appendix A).

The following are the thresholds we have used:

NOT_COMPLETELY_FALSE	0.0,
RATHER_TRUE	0.3,
TRUE	0.5,
VERY_TRUE	0.7,
VERY_VERY_TRUE	0.9.

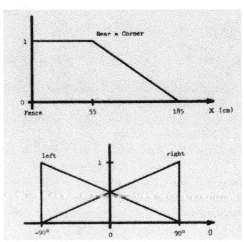

Fig. 9. Fuzzy sets for "Turning point?".

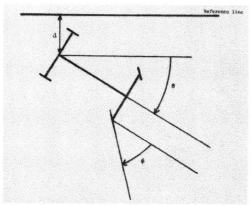

Fig. 10. The variables in the control rules for "go straight".

The logical connectives used in our system are only 1_AND, b_OR, and NOT. The fuzzy sets are represented as L-R type flat fuzzy numbers [1] $(m_1, m_2, \alpha, \beta)_{LR}$ with $L(x) = R(x) = \max(0, 1 - |x|)$ (see Appendix B). The parameters for the membership functions and the thresholds are determined by modifying those obtained in computer simulation.

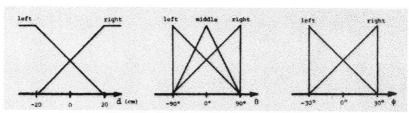

Fig. 11. Fuzzy sets for "go straight".

3. Results and conclusion

Some results of experiments are shown in Figures 12(a)-(e). Occasionally the control failed because of a sensing error, communication error, or speech recognition error. The design of a fuzzy (algorithmic) controller is not difficult due to the usage of linguistic expressions in an algorithm. This fact should be pointed out also in our work. Experiences in designing fuzzy controllers show the realization of this advantage.

Linguistic expressions based on fuzzy sets are useful to branching conditions as well as control rules. In the case of branching conditions, their advantages come out when fuzzy sets are connected by compensatory operations like $b\_OR(X, Y) = \min(1, x + y)$. See the program of the branching condition "Turning Point?" mentioned in Section 2.3 and its results (Figures 12(a) and (b)). If the direction of the car is in the middle between "*right*" and "*left*", the car turns at the middle point between "*near*" and "NOT *near*". In the case $b\_OR(x, y) = \max(x, y)$, which is not compensatory, the car cannot turn if the direction of the car is in the middle.

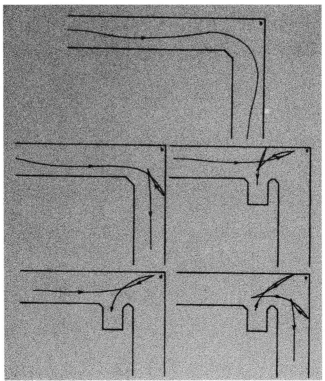

Fig. 12. Results of fuzzy control. (a), (b): "turn right" (c), (d): "enter garage"; (e): "go out of garage".

An obvious drawback in fuzzy algorithmic control is that a person who cannot manually control a system cannot make a good controller. One of the authors who first took charge of the algorithm *"enter garage"* could not make it because he could not drive a car and thus had never put a car into a garage.

In this study we have determined the parameters for membership functions by computer simulation. We should investigate a parameter-determining method by learning. For example, we can develop a system which learns parameters by linguistic teaching such as "Turning point is too far from the corner" or even a self-learning system with learning rules such as "in the case of *'turn right'* if the car cannot finish the turn, set the turning point a little bit apart from the corner".

Our fuzzy algorithmic controller has been designed by an operator's experience and knowledge, and it has shown good performance. It is, however, often the case that an operator cannot verbalize his/her knowledge. In such cases it is useful to derive fuzzy control rules from an operator's actions, and we designed the rule-based controllers by such derivation in past studies [2, 3]. Summing up the past experiences, we now think that a fuzzy algorithm can hardly be derived only by a purely numerical method like [2, 3] and that an operator's linguistic support is necessary for its derivation.

Appendix A

Here we explain the process of reasoning of the fuzzy controller for the instruction "*go straight*". The control rules and the fuzzy sets used for them are shown in Section 2.3. Let the distance and the angle between the reference line and the car be d_0 cm and θ_0°, respectively. Then the weights, i.e., the compatibilities w_1, w_2, w_3 and w_4 of the premises, are calculated by

$$w_1 = \min(\mu_{left}(d_0),\ \mu_{middle}(\theta_0)),$$
$$w_2 = \min(\mu_{right}(d_0),\ \mu_{middle}(\theta_0)),$$
$$w_3 = \mu_{left}(\theta_0),$$
$$w_4 = \mu_{right}(\theta_0).$$

Next we find the outputs of the rules ψ_1, ψ_2, ψ_3 and ψ_4 for which
$$w_1 = \mu_{right}(\psi_1),\quad w_2 = \mu_{left}(\psi_2),\quad w_3 = \mu_{right}(\psi_3),\quad w_4 = \mu_{left}(\psi_4)$$
Then the output ψ_0 of the controller is inferred:

$$\psi_0 = \frac{w_1\psi_1 + w_2\psi_2 + w_3\psi_3 + w_4\psi_4}{w_1 + w_2 + w_3 + w_4}$$

Appendix B

Here we explain L-R type flat fuzzy numbers. An L-R type flat fuzzy number $M = (m_1, m_2, \alpha, \beta)_{LR}$ $(m_1 \leq m_2 \leq \alpha > 0, \beta > 0)$ is a fuzzy set of the real line with the following membership function μ_M:

$$\mu_M(x) = \begin{cases} L((m_1 - x)/\alpha), & x \leq m_1, \\ 1, & m_1 < x < m_2, \\ R((x - m_2)/\beta), & m_2 \leq x, \end{cases}$$

where L and R are non-increasing functions defined on $[0, \infty)$ satisfying
(1) $L(x) = 1$ if $x = 0$,
(2) $L(-x) = L(x)$.

Figure 13 shows an example of an L-R type flat fuzzy number $(m_1, m_2, \alpha, \beta)_{LR}$ with $L(x) = R(x) = \max(0, 1 - |x|)$.

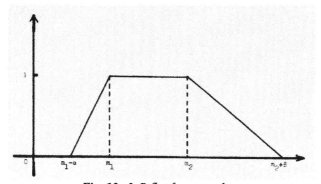

Fig. 13. L-R flat fuzzy number.

Acknowledgment

The authors wish to thank Mr. K. Katayama (Tokyo Gas Co. Ltd.) and Mr. Y. Nagai (University of Electro-Communications) for their help.

References

[1] D. Dubois and H. Prade, *Fuzzy Sets and Systems: Theory and Applications* (Academic Press, New York, 1980).
[2] M. Sugeno and M. Nishida, Fuzzy control of model car, *Fuzzy Sets and Systems* **16** (1985) 103-113.
[3] M. Sugeno and K. Murakami, An experimental study on fuzzy parking control using a model car, in: M. Sugeno, Ed., *Industrial Applications of Fuzzy Control* (North-Holland, Amsterdam, 1985) 125-138.
[4] M. Sugeno and K. Katayama, Linguistic control of model car, *First IFSA Congress* (Mallorca, 1985).
[5] M. Sugeno, An introductory survey of fuzzy control, *Inform. Sci.* **36** (1985) 59-83.
[6] L.A. Zadeh, Fuzzy algorithm, *Inform. Control* **12** (1968) 94-102.
[7] L.A. Zadeh, Outline of a new approach to the analysis of complex systems and decision processes, *IEEE Trans. Systems Man Cybernet.* **3** (1973) 28-44.

Application of Fuzzy Reasoning to the Water Purification Process[*]

0. Yagishita
Technology Department II, Public Utility Division
Fuji Electric Co., Ltd.
12-1 Yurakucho 1-chome, Chiyoda-ku, Tokyo 100, Japan

0. Itoh
Systems Research Section I, Systems Development Department
Fujifacom Corporation
Fujimachi 1-chome, Hino 191, Japan

M. Sugeno
Department of Systems Science
Tokyo Institute of Technology
4259.Nagatsuta, Midori-ku, Yokohama 227, Japan

Abstract

In the water purification plant, raw water is promptly purified by injecting chemicals. The amount of chemicals is directly related to the water quality such as turbidity and alkalinity of raw water. At present, however, the process of the chemical reaction to the turbidity has not been clarified as yet, so the amount of chemical cannot be calculated from the data of water quality only. Accordingly, it has to be judged and determined by skilled operators according to the data obtained from past experience. In connection with this, we studied a method to determine the amount of chemicals automatically by reproducing the judgements of the operators by means of fuzzy implication, and performed a field test in the water purification plant using a fuzzy controller. This report deals with the construction and functions of the fuzzy controller and the result of the field test.

[*] © 1985 North-Holland. Retyped with written permission from
Industrial Applications of Fuzzy Control (M. Sugeno, Ed.), (1985), 19-39.

1. Introduction

The role of a purification plant is to supply high-quality safe drinking water, and water for factories without fail. The source of water for this purpose is underground water and surface water of rivers, lakes, etc. Recently, surface water has been mainly used in order to protect resources. Surface water is available in relatively abundant quantities but contains a large amount of impurities, and that amount fluctuates widely. A coagulant dosage is fed to remove these impurities fully and effectively. The coagulant feed rate has to be changed according to a change in quality of a raw water taken into the purification plant. However, the process of coagulating the impurities with a coagulant is complicated, and only limited items can be measured with on-line water-quality sensors. For these reasons, no practical method of fixing a feed rate has been established yet, and currently, feed rate is fixed by a skilled operator's judgement based on his rich experience.

This paper describes the method of expressing an operator's judgement based on his empirical knowledge by fuzzy reasoning and a coagulant addition fuzzy controller using this method. This paper also reports the result of a field test of this controller at an actual water purification plant.

The feed rate is estimated by the conventional regression model plus fuzzy reasoning, which is used in respect of those areas not handled by the regression model, to estimate the required amount of compensation [2]. A fuzzy implication or the amount of compensation is a combination of the operator's grasp of states (e.g., rise in turbidity and formation of flocs) and the appropriate rate of increase/decrease of the feed rate. Hence, the fuzzy controller makes control calculations with the operator's recognition as an input value in addition to process measurements.

The output of the fuzzy controller can be inferred according to the compositional rule of inference [1]. Tong and others incorporated unmeasurable factors of a state into fuzzy reasoning and simulated the active sludge process to demonstrate the effectiveness of the fuzzy controller [3]. The study presented in the paper used a reasoning method based on compositional rule of inference, which was partially improved so as to reduce calculations and thereby make the reasoning readily understandable. This method has already been put into practical use for control of a cement kiln by fuzzy logic [4].

The field test was carried out for two months. Fuzzy implications on the amount of compensation were made on the basis of the operation data for three months including the first one month of the test period. They were verified with the data for the last month of the test period. As a result, it was discovered that the fuzzy controller is effective even in such situations that cannot be handled by the conventional system -- especially a case where the state of flocs must be frequently observed at the time the turbidity of raw water is changing.

Chemicals Addition Control for Purification Plant

The purpose and the method of chemicals addition control at a purification plant will be described here, and the present control method will be explained.

An alkaline agent, coagulant and chlorine agent are fed in the purification plant. Lime or caustic soda is used as an alkaline agent for pH adjustment. Aluminum sulfate or PAC (polymerized aluminium chloride) is used as a coagulant. All of these are used to promote coagulation of the impurities to raise the efficiency of sedimentation.

A suitable chemical agent is selected according to the trend in water quality at the time a purification plant is constructed. Aluminium sulfate is economical but is not very effective in low-temperature water. The effect of PAC is not decreased in low-temperature water, but it is expensive. A chlorine agent is fed for sterilization.

A chlorine agent is called a pre-chlorination or post-chlorination agent according to the feed point. A pre-chlorination agent is used to prevent generation of algae or the like in a sedimentation basin, filtration basin, etc., and a post-chlorination agent is used to turn the water into hygienic drinking water.

An outline of the purification system and chemical feed points is shown in Fig. 1. The sedimentation and filtration processes are most fundamental and represent a large portion of the purification plant. The sedimentation basin is designed to precipitate the impurities in water. A coagulant is fed to quicken the sedimentation process. The coagulant aggregates the impurities in water (into flocs) to make them gain weight and sink faster. Thus formation of good flocs is essential for smooth sedimentation, and this gives a reference for coagulant addition. The turbidity of raw water is 10-odd mg/l normally and may rise to even hundreds of mg/l. The impurities are largely removed by a sedimentation basin, and the effluent at its outlet has a turbidity of only several mg/l. The fact that the coagulant plays a very important role can be inferred from this. The residual impurities are then removed by a sand filtration basin to reduce the turbidity further to 1 mg/l or less. If impurities inadequately precipitated are transferred to the filtration basin, that basin is promptly clogged; thus it cannot perform its filtration function.

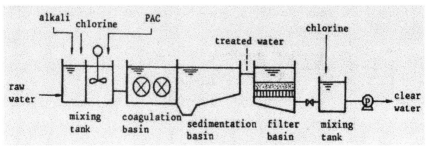

Fig. 1. Water purification process.

The method of feed rate control or a chemical agent will be described. A suitable feed rate is prefixed for any particular chemical agent, and the quantity fed is kept in proportion to the flow of the raw water. An alkaline agent and a chlorine agent need to be fed at a constant rate despite some fluctuation of water quality; thus no difficulty is involved in the fixing of a feed rate for such chemicals. On the other hand, a coagulant dosage reacts to the impurities in water through a process so complicated that a process model is hard to work out. For this reason, there is no widely accepted method of fixing a feed rate. Some of the methods currently used will be explained. One method involves a jar test. Under this method, specimens of raw water are collected in five to six beakers, a coagulant is fed in these beakers at different rates, and the mixtures are stirred. A feed rate is fixed after analysis of the sedimentation of the impurities in these beakers. This is an effective method, but the test takes too much time. The test is especially time-consuming work if the turbidity of the raw water is changing. For this reason, studies have been made to find out a way of fixing a feed rate on the basis of the quality of raw water. Under this method, a regression model for historical plant data is obtained by regression analysis and is used as a feed rate equation. As measurements with on-line water-quality sensors are employed, this method permits automated coagulant addition control. Only limited measurements, however, can be made with on-line water-quality sensors. Since the turbidity is measured by use of transmitted light, particle size and the chemical composition cannot be found out. Thus only gross measurements can be made, and therefore, this method cannot respond to slight changes in water quality.

As can be seen from the forgoing, the main difficulty in coagulant addition control lies in the fixing of an optimum feed rate in response to a change in quality of raw water. An especially important factor in the connection is the coagulant which directly affects the sedimentation process. Hence an analysis will be made in this paper with primary attention to the coagulant.

2. Application of Fuzzy Reasoning

2.1 *Basic Configuration*

The basic configuration for estimating the feed rate is shown in Fig. 2. The value of the feed rate equation is used as a reference, and that portion that cannot be handled with that value alone is inferred by fuzzy implication and added as an amount of compensation.

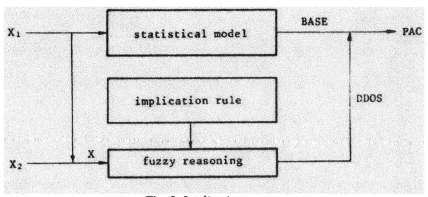

Fig. 2. Implication system.

X_1: water quality data. X_2: keyboard input data.
PAC: amount of PAC.   BASE: amount of PAC basis.
DDOS: amount of PAC modification

X_1 is the universe of discourse of water quality measurements regularly made and X_2 is the universe of discourse of measurements that cannot be regularly made, such as the state of flocs. The latter measurements are entered by the operator as a "large", "medium" or "small" value. X is the universe of discourse. A fuzzy implication on the amount of compensation is represented by (1) which is formed from the proposition P that declares the input X to be "large" or "small" and the similar proposition Q for the amount of compensation.

if P then Q
(if P is valid then Q is valid)   (1)

How P and Q are expressed will be first explained. The proposition P regarding certain input $x \in X$ is represented by (2) by using the fuzzy variable f which generally declares x to be "*large*", "*small*", etc. and the modifier m. F is the universe of discourse of f.

$$P : X =^m f \quad x \in X, \quad f \in F \tag{2}$$

Four cases are considered -- absence of a modifier, excess, ">", falling short, "<", and negation A-A. These are represented by the expressions in (3) below.

$$\begin{aligned} x &= f \; : x \text{ is equal to } f. \\ x &=> f : x \text{ exceeds } f. \\ x &=< f : x \text{ is smaller than } f. \\ x &= \neg f : x \text{ is not } f. \end{aligned} \tag{3}$$

Examples of propositions are given in Fig. 3. As can be seen from Fig. 3, the value exceeding the middle, i.e., the value to the left of the symbol, is of Grade 1. If the fuzzy variable f is "*medium*" (MM) in ① in Fig. 3 and x is the water temperature (TEMP), the proposition TEMP = ^{m}MM is one among ② to ⑤ in Fig. 3. Each of these is a fuzzy set defined in relation to water temperature.

The fuzzy variable f is represented by a fuzzy set in a unit interval. The membership function $hf(x)$ is represented by (4), and the fuzzy variables shown in Table 1 are provided. The fuzzy variables "*large*", "*small*", etc. are relative magnitude descriptions. The scope indicated by such a description changes if the scope of the variable changes. Conversely, these descriptions can be generalized by normalizing the range of scope. The variables so obtained are called fuzzy variables. Various inputs can be accommodated only if a number of basic variables are provided.

$$h_f(x) = \exp\left\{-\frac{(x-\mu_f)^2}{2\sigma_f^2}\right\} x, \mu_f \in I \tag{4}$$

The proposition Q regarding the amount of compensation, DDOS, is represented by the expression (5) after the style of handling for (1).

$$Q; \text{DDOS} = f \in F \tag{5}$$

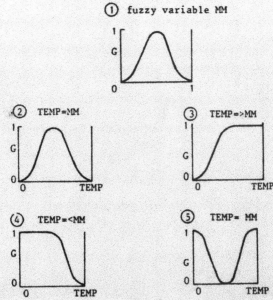

Fig. 3. Examples of membership function of modified fuzzy variable.

Table 1. Fuzzy variables

Variables	Support Set	μ	σ	Meaning
SS	0, 1	0.0	0.15	Small Small
SA	0, 1	0.0	0.3	Small
SM	0, 1	0.3	0.15	Small Medium
MM	0, 1	0.5	0.15	Medium
ML	0, 1	0.7	0.15	Medium Large
LA	0, 1	1.0	0.3	Large
LL	0, 1	1.0	0.15	Large Large
NB	-1, 1	-1.0	0.4	Negative Big
NM	-1, 1	-0.5	0.2	Negative Medium
NS	-1, 1	-0.2	0.2	Negative Small
ZE	-1, 1	0.0	0.2	Zero
PS	-1, 1	0.2	0.2	Positive Small
PM	-1, 1	0.5	0.2	Positive Medium
PB	-1, 1	1.0	0.4	Positive Big

The fuzzy variable used in this expression is one with the interval [-1, 1] on Table 1.

The fuzzy implications on the amount of compensation are given by (6) by use of P represented by (1) and Q represented by (5). The number of inputs is n, and the number of implications m.

$$\begin{matrix} \text{if } P_{11}, & \cdots & \cdots & \cdots & P_{1n} & \text{then} & Q_1 \\ \vdots & \ddots & & & \vdots & & \vdots \\ \vdots & & P_{ij} & & \vdots & & \vdots \\ \vdots & & & \ddots & \vdots & & \vdots \\ \text{if } P_{11}, & \cdots & \cdots & \cdots & P_{mn} & \text{then} & Q_m \end{matrix} \qquad (6)$$

2.2 Fuzzy Reasoning

The amount of compensation is inferred by the fuzzy implications on the amount of compensation given by (6). The propositions P are in AND relation with one another, and the implications are in OR relation with one another. The fuzzy set DDOS (amount of compensation) is calculated by using (7). This expression simplifies the composition rule of inference. The amount of calculations is small, and the relation between the grade p_{ij} of implication and the output $p_{ij}*Q$ corresponding to it can be easily grasped. p_{ij} represents the grade of the proposition p_{ij}, c union of fuzzy subsets, and * multiplication by a constant of the grades of proposition Q. Q_0 indicates that the amount of compensation is approximately "zero". That is, DDOS = ZE. The amount of compensation is zero in a state undefined by the fuzzy implication. This means that the value of the feed rate equation is to be used.

$$\widetilde{\text{DDOS}} = U\{\text{Min } p_{ij}*Q_i, p0*Q_0\}$$
$$\quad 1 \leq i \leq m \quad 1 \leq j \leq n$$
$$p0 = 1 - \text{Max}\{\text{Min } p_{ij}\}$$
$$\quad 1 \leq i \leq m \quad 1 \leq j \leq n \qquad (7)$$

Examples are given in Fig. 4. Control rules expressed by fuzzy implications are indicated on the left side.
Their meanings are explained below:
(1) If the rise of the turbidity of the raw water is medium, set the amount of compensation at some positive level.
(2) If the turbidity of the raw water is medium and the water temperature is not low, set the amount of compensation at a medium negative level.

If the measurement and the keyboard entry are at ▲ in the figure, the grade of the proposition TUUP = MM holding true (grade) is 0.75 under the rule (1), and the solid line ② in Fig. 4 is obtained. Under the rule (2), the grades of the propositions are calculated, and the solid line ⑤ in Fig. 4 is obtained. If there are two or more propositions the lowest grade is selected. The functions are combined. The largest function is represented by the solid line ⑥ in Fig. 4. It is a membership function representing a fuzzy set of feed rate compensations. It is abbreviated to h_D. The amount of compensation is derived from h_D by using the expression (8). "d" is a variable within the amount of compensation range. That is, the value which divides the area under the membership function represented by ⑥ in Fig. 4 is defined as the amount of compensation. This is because consideration of only the rule having the highest grade is inadequate. The amounts of compensation of the rules have to be reflected according to the grade.

$$DDOS = \int_d \frac{\Sigma_d \, d.h_D}{\Sigma_d \, h_D} \tag{8}$$

In the example given in Fig. 4, an extra amount of 1 mg/l is provided.

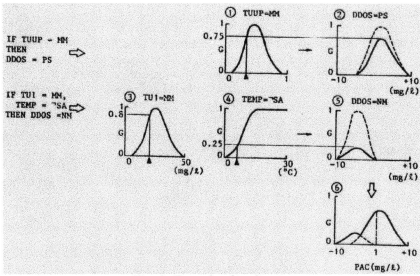

Fig. 4. Example of implication

Composition of Fuzzy Controller

The features of a coagulant addition fuzzy controller include its ability to recognize a process measurement and the operator's recognition of a state such as "large" or "small" flocs, as input and make a control calculation. Such recognition of a state is based on the operator's long experience, and mere quantification of such recognition may cause loss of an important factor. Hence such recognition by the operator has to be treated as an approximate quantity ("large", "small", etc.), i.e., a fuzzy variable. Moreover, the empirical rule should preferably be related directly to definitions of fuzzy variables, selection of operation instruction variables (turbidity, alkalinity, etc.) and the composition of the implications. It is desirable to make an addition, or change or effect deletion according to the situation so as to facilitate renovation of a fuzzy reasoning. The hardware taken into account is shown in Fig. 5, and the software so designed in Fig. 6.

Application of Fuzzy Reasoning to the Water Purification Process 97

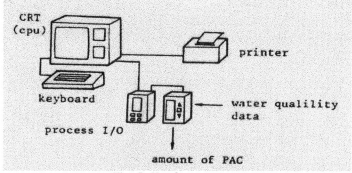

Fig. 5. Hardware system.

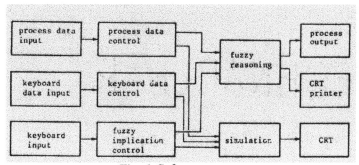

Fig. 6. Software system.

The functions of the components are described below:

CPU   : Calculation of fuzzy logic, simulation.
CRT   : Displaying of a feed rate, the state of reasoning by control rule, the states of input variables and the result of simulation.
Keyboard   : Input of the operator's recognition of a state, fuzzy variables and a change of the implication.
Process I/O : Input of process measurements and output of the reasoning, i.e., the feed rate.
Printer   : Output of input data, the state of reasoning by control rule and output of reasoning.

Among these components, the CRT plays an especially important role as a man-to-machine interface. Examples of presentations on the display are given in Fig. 7.

A presentation on the display of a state is shown in ① in Fig. 7. The estimate and the action under the major control rule are displayed. The control rule of the highest grade, that is, the effective rule at that point, is selected and displayed. The grades of the individual proposition of the rule and the grade of that rule are graphically presented in such a way as to permit ready matching with the fuzzy set of the amounts of compensation for reasonings.

The result of a simulation to find out the estimate after a change of the implication rule is shown in ② of Fig. 7. The operator looks at this and decides whether to change the rule, according to the feed rate after that change. The rule can be changed by answering the question "FIX?" with YES.

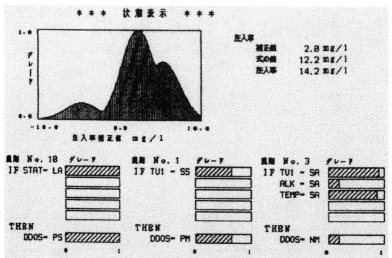

Fig. 7. Example of CRT output. ① *Result of reasoning.*

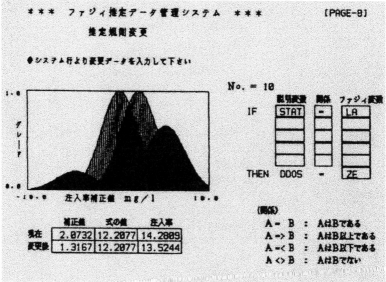

Fig. 7. Example of CRT output. ② *Simulation and modification of reasoning.*

The section ③ of Fig. 7 is an area where keyboard entries, that is, the operator's recognitions of states as "*small*", "*large*", etc., are displayed. In this example, the rise of temperature is small, the flocs are somewhat larger (good) than the average, and the start-up speciality is large.

Fig. 7. Example of CRT output. ③ *Keyboard input data.*

An addition to and an alternation of fuzzy variables are made in ④ of Fig. 7. General fuzzy variables are listed in Table 1. An addition and an alternation can be made to adapt them to the peculiarities of a place and an individual's peculiar criterion of judgment. A fuzzy set is graphically presented in accordance with given parameters on the left side so that the meanings of those parameters may be readily understood.

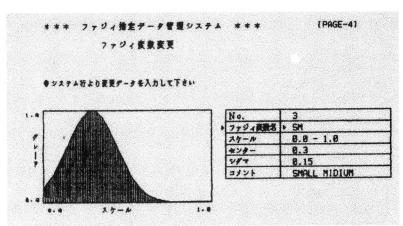

Fig. 7. Example of CRT output. ④ *Fuzzy variable modification.*

In addition to the presentations shown here, there are a total of 11 varieties including presentations of changes in input variables and the feed rate equation and lists of implications, input variables and fuzzy variables.

3. Field Test

A field test was made at the Toyoiwa Water Purification Plant in Akita City in October and November in 1983. The test procedure, method and results will be described. PAC was employed at this purification plant as a coagulant. The plant was in intermittent operation in the daytime only because of scanty raw water supply.

First, the test procedure will be explained. The procedure can be divided into three major parts:

(1) Personal inquiry into empirical rules and drafting of implication.
(2) Verification of implication with actual data and its expansion.
(3) Field test.

(1) The results of the personal inquiry will be summarized here. In the directions given below, "addition" means the setting of the feed rate at a level higher by 3 to 4 mg/l, and "subtraction" the setting of the feed rate at a level lower by 3 to 4 mg/l.
- Addition if the turbidity of raw water is rather low.
- Subtraction if the alkalinity is low.
- Addition if the supernatant liquid of the sedimentation basin has a high turbidity.
- Addition if the turbidity of raw water rises (or such rise is anticipated).
- Addition if the flocs are small.
- Slight addition at the time the operation (intermittent) is started.

In Table 3, initial implications are made out of the results of the personal inquiry by using output variable symbols in Table 2 and fuzzy variables in Table 1.

(2) The field test in October was made under the implications in Table 3. The implications were then expanded by use of the operation data for this period and the data for July and August. At that time, a study was made of various factors including the turbidity of treatment water and the water temperature which were generally believed to affect the feed rate although no such effect had been noted through the operator's personal (firsthand) inquiry. The final implications after the study are shown in Table 4. Fuzzy implications for a case where the raw water has medium turbidity have been added to Table 3. The fuzzy implications for a case where the alkalinity is low were divided into two fuzzy subspaces, and those for a case where the turbidity rises were divided into three fuzzy subspaces.

A mathematical model for the feed rate was worked out by statistical analysis using the data for July, August and October. The equation is given below.

$$\text{BASE} = 18.572 + 2.375\sqrt{\text{TUI}} - 0.93\text{ALK} \tag{9}$$

Application of Fuzzy Reasoning to the Water Purification Process 101

where, BASE is the reference feed rate (mg/l), TUI the turbidity of the raw water (mg/l), and alkalinity.

Table 2. Variables

	Variables	Symbol	Span	Dimension
	Turbidity of raw water	TUI	0, 50	mg/l
	Alkalinity	ALK	8, 18	"
Input data	water temperature	TEMP	0, 30	°c
	Turbidity of treated water	TUSE	0, 3	mg/l
	Turbidity increase	TUUP	0, 1	-
	floc formation	FLOC	0, 1	-
	Operation start	STAT	0, 1	-
Output data	Amount of PAC modification	DDOS	-10, 10	mg/l

Table 3. Implication rule (primary)

1	IF TUI = SS	THEN DDOS = PM
2	IF ALK = SA	THEN DDOS = NM
3	IF TUSE = LA	THEN DDOS = PM
4	IF TUUP = LA	THEN DDOS = PM
5	IF FLOC = SA	THEN DDOS = PM
6	IF STAT = LA	THEN DDOS = PM

Table 4. Implication rule (final)

No.		
1	IF TUI = SS	THEN DDOS = PM
2	IF TUI = MM TUSE =¬ LA TEMP =¬ SA	THEN DDOS = NM
3	IF TUI = SA ALK = SA TEMP = SA	THEN DDOS = NM
4	IF TUI = LA ALK = SA	THEN DDOS = NM
5	IF TUSE = LA	THEN DDOS = NM
6	IF TUUP = LL	THEN DDOS = PB
7	IF TUUP = ML	THEN DDOS = PM
8	IF TUUP = MM	THEN DDOS = PS
9	IF FLOC = SA	THEN DDOS = PM
10	IF STAT = LA	THEN DDOS = PS

(3) The test for November was carried out with the implications (3) and (2) and the feed rate equation stored in the controller. Details of the installation of the controller are given in Fig. 8.

Fig. 8. Outside view of fuzzy controller installed in the control center of the water purification plant.

Next the test method will be described. The test was made by comparison with the value preset by the operator. At a point in the test period, the operator preset feed rate by a jar test or by using his judgment based on experience. It was decided that the validity of fuzzy reasoning could be established by demonstrating conformance of the output of fuzzy reasoning to the presetting by a skilled operator and improvement beyond the result obtained with the regression model. The fuzzy reasoning included information regarding the state of flocs and other states unfit for automatic measurement and is expected to provide a better result than the regression mode. By demonstrating that the output of fuzzy reasoning is closer to the result obtained by a skilled operator than to the result obtained with a regression model, it can be proven that the information on non-automatic measurements which is used for fuzzy reasoning plays an important role in fixing feed rate.

Moreover, it can be demonstrated that the experience and knowledge of a skilled operator can be fully utilized, which is the objective of the fuzzy reasoning in this paper, and that by so utilizing a skilled operator's experience and knowledge, a system which can be run by even an unskilled operator with the cleverness of a skilled operator can be realized. Non-automatic measurement information is entered from a keyboard, but during the test, input data were entered when the preset feed rate was changed. This was due to certain judgment of the situation which had led to change of the presetting, and the above set-up was designed to reflect that judgment. There were three operators, and it was discovered as a result of the inquiry that a slight difference in judgment existed among these operators. However, the test period was relatively short, and consequently, such difference for operator to operator could not be verified on the basis of the test results.

Now the test results will be analyzed. As already mentioned, the fuzzy implications established on the basis of the data for July, August and October were applied to the data for November. The effect was verified by comparing the estimate with the number of data collected in November which was 122. The value of the feed rate equation was obtained for comparison. Part of the test results are shown in Fig. 9. It will be seen that the fuzzy reasoning affected close follow-up even in that area which could not be matched with the operator's presetting under the method using a regression model. This can also be demonstrated with the sum of squares of the residuals in Table 5 or the standard deviation of the residual. The residual here refers to the difference between the actual feed rate and the value obtained by fuzzy reasoning and the difference between the actual feed rate and the value of the feed rate equation.

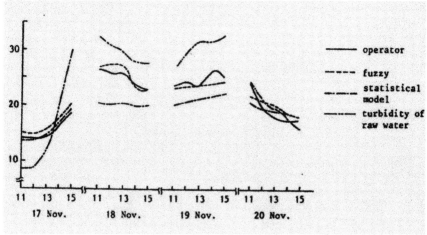

Fig. 9. Example of fieldtest.

Table 5. Example of result of fieldtest

	Statistical model	Fuzzy
Residual square sum	628	224
Standard deviation of residual	2.28	1.36
Number of plus Y	33	41
Number of minus Y	41	33

Furthermore, the criterion for judgment given by the expression (10) should be considered in order to analyze coagulant addition and its effect. The residual is the value defined already. TUSE is the deviation of the turbidity of the effluent of the sedimentation basin (TUSE) from the average turbidity. To allow for the detention time in the sedimentation basin, ᵃTUSE data after time lag of two hours should be used instead of the feed rate.

$$Y = \text{Residual} \times \Delta \text{TUSE} \qquad (10)$$

If Y is positive,
    (1) the residual and ΔTUSE are both positive, or
    (2) the residual and ΔTUSE are both negative.

If ΔTUSE is positive, the feed rate has to be higher than the actual feed rate. If the residual is positive, this value should be a lowered one. If ΔTUSE is negative, the feed rate should be lowered. In either case, the residual should be a narrowed one. Conversely, if Y is negative, the residual is widened. That is, the greater the number of positive symbols, the better the feed rate. As can be seen from this, fuzzy reasoning is superior in this respect also to the method using a regressions model only.

Conclusion

This paper deals with the application of a fuzzy theory to coagulant addition control for a water purification plant. The paper analyzes feed control of a coagulant which is used in larger quantity than any other chemical, and reacts in a complicated process and for which a model cannot be easily worked out. The fixing of a feed rate for the coagulant depends heavily on the operator's judgment based on his experience and knowledge. This judgement is expressed by using fuzzy reasoning so that a feed rate may be fixed by even an unskilled operator. The feed rate was estimated by use of a regression model under the conventional method. The compensation required was calculated and actually made. The necessary amount of compensation was found out by fuzzy reasoning by use of the operator's recognition of a state.

A coagulant addition fuzzy controller incorporating this fuzzy reasoning method was fabricated, and a field test was made at the water purification plant. As a result of this test, it was demonstrated that the output of the fuzzy reasoning is matched with a skilled operator's presetting and is better than the result obtained with a regression model.

A task to be accomplished in the future is addition of a learning system by which the value of feed rate of a chemical agent can be evaluated and the fuzzy implication can be improved as required. Also generalization of rules and development of a general-purpose controller will be in increasing demand in the future.

Before concluding the paper, we wish to express our sincere gratitude to the staff of the Toyoiwa Water Purification Plant in Akita City in Japan for their kind cooperation in the field test.

References

[1]. Michio Sugeno: "Application of Fuzzy Set and Logic To Control", *"Measurement and Control"*, Vol. 18, No. 2, pp. 150-160 (1975).
[2]. Yagishita and Tanaka : "Estimation of Chemicals Feed Rate By Fuzzy Reasoning", *34th National Water Service Research Presentation and Lecture Meeting*, pp. 365-367 (1983).
[3]. R. M. Tong et al: Fuzzy Control of the Activated Sludge Water Treatment Process; *Automatics*, Vol. 16, pp. 695-701 (1980).
[4]. L. P. Holmblod and J. J. Ostergaard: Control of a Cement Kiln by Fuzzy Logic, in *Fuzzy Information and Decision Processors* (M. M. Gupta and E. Sanchez, Eds.), North-Holland, pp. 389-399 (1982).

A Microprocessor Based Fuzzy Controller For Industrial Applications[*]

Tsukasa Yamazaki
Systems and Control Design Department
Systems Engineering Division
JGC Corporation
1-14-1, Bessho, Minami-ku
Yokohama 232, JAPAN

Michio Sugeno
Department of Systems Science
Tokyo Institute of Technology
4259, Nagatsuta, Midori-ku
Yokohama 227, JAPAN

1. Introduction

Various fuzzy control algorithms have been proposed since Mamdani and Assilian [1] first applied fuzzy logic to the control of a pilot steam engine. Most of them are specialized to the control of specific processes and implemented by a rather versatile computer with adequate peripherals.

Because of a growing interest in fuzzy control among the process industries, it was thought that the construction of a microprocessor based fuzzy controller for general purposes would be a valuable step to explore the potential demands for fuzzy control. Designing a general purpose algorithm and implementing it with a microprocessor necessitate careful consideration of a new angle due to the required flexibility of control performance and the constraints from its hardware. As a first step, a prototype controller was developed for single input-single output control. The paper discusses some important results obtained from this study.

2. Algorithm

2.1. *Inference method*

The basic fuzzy control algorithm adopted, including the inference method, is based on the control part of SOC (Self-Organizing fuzzy Controller) investigated by Yamazaki [2]. The i-th fuzzy control rule is expressed with three fuzzy variables E (Error), CE (Change in Error), and CO (Change in Output) in the following manner.

If E is E_i and CE and CE_i then CO is CO_i (1)

where, E_i, CE_i and CO_i belong to one of fuzzy labels of E, CE and CO, respectively. They are normalized in common over the same discrete support set as shown in Fig. 1.

	Universes Of Discourse E, CE, CO												
	-6	-5	-4	-3	-2	-1	0	1	2	3	4	5	6
NB	10	7	3	0	0	0	0	0	0	0	0	0	0
NM	3	7	10	7	3	0	0	0	0	0	0	0	0
NS	0	0	3	7	10	7	3	0	0	0	0	0	0
ZO	0	0	0	0	3	7	10	7	3	0	0	0	0
PS	0	0	0	0	0	0	3	7	10	7	3	0	0
PM	0	0	0	0	0	0	0	0	3	7	10	7	3
PB	0	0	0	0	0	0	0	0	0	0	3	7	10

NB : Negative Big
NM : Negative Medium
NS : Negative Small
ZO : Zero
PB : Positive Big
PM : Positive Medium
PS : Positive Small

Fig. 1. Fuzzy labels.

Therefore parameters GE, GCE and GCO are introduced to map measured variables onto the values inside the algorithm, i.e., the support sets. This is a widely used technique and is very convenient for designing a general purpose controller because various processes can be controlled by just adjusting the parameters.

The fuzzy relation R_i of the conditional statement (1) is defined as the Cartesian product by multiplication instead of min operation, and the connective AND is also treated as the Cartesian product, which is interpreted as an interactive conjunction. The mathematical expression of the above is given below.

$$R_i \triangleq (E_i(e) \cdot CE_i(ce)) \cdot CO_i(co) \qquad (2)$$

$$e \in E, \quad ce \in CE, \quad co \in CO$$

The compositional rule of inference infers a control action $CO^0{}_i$ from R_i when a control situation (E^0, CE^0) is given. With a simplification the fuzzy subsets of E^0 and CE^0 are treated as crisp sets, that is, E^0 and CE^0 have the membership value of 1.0 only at e^0 and ce^0, respectively, and 0.0 elsewhere. CO^0 inferred from all rules is given as follows.

$$CO^\circ(co) = \max_i \{E_i(e^\circ) \, CE_i(ce^\circ) \, CO_i(co)\} \qquad (3)$$

The use of the center of gravity procedure, as the interpretation of a fuzzy subset CO^0, gives the following in the membership from:

$$CO^\circ = \sum_{j=1}^{13} co_j \cdot \mu CO^\circ(co_j) / \sum_{j=1}^{13} \mu CO^\circ(co_j) \qquad (4)$$

The inference method mentioned above is quite simple but the usefulness of it has been shown in a simulation study [3].

2.2. Nest arrangement of control rules

For a general use, especially industrial, it is vital that a controller can produce non-abruptly rigorous or fine control actions corresponding to a control situation given. Though it is possible to meet the above requirement by using a large support set for CO, this leads to a huge increase of memory space or calculation time, or even both, whichever the support set is discrete or continuous. This is not a desirable situation.

For this reason, the control algorithm is arranged to have a "nested control window", in which the same set of control rules is used recursively and a gain for CO is changed according to the degree of the nest. By adopting the control rule table representation, the procedure of nesting a set of control rules is illustrated in Fig. 2. When the control situation goes into the region specified on the table, for example, $E = ZO$ and $CE = ZO$, the same set of control rules being used is mapped to this region together with the decrease of the controller gain GCO. This means that the control situation corresponding to the region is subdivided further and hence finer control output can be produced. Conversely, when the control situation goes out of the table, the set of control rules is mapped inversely accompanying the increase of the controller gain GCO accordingly. Thus, both the fine and rigorous control actions are achieved without the abovementioned problems.

The idea of "nested control window" is a natural extension of the human operators behavior because an operator first tries to bring the process value within a small deviation band around the setpoint, and then starts near the setpoint when a large deviation of the process value occurs. In other words, operator's consciously or unconsciously try first to bring the process value to a roughly desirable situation and then to a precisely desirable one.

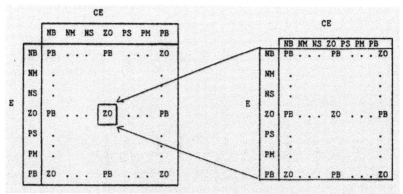

Fig. 2. Scheme of nesting control rules.

2.3. Selection of operational control rules

With a view to easy operation and saving of memory space, the controller is designed so that appropriate control rules, as an operational set of rules, can be chosen from prospective control rules stored in a nonvolatile memory (PROM). The rule selection is specified by the rule numbers, which are assigned to all prospective rules stored in the PROM. This is a practical method because the controller, like an ordinary industrial controller, has limited man-machine interface devices such as a digital display and a keyboard. It should be noted that the prospective control rules stored in the PROM were determined after a number of control simulations on various processes.

3. Hardware and Operation

The hardware of the controller comprises a microcomputer board, an A/D and D/A converters board and an operational interface board as shown in Fig. 3.

The microcomputer board consists of a Z-80 microprocessor, two PROMs (2k byte), a RAM (1k byte) and auxiliary ICs. Both the A/D and D/A converters have a resolution of 8 bits. The operational interface board is mounted with two digital display sections, which indicate rule numbers and the other data separately, and with numerical and functional keys. Fig. 4 depicts the arrangements of such components on the board.

The controller needs several items to be input: controller parameters GE, GCE and CO, operating parameters SP (set point) and TS (sampling period), and operational control rules. In the sequel, the operation of the controller is described by referring to Fig. 4.

3.1. Setting and read-out of rules

The selection of operational control rules is initiated by the RIN key. Control rules can be stored in the RAM by repeating the procedure of pressing the ENT key after entering the rule number to be used until all the necessary rules are stored. The read-out of rule numbers stored in the RAM is started by the ROUT key, and then every time the ENT key is pressed, a rule number is shown sequentially on the rule number display section until all the stored rules are shown. The deletion of operational rules can be performed by entering the RDEL key, the rule number to be deleted and the ENT key in this sequence. All the operation for rule manipulation can be terminated by the END key.

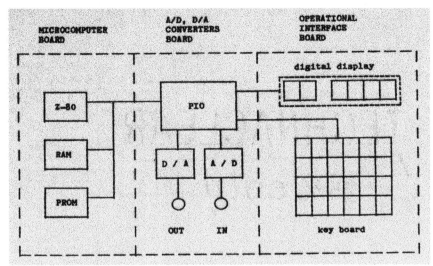

Fig. 3. Hardware configuration of prototype fuzzy controller.

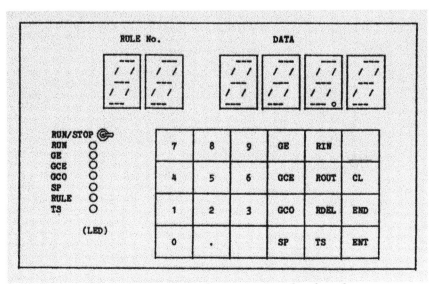

Fig. 4. Arrangement of operating interface board.

3.2. Setting and read-out of parameters

The operation for GE parameter is initiated by pressing the GE key, by which the value of GE stored in the RAM is shown on the data display section. When setting of GE is required, the CL key needs be pressed to clear the display instead of the END key. A value of GE can be stored in the RAM by the ENT key after entering the desired value with the numerical keys. The operation for other parameters GCE, GCO, TS and SP are performed likewise.

3.3. Indicating function

Apart from the indication on the digital displays which is associated with the key operations mentioned above, the controller has two indication functions. One is the indication of a process value being controlled when the controller is at the RUN mode which is selected by a RUN / STOP switch. The other is the indication of the control rule in use, which is one of essential functions for a rule-based controller. On the rule number display section, it indicates the rule number which contributes most to the controller output inferred every sampling period. This function is a useful remedy for improving the operational control rule and making its procedure efficient.

4. Performance of the Controller

A full investigation of the performance of the controller has not been done yet due to the shortage of time. The results obtained so far are described briefly here. To examine to what extent the controller can control difficult processes, an experiment was conducted by connecting it with an Apple-II microcomputer, in which a process model was programmed, as shown in Fig. 5. The operational set of control rules is illustrated as a control rule table in Fig. 6. The transfer function of the process is given by $G(s) = e^{-Ls}/(1+Ts)$. Varying L and T, the performance of the controller was evaluated.

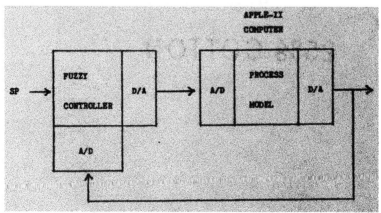

Fig. 5. Control flow of the experiment.

		CONTROL DECISION TABLE CHANGE IN ERROR (CE)						
		NB	NS	NS	ZO	PS	PM	PB
	NB	(- -) **	(2) PB	(- -) **	(8) PM	(10) PM	(13) PS	(- -) **
E	NM	(16) PB	(- -) **	(19) PM	(- -) **	(22) PS	(- -) **	(25) NS
R	NS	(- -) **	(27) PM	(28) PS	(29) PS	(30) ZO	(13) NS	(34) NM
R	ZO	(35) PB	(- -) **	(37) PS	(38) ZO	(39) NS	(- -) **	(41) NB
O	PS	(42) PM	(44) PS	(46) ZO	(47) NS	(48) NS	(49) NM	(- -) **
R	PM	(51) PS	(- -) **	(53) NS	(- -) **	(57) NM	(- -) **	(60) NB
(E)	PB	(- -) **	(63) NS	(66) NM	(68) NM	(- -) **	(74) NB	(- -) **

Fig. 6. Operational set of control rules used in the experiment.

Fig. 7 shows one of the control responses which was obtained by a step change of the setpoint from 0 to 50% and then to 40%.

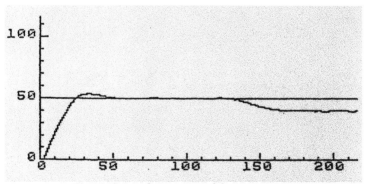

Fig. 7. Control response of the experiment.

In view of the results, the performance of the controller is in general satisfactory as far as a ratio of L/T is less than 0.4. This fact indicates that this prototype controller compares to an optimally tuned PID controller in terms of ability to control difficult processes if one takes the ratio (L/T) as its index. This limitation is quite conceivable since the control rule form (1) does not include the element of time.

A human operator sometimes uses novel tactics, such as "Wait and See". Therefore, it is our belief that the function of variable interval control action should overcome the limitation found above. During the course of the experiment, two hardware problems were detected, though these are minor as can be seen from Fig. 7. One comes from the low resolution of A/D and D/A converters, and the other is noise into the analog circuit of the controller.

However, these problems are not particular to this controller, and hence this point would not be discussed further just by stating that the improvement is under way.

5. Conclusion

The objective of this study is not to investigate a new fuzzy control algorithm but to demonstrate the possibility of constructing a microprocessor based fuzzy controller for industrial use. The study showed that the prototype fuzzy controller with the simple algorithm was quite capable, and that there were not many difficulties in constructing a microprocessor based fuzzy controller if careful considerations were given to the man-machine interface. It is hoped that the study reported here will encourage industrial people to explore the potential of fuzzy control.

6. Acknowledgment

The authors are grateful to JGC corporation for the financial support which made this study possible. Many thanks are also given to Mr. Kawamoto and Mr. Tominaga for their assistance.

References

[1] Mamdani, E. H. and Assilian, S.: An Experiment in Linguistic Synthesis with a Fuzzy Logic Controller, *Int. J. of Man-Machine Studies*, 7, 1-13, 1974
[2] Yamazaki, T.: An Improved Algorithm for a Self-Organising Controller, and Its Experimental Analysis, *Ph.D Thesis*, London University, 1982
[3] Yamazaki, T. and Mamdani, E. H.: On the Performance of a Rule-Based SelfOrganising Controller; *Proc. of IEEE Conf. on Applications of Adaptive and Multivariable Control*, Hull, U.K. 50-55, 1982

Fuzzy Modeling and Control of Multilayer Incinerator[*]

M. Sugeno and G.T. Kang
Department of Systems Science,
Tokyo Institute of Technology,
4259 Nagatsuta, Midori-ku,
Yokohama 227, Japan

Received January 1985
Revised March 1985

Abstract

A method to design a fuzzy controller based on a fuzzy model is suggested. The identification method of a fuzzy model is also discussed. The suggested method is applied for the modeling and control of a multilayer incinerator. The performance of the designed fuzzy controller is tested by computer simulation using the identified process model. The results are satisfactory.

Keywords: fuzzy model, identification, fuzzy control, model-based controller design, multilayer incinerator.

1. Introduction

Since Mamdani's pioneering work [2] on fuzzy control motivated by Zadeh's approach to inexact reasoning [7], there have been reported more than 100 works in this research field so far. Recently we have had some industrial applications of fuzzy control; control of cement kiln [3] (in Denmark, 1981), control of a water purification process [4] (in Japan, 1983), and automatic train operation [6] (in Japan, 1984).

---

[*] © 1986 North-Holland, Retyped with written permission from
Fuzzy Sets and Systems, **18** *(1986) 329-346.*

As far as fuzzy control is concerned, our main interest is now towards its applications to industrial processes. There are, however, still some problems to be solved. One of the most important problems is a method to design a fuzzy controller, i.e., to find fuzzy control rules. There are three applicable methods in a practical situation:
- (1) based on human operator's experience and/or control engineer's knowledge.
- (2) based on fuzzy modeling of human operator's control actions.
- (3) based on the fuzzy model of a process.

The first method is nothing but one used in an expert system. Most fuzzy controllers have been designed by this method so far. This method is effective in cases where human experience and knowledge play an important role. However, the situations in control are often different from, for example, those in medical diagnosis. There are many situations in process control where we need skill rather than knowledge. Further it is often the case that an operator cannot explain linguistically what control actions he takes.

Then the second method becomes very important. For example, suppose that we drive a car. We know the driving techniques by our arms and legs rather than by our brains. We can realize this sort of technique by a computer if we build a fuzzy model of human control actions by using this input-output data. By a fuzzy model what we mean here is the description of an input-output relation using fuzzy implications which are of the same type as fuzzy control rules.

However, the second method is still of the same kind as the first one. We can apply the first and second methods only in the case where we can believe that an operator performs an excellent control. It should be noted that we can only do, by building a model of an operator, at most the same control as he does, but never more.

When a process is complex, there may exist no expert who can well control it. In this situation we cannot use operator's experience and/or knowledge. Instead of those, what is desired is to build a mathematical model of a process which enables us to derive control rules theoretically. It is expected that we can design a fuzzy controller based on a fuzzy model of a process as we do in ordinary control theory. Here we have to develop two methods: one is a method of fuzzy identification and the other is a method of the design of a fuzzy controller based on a fuzzy model.

This paper uses the method of fuzzy systems identification developed by Takagi and Sugeno [5], and suggests a method for the model-based design of a fuzzy controller. These methods are applied for the control of an incinerator.

2. Fuzzy modeling

2.1. *Fuzzy process laws*

A fuzzy model is one that expresses a complex system in the form of fuzzy implications. In fuzzy modeling of a process, a fuzzy implication is particularly called a fuzzy process law in contrast to a fuzzy control rule. In general a fuzzy model is built by using the physical properties of a system, the observation data, the empirical knowledge and so on.

Here we use the fuzzy reasoning method suggested by Takagi and Sugeno [4] where a fuzzy implication is of the following form:

$$L^i : \text{if } x_1 \text{ is } A_1^i, x_2 \text{ is } A_2^i, \cdots, x_m \text{ is } A_m^i,$$
$$\text{then } y^i = c_0^i + c_1^i x_1 + c_2^i x_2 + \cdots + c_m^i x_m \quad (1)$$

where L^i is the i-th process law, A_j^i is a fuzzy set, x_j is an input variable and y^i is the output from the i-th process law. The above process law describes a relation between multi-inputs and a single output. The superscript i of y^i does not mean the number of a process output but means the number of a process law. In the sequel, the membership function of a fuzzy set A will be written as $A(x)$, which is formed by straight lines. The fuzzy implication in (1) will be simply written as follows:

$$L^i : \text{if } x = A^i \text{ then } y^i = (C^i)^T x^* \quad (1a)$$

where T stands for the transpose of a vector and

$$x = (x_1, x_2, \ldots, x_m), \quad x^* = (1, x_1, x_2, \ldots, x_m),$$
$$A^i = (A_1^i, A_2^i, \ldots, A_m^i), \quad C^i = (c_0^i, c_1^i, \ldots, c_m^i). \quad (2)$$

Given an input $x^0 = (x_1^0, x_2^0, \ldots, x_m^0)$, the output y^0 is inferred by taking the weighted average of y^i's.

$$y^0 = \frac{\sum w^i y^i}{\sum w^i} \quad (3)$$

where a weight w^i implies the truth value of the premise of the i-th process law calculated as

$$w^i = \prod_{j=1}^{m} A_j^i(x_j^0). \quad (4)$$

Here we show an example to see the performance of the fuzzy reasoning as well as fuzzy implications. Suppose we have four fuzzy implications as are shown in Figure 1. Figure 2 shows fuzzy subspaces specified in the premises of the implications, and Figure 3 shows the input-output relation expressed by those fuzzy implications. We can see that they express a highly nonlinear functional relation.

2.2. Process identification

Let us discuss a way to identify a fuzzy model composed of fuzzy process laws. The process identification using input-output data consists of two parts: structure identification and parameter identification. By structure identification what we first mean is to find the input variables which affect the outputs. There is no general way to solve this problem. We can use some methods in ordinary systems theory. For example the following criteria of finding variables are considered:

  (1) to use the physical properties of a system and the empirical knowledge,

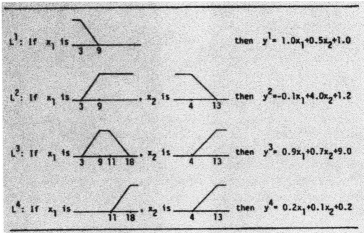

Fig. 1. *Fuzzy implications.*

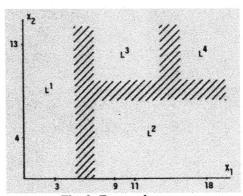

Fig. 2. *Fuzzy subspaces.*

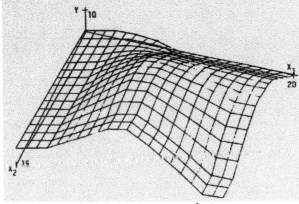

Fig. 3. *Input-output relation.*

(2) to choose a variable with a big partial correlation coefficient,
(3) to choose a variable with a small PSS (Prediction Sum of Square) value.

There is another kind of structure identification problem which appears particularly in fuzzy modeling. We have to find which variables are necessary in the premise of an implication; there do not always appear all the input variables in a premise. We also have to find the fuzzy partition of the inputs space, as is seen in Figure 2, which is directly related to the number of necessary implications

For example suppose that the system with three input variables and one output is expressed by two implications where the first implication is:

if x_1 is *small*, x_2 is *small* then y = 2.0 + 1.0x_1 + 3.0x_2 + 0.5x_3,
and the second one is:

if x_1 is *big* then y = 3.0 + 0.8x_1 + 2.0x_2 + 1.0x_3,

We see that the variable x_3 does not appear in the premises. There is only one input variable x_1 in the premise of the second implication. It is of course possible to put the term 'x_2 is *any*' as a dummy in the premise of the second implication. But it does not help us since to identify the membership function of '*any*' is to the same thing to find that x_2 is not necessary. As is seen in the premises, the input space is partitioned into two fuzzy subspaces.

The next problem is the parameter identification: the parameters of the fuzzy variables in the premises and the parameters in the consequences, in this case, are the coefficients of linear equations. These parameters are identified so that the output error is minimized. We omit the details in this study (see [5]).

The algorithm of the identification used in this study is shown in Figure 4 where the parameters of the premises and the consequences are identified by using the observation data.

3. Model-based design of fuzzy controller

Since a fuzzy model is essentially a nonlinear model composed of fuzzy process laws, the model-based design of a controller seems to be very difficult. However we can use the fact that the consequence of a fuzzy process law is a linear equation. As it has been suggested [3], the basic idea of a model-based design is to set one control rule for one process law and then to combine control rules to form a controller. In this paper we suggest the following approach by taking notice of the linearity of the consequences.

First apart from the premises, we focus our attention on the consequences of process laws where linear system equations are written. We form some linear subsystems from the consequences as later explained in details and we design a controller for each linear subsystem. Next we make a fuzzy control rule by putting the original premise to the designed controller. Here the premise of a fuzzy control rule is that of a process law from which a linear subsystem is formed.

When the process has only one output, the consequence of each fuzzy process law becomes a linear subsystem, and its premise becomes that of a fuzzy control rule derived from this linear subsystem. So there are the same number of control rules as that of the process laws. On the other hand when the process has multi-outputs, the method to form subsystems is different and becomes complicated.

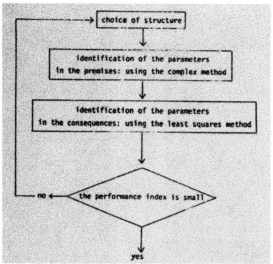

Fig. 4. Outline of identification.

Let us assume without loss of generality that the process has two inputs and two outputs, and also assume that there are two process laws for each output. For example we have:

$$
\begin{aligned}
L_1^1 &: \text{if } Z = A_1^1 \text{ then } y_1^1 = (C_1^1)^T Z^*, \\
L_1^2 &: \text{if } Z = A_1^2 \text{ then } y_1^2 = (C_1^2)^T Z^*, \\
L_2^1 &: \text{if } Z = A_2^1 \text{ then } y_2^1 = (C_2^1)^T Z^*, \\
L_2^2 &: \text{if } Z = A_2^2 \text{ then } y_2^2 = (C_2^2)^T Z^*,
\end{aligned}
\quad (5)
$$

where $Z = (x_1,...x_m, u_1, u_2)$, with x state variable and u input variable, and where

$$A_j^i = (A_j^{i1}, A_j^{i2}, ..., A_j^{im}, A_j^{i(m+1)}, A_j^{i(m+2)}).$$

Note that the i of L_j^i and y_j^i is the number of a process law and the j is the number of an output. By taking one of $L^1{}_1$, and $L^2{}_1$, and one of $L^1{}_2$, and $L^2{}_2$, we have four linear subsystems S_1 - S_4, with two outputs in this case. For example,

$$S_1 : \begin{aligned} y_1 &= (C_1^1)^T Z^* \\ y_2 &= (C_2^1)^T Z^* \end{aligned} \quad (6)$$

$$S_2 : \begin{aligned} y_1 &= (C_1^2)^T Z^* \\ y_2 &= (C_2^2)^T Z^* \end{aligned} \quad (7)$$

where S_1 comes from L^1_1, and L^1_2 and S_2 from L^1_1, and L^2_2. Now we put the following premise PS_1 to the subsystem S_1:

$$PS_1 : Z = A^1_1 \cap A^1_2,$$

where,

$$A^1_1 \cap A^1_2 = (A^{11}_1 \cap A^{11}_2, A^{12}_1 \cap A^{12}_2, \ldots A^{1(m+2)}_1 \cap A^{1(m+2)}_2), \tag{8}$$

and also PS_2 to S_2:

$$PS_2 : Z = A^1_1 \cap A^2_2,$$

It is easily seen that PS_1 is the '*and*' connection of the premises of L^1_1 and L^1_2. Next as for S_1 if all the fuzzy sets $A^{1j}_1 \cap A^{1j}_2$ ($j = 1, \ldots, m+2$) in PS_1 are normal, then we select S_1 as a subsystem. If not, we delete it. Note that $A^{1j}_1 \cap A^{1j}_2$ may be empty for some j. So in general the number of subsystems is smaller than that of all the combinations.

Let us further assume that the subsystems S_1 and S_2 are selected. We can then design two controllers for S_1 and S_2 according to linear control theory. For example as for $S1$ we may have the following controller C_1:

$$C_1 : \begin{aligned} u_1 &= f^1_1(x_1, \ldots, x_m, y_1, y_2), \\ u_2 &= f^1_2(x_1, \ldots, x_m, y_1, y_2), \end{aligned} \tag{9}$$

and C_2 as for S_2:

$$C_2 : \begin{aligned} u_1 &= f^2_1(x_1, \ldots, x_m, y_1, y_2), \\ u_2 &= f^2_2(x_1, \ldots, x_m, y_1, y_2), \end{aligned} \tag{10}$$

Then putting the premise PS_1 to C_1 and PS_2 to C_2, we obtain two control rules such that

$$R^1 : \text{if } Z = A^1_1 \cap A^1_2 \text{ then } u_1 = f^1_1(x_1, \ldots, x_m, y_1, y_2),$$
$$u_2 = f^1_2(x_1, \ldots, x_m, y_1, y_2), \tag{11}$$

$$R^2 : \text{if } Z = A^1_1 \cap A^2_2 \text{ then } u_1 = f^2_1(x_1, \ldots, x_m, y_1, y_2),$$
$$u_2 = f^2_2(x_1, \ldots, x_m, y_1, y_2). \tag{12}$$

As an illustrative example, let us consider a fuzzy model of a process as is shown in Figure 5 where a process has two inputs and two outputs. In this case we have 16 subsystems altogether. However, 8 subsystems are selected according to the criterion mentioned above. For example we have a subsystem from L^1_1 and L^1_2 such that:

$$y_1(t+1) = 1.2y_1(t) - 0.8y_1(t-1) + 0.2y_2(t) + 0.7u_1(t) + 0.2u_2(t-1),$$
$$y_2(t+1) = 0.13y_1(t-1) + 1.5y_2(t) - 0.8y_2(t-1) \tag{13}$$
$$- 0.1u_1(t) + 1.3u_2(t) - 0.5u_2(t-1)$$

The premise concerned with this subsystem is that

$y_1(t)$ is A_1, $\quad y_2(t-1)$ is $B_1 \cap B_3$ $\quad y_2(t)$ is C_1.

Given the set points r_1, and r_2 for y_1 and y_2, an optimal controller is designed under a quadratic performance index. Putting the premise to this controller, we obtain the following fuzzy control rule:

if $y_1(t)$ is A_1, $y_2(t-1)$ is $B_1 \cap B_3$, $y_2(t)$ is C_1
then $u_1(t) = 0.43e_1 - 0.59e_2 + 0.37e_3 - 0.89e_4$
$\qquad +0.12u_1(t-1) + 0.23u_2(t-1) + 0.4r_1 - 0.033r_2$ \hfill (14)
$u_2(t) = 0.41e_1 - 3.3e_2 - 1.7e_3 + 0.98e_4$
$\qquad +0.95u_1(t-1) + 1.1u_2(t-1) + 4.7r_1 - 2.44r_2$

where

r_1 = set point for y_1, r_2 = set point for y_2,
$e_1 = y_1(t-1) - r_1$, $e_2 = y_1(t) - r_1$, \hfill (15)
$e_3 = y_2(t-1) - r_2$, $e_4 = y_2(t) - r_2$.

Similarly we obtain seven other control rules. Figure 6 shows the results of the fuzzy control where the set points r_1 and r_2 for y_1 and y_2 are changed from $(r_1, r_2) = (3, 6)$ to $(r_1, r_2) = (8, 10)$ at $t = 15$. We can see that the designed fuzzy controller works well.

(a) fuzzy implications for output $y_1(t+1)$

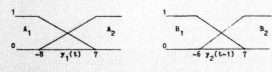

fuzzy variables in premises		parameters of linear equations in consequences				
$y_1(t)$	$y_2(t-1)$	$y_1(t)$	$y_1(t-1)$	$y_2(t)$	$u_1(t)$	$u_1(t-1)$
A_1	B_1	1.2	-0.8	0.2	0.7	0.2
A_1	B_2	0.7	-0.3	0.15	0.5	-0.4
A_2	B_1	0.9	-0.4	-0.1	0.8	-0.3
A_2	B_2	1.5	-0.9	0.17	0.6	-0.35

(b) fuzzy implications for output $y_2(t+1)$

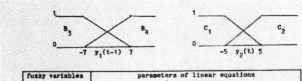

fuzzy variables in premises		parameters of linear equations in consequences					
$y_2(t-1)$	$y_2(t)$	$y_1(t-1)$	$y_2(t)$	$y_2(t-1)$	$u_1(t)$	$u_2(t)$	$u_2(t-1)$
B_3	C_1	0.13	1.5	-0.8	-0.1	1.3	-0.5
B_3	C_2	0.2	1.2	-0.6	0.12	1.5	-0.8
B_4	C_1	-0.1	0.9	-0.5	-0.15	0.9	-0.6
B_4	C_2	0.3	1.3	-0.9	-0.2	1.6	-0.8

Fig. 5. Fuzzy model.

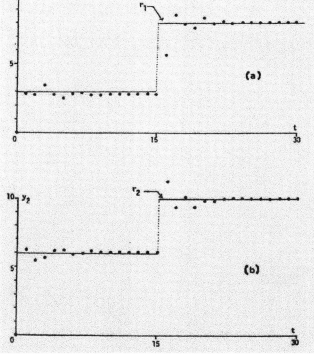

Fig. 6. Results of fuzzy control (a) output y_1, (b) output y_2.

4. Fuzzy controller for multilayer incinerator

4.1. *Incineration process*

The multilayer incinerator discussed in this study is a furnace for burning the sludge produced at a sewage treatment plant. The incinerator is of a cylindrical type, and with about 20 m in height and about 4 m in diameter as is shown in Figure 7. It consists of 14 layers. Before its treatment in the incinerator, the sludge is dried and cut into small pieces which are called sludge cakes. The sludge cakes are first put into the incinerator from the top and then fall down to the lower layers as the axis of the incinerator rotates. The cakes are heat-dried at the upper layers (1st - 4th layers), burned at the middle layers (5th-10th layers), and become ashes at the lower layers. The treatment of the cakes takes about 70 min depending on the rotation speed of the axis. The heat source of burning is hot air warmed by oil burners.

If the cakes begin to burn under a good condition, those burn continuously without hot air supply. So the control problem is to keep a desired temperature distribution for self-burning. Usually the temperature of the 7th layer is the highest at a steady state. As seen in Figure 7, the incinerator is at present controlled by adjusting three cold air inputs and one hot air input, and also by changing the rotation speed of the axis. Cold air is put into the 4th and 6th layers at a fixed ratio and also put independently into the 10th and 12th layer. Cold air to the 10th layer is input when hot air to the 9th layer is stopped.

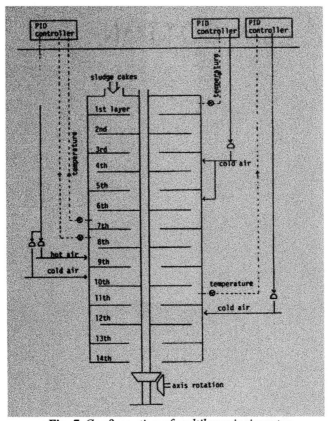

Fig. 7. Configuration of multilayer incinerator.

The incinerator is equipped with PID controllers. However those do not work well, and actually human operators play an important role in keeping the process at a steady state. It is easily seen that the process is dynamical and of distributed constants. Its input-output relation is nonlinear and very complex. So the model-based design of a controller becomes necessary.

4.2. Fuzzy model

In order to identify the fuzzy model of the process, we have performed two kinds of experiments. One is to take step responses where an input is changed in a stepwise manner, whereas the other inputs are kept the same. The other is to measure the temperature change at each layer by putting random inputs according to M series signal. Figure 8 shows some results of the experiments. In building a process model and designing a controller, we chose two inputs: cold air to both the 4th and 6th layers, and cold air to the 10th layer. The amount of the cakes is treated as a disturbance input. We keep the rotation speed constant and assume that the incinerator is at a self-burning condition; the data under this condition are used for the process identification.

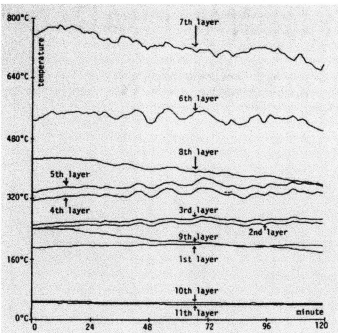

Fig. 8a. *Inputs – outputs data for identification: Temperature changes at layers.*

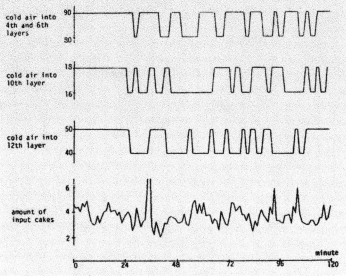

Fig. 8b. *Inputs-outputs data for identification: Random inputs.*

We first made the time series analysis of the temperature changes and decided that the sampling period is 2 min. This gives us 143 data of time series. Let us define variable names used in the model as follows:

$n(t)$: temperature of n-th layer at time t,
$u_1(t)$: amount of cold air to 4th and 6th layers at time t,
$u_2(t)$: amount of cold air to 10th layer at time t.

Then the structure of the model is identified as explained in Section 2. For example, the temperature of the n-th layer at time $(t + 1)$ is determined by the variables as shown in Figure 9. We can see that the variables affecting $6(t + 1)$ are marked by white squares and those are $5(t)$, $6(t)$, $6(t - 1)$, $7(t)$, $u_1(t)$, $u_2(t-1)$.

Fig. 9. *Variables affecting the temperature of the n-th layer at time $(t + 1)$.*

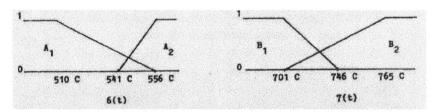

	Fuzzy variables in premises		Parameters of linear equations in consequences						
	$6(t)$	$7(t)$	const.	$5(t)$	$6(t)$	$6(t-1)$	$7(t)$	$u_1(t)$	$u_1(t-1)$
1	A_1	B_1	22.7	-0.133	1.075	-0.124	0.059	-0.356	0.454
2	A_1	B_2	46.4	0.463	0.987	-0.570	0.181	-0.945	0.621
3	A_2	B_1	-257.5	0.733	2.737	-2.558	0.641	-0.698	0.473
4	A_2	B_2	410.2	0.276	0.144	-0.264	0.244	-0.677	-0.226

$6(t)$: temperature of 6th layer at time t.
$u_1(t)$: cold air input into 4th and 6th layers at time t.

Fig. 10. *Process laws for the temperature of the 6th layer at time $(t + 1)$.*

Taking into account that the cakes pass a layer in 5 min, this structure seems very reasonable from a physical point of view.

As an example, Figure 10 shows the identified four process laws giving $6(t+1)$ where we find only two variables $6(t)$ and $7(t)$ in the premises. As the result of the identification, we obtain 36 process laws altogether giving $1(t+1)$, ..., $11(t+1)$. Figure 11 shows the accuracy of the fuzzy model by comparing the outputs of the model and those in the experiments. It is seen that the obtained model is satisfactory. Table 1 also compares the performance of the fuzzy model with the ordinary statistical model where we can find the error of the fuzzy model is much less than that of the statistical model.

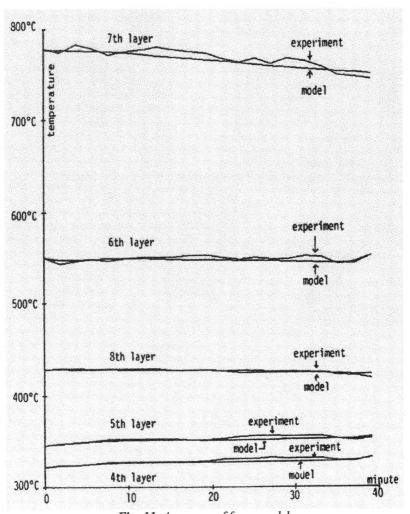

Fig. 11. Accuracy of fuzzy model.

Table 1. Comparison of the fuzzy model and the statistical model. (N: the number of fuzzy process laws PI = $\Sigma(y_i - y_0^i)^2 / n$, E = $\Sigma |y_i - y_0^i| / ny$ n: the number of data, y_i: actual temperature, y_i^0: model temperature, y: average of y_i's.)

Layer	N	Fuzzy model PI	Fuzzy model E(%)	Statistical PI	Statistical E(%)
1	2	0.69	0.24	0.73	0.26
2	2	0.80	0.23	0.87	0.26
3	2	0.78	0.22	0.83	0.24
4	4	1.14	0.26	1.51	0.36
5	4	1.17	0.25	1.42	0.32
6	4	2.60	0.37	3.16	0.46
7	4	3.46	0.37	4.43	0.47
8	2	1.12	0.22	1.22	0.24
9	4	0.64	0.22	0.81	0.31
10	4	0.36	0.48	0.48	0.68
11	4	0.34	0.37	0.4	0.55
Total 36					

4.3 Fuzzy Controller

It is necessary to control the temperature distribution in such a way that the temperatures of the middle layers are maintained high enough for self-burning while the temperatures of the other layers are kept not too high for safety.

Here let us design a fuzzy controller for the control of the temperatures of the 4th, 6th, 7th, and the 8th layers using the fuzzy model. We use two control inputs $u_1(t)$ and $u_2(t)$ to the purpose. We have 14 process laws for $4(t + 1)$, $6(t + 1)$, $7(t + 1)$ and $8(t + 1)$. From these we obtain 8 subsystems according to the previously mentioned criterion. Then we can derive 8 fuzzy control rules.

Figure 12 shows the premises of the obtained fuzzy control rules. For example, the first control rule is written as

if $4(t-2)$ is S_4, $6(t)$ is L_6, $7(t)$ is S_7, then

$$u_1(t) = -0.003e_1 + 0.012e_2 - 0.045e_3 + 1.54e_4 - 2.51e_5 + 0.43e_6$$
$$- 0.04e_7 - 0.023e_8 - 0.2e_9 + 0.76e_{10} + 0.27e_{11} - 0.23e_{12}$$
$$- 0.48e_{13} + 1.55e_{14} - 0.14r_4 - 2.2r_6 + 0.77r_7 + 0.21r_8 - 199, \quad (16)$$

$$u_2(t) = -0.009e_1 + 0.019e_2 - 0.074e_3 + 0.22e_4 - 0.51e_5 + 0.014e_6$$
$$- 0.051e_7 - 0.03e_8 - 0.32e_9 + 0.26e_{10} - 0.32e_{11} - 0.14e_{12}$$
$$- 0.2e_{13} + 0.79e_{14} - 0.09r_4 - 0.31r_6 + 0.12r_7 + 0.14r_8 - 10.25,$$

where

$u_1(t+1)$ = amount of cold air to 4th and the 6th layers at time $(t+1)$,
$u_2(t+1)$ = amount of cold air to 10th layer at time $(t+1)$,
r_k = set temperature of k-th layer, k=4, 6, 7, 8,
$e_1 = 4(t-2) - r_4$, $e_2 = 4(t-1) - r_4$, $e_3 = 4(t) - r_4$,

$$e_4 = 6(t-1) - r_6, \quad e_5 = 6(t) - r_6, \quad e_6 = 7(t) - r_7,$$
$$e_7 = 8(t-2) - r_8, \quad e_8 = 8(t-1) - r_8, \quad e_9 = 8(t) - r_8, \qquad (17)$$
$$e_{10} = u_1(t-3), \quad e_{11} = u_1(t-2), \quad e_{12} = u_1(t-1),$$
$$e_{13} = u_2(t-2), \quad e_{14} = u_2(t-1),$$

Rule	Premise
1	$4(t-2)$ is S_4, $6(t)$ is S_6, $7(t)$ is S_7
2	$4(t-2)$ is S_4, $6(t)$ is S_6, $7(t)$ is L_7
3	$4(t-2)$ is S_4, $6(t)$ is $L6$, $7(t)$ is S_7
4	$4(t-2)$ is S_4, $6(t)$ is L_6, $7(t)$ is L_7
5	$4(t-2)$ is L_4, $6(t)$ is S_6, $7(t)$ is S_7
6	$4(t-2)$ is L_4, $6(t)$ is S_6, $7(t)$ is L_7
7	$4(t-2)$ is L_4, $6(t)$ is L_6, $7(t)$ is S_7
8	$4(t-2)$ is L_4, $6(t)$ is L_6, $7(t)$ is L_7

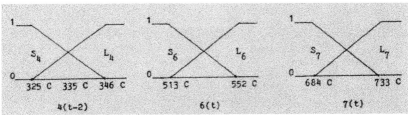

Fig. 12. Premises of fuzzy control rules.

Figure 13(a) shows the results of the fuzzy control and (b) shows the control inputs. These are obtained by computer simulation using the fuzzy model of the process. We can see in Figure 13 that the process is well controlled by the fuzzy controller.

Editor's Note: The Reference listing in the original paper was inadvertently omitted and hence does not appear in this reproduction. However, most of the references cited in this chapter can be obtained from the reference listings in Chapters 4, 5, 7, 10, and 13.

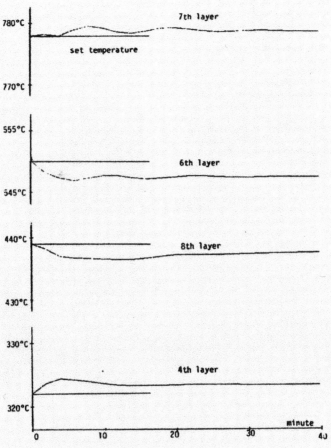

Fig. 13a. Results of fuzzy control: Output.

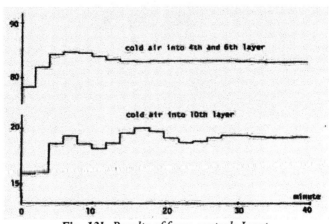

Fig. 13b. Results of fuzzy control: Input.

Fuzzy Identification of Systems and its Applications to Modeling and Control*

Tomohiro Takagi And Michio Sugeno
Department of Systems Science,
Tokyo Institute of Technology,
4259 Nagatsuta, Midori-ku, Yokohama 227, Japan

Received September 30, 1983
Revised April 6, 1984

Abstract

A mathematical tool to build a fuzzy model of a system where fuzzy implications and reasoning are used is presented in this paper. The premise of an implication is the description of fuzzy subspace of inputs and its consequences is a linear input-output relation. The method of identification of a system using its input-output data is then shown. Two applications of the method to industrial processes are also discussed: a water cleaning process and a converter in a steel-making process.

1. Introduction

The main purpose of this paper is to present a mathematical tool to build a fuzzy model of a system. There has been a considerable number of studies [1]-[3] on fuzzy control where fuzzy implications are used to express control rules. Most of those implications contain fuzzy variables with unimodal membership functions since those are linguistically understandable and thus called linguistic variables. As for reasoning, the so-called compositional rule of inference or its simplified version is used. However, when we use this type of reasoning together with unimodal fuzzy variables for multivariable control, we have much difficulty because we need many fuzzy variables, i.e., many implications; it is usual to use five variables in each dimension of input space.

© 1985 IEEE. Retyped with written permission from
IEEE Trans. on Systems, Man and Cybernetics, 15, No 1
(1985), 116-132.

The authors have suggested multidimensional fuzzy reasoning [6] where we can surprisingly reduce the number of implications. The study in this paper is related to the above idea of reasoning, where a fuzzy implication is improved and reasoning is simplified.

Recently some studies [4], [5] have also been reported on fuzzy modeling of a system. Fuzzy modeling based on fuzzy implications and reasoning may be one of the most important fields in fuzzy systems theory. Here we have to deal with a multivariable system in general and so therefore have to consider a multidimensional reasoning method.

Generally speaking, model building by input-output data is characterized by two things; one is a mathematical tool to express a system model and the other is the method of identification. A mathematical tool itself is required to have simplicity and generality. The fuzzy implication presented as a tool in the paper is quite simple. It is based on a fuzzy partition of input space. In each fuzzy subspace, a linear input-output relation is formed. The output of fuzzy reasoning is given by the aggregation of the values inferred by some implications that were applied to an input.

This paper also shows the method of identification of a system using its input-output data. As is well-known, identification is divided into two parts: structure identification and parameter identification.

In its nature, structure identification is almost independent of a format of system description. We omit this part in the paper, so by identification we mean parameter identification in fuzzy implications. However, a kind of structure problem partly appears.

Finally this paper shows two applications to industrial processes. One is a water cleaning process where an operator's control actions are fuzzily modeled to design a fuzzy controller. The other is a converter in the steel-making process where the conversion process is fuzzily modeled and model-based fuzzy control is considered.

Most fuzzy controllers have been designed based on human operator experience and/or control engineer knowledge. It is, however, often the case that an operator cannot tell linguistically what kind of action he takes in a particular situation. In this respect it is quite useful to give a way to model his control actions using numerical data. Further, if there is no reason to believe that an operator's control is optimal, we have to develop model-based control just as in ordinary control theory. To this aim it is necessary to consider a means for fuzzy modeling of a system.

2. Format of Fuzzy Implication and Reasoning Algorithm

In this paper we denote the membership function of a fuzzy set A as $A(x)$, $x \in X$. All the fuzzy sets are associated with linear membership functions. Thus, a membership function is characterized by two parameters giving the greatest grade 1 and the least grade 0. The truth value of a proposition "x is A and y is B" is expressed by

$$|x \text{ is } A \text{ and } y \text{ is } B| = A(x) \wedge B(y).$$

A. Format of Implications

We suggest that a fuzzy implication R is of the format:

$$R: \text{If } f(x_1 \text{ is } A_1, \cdots, x_k \text{ is } A_k) \text{ then } y = g(x_1, \cdots, x_k) \quad (1)$$

where

- y Variable of the consequence whose value is inferred.
- $x_1 - x_k$ Variables of the premise that appear also in the part of the consequence.
- $A_1 - A_k$ Fuzzy sets with linear membership functions representing a fuzzy subspace in which the implication R can be applied for reasoning.
- f Logical function connects the propositions in the premise.
- g Function that implies the value of y when $x_1 - x_k$ satisfies the premise.

In the premise if A_i is equal to X_i for some i where X_i is the universe of discourse of x_i, this term is omitted; x_i is unconditioned.

Example 1:

$$R: \text{If } x_1 \text{ is small and } x_2 \text{ is big then } y = x_1 + x_2 + 2x_3.$$

This implication states that if x_1, is small and x_2 is big, then the value of y would be equal to the sum of x_1, x_2 and $2 x_3$, where x_3 is unconditioned in the premise.

In the sequel we shall only use "and" connectives in the premise and adopt a linear function in the consequence as is seen in the above example. So an implication is written as:

$$\begin{aligned} R: &\text{If } x_1 \text{ is } A_1 \text{ and } \cdots \text{ and } x_k \text{ is } A_k \\ &\text{then } y = p_0 + p_1 x_1 + \cdots + p_k x_k. \end{aligned} \quad (2)$$

B. Algorithm of Reasoning

Suppose that we have implications R^i ($i = 1, \ldots, n$) of the above format. When we are given

$$(x_1 = x_1^0, \cdots, x_1 = x_k^0)$$

where $x_1^0 - x_k^0$ are singletons, the value of y is inferred in the following steps.

(1) For each implication R, y^i is calculated by the function g^i in the consequence

$$\begin{aligned} y^i &= g^i(x_1^0, \cdots, x_k^0) \\ &= p_0^i + p_1^i x_1^0, \cdots, + p_k^i x_k^0. \end{aligned} \quad (3)$$

(2) The truth value of the proposition $y = y^i$ is calculated by the equation

$$\begin{aligned} |y = y_i| &= \left| x_1^0 \text{ is } A_1^i \text{ and } \cdots \text{ and } x_k^0 \text{ is } A_k^i \right| \wedge \left| R^i \right| \\ &= \left(A_1^i(x_1^0) \wedge \cdots \wedge A_k^i(x_k^0) \right) \wedge \left| R^i \right| \end{aligned} \quad (4)$$

where $|*|$ means the truth value of proposition $*$ and $\wedge$ stands for min operation, and $|x^0 \text{ is } A| = A(x^0)$, i.e., the grade of the membership of x^0.

For simplicity we assume

$$\left| R^i \right| = 1 \quad (5)$$

so the truth value of the consequence obtained is
$$|y = y_i| = A_1^i(x_1^0) \wedge \cdots \wedge A_k^i(x_k^0). \tag{6}$$

3) The final output y inferred from n implications is given as the average of all y^i with the weights $|y = y^i|$

$$y = \frac{\sum |y = y^i| \times y^i}{\sum |y = y^i|}. \tag{7}$$

Example 2: Suppose that we have the following three implications:
R^1: If x_1 is small$_1$ and x_2 is small$_2$ then $y = x_1 + x_2$
R^2: If x_2 is big$_1$ then $y = 2 \times x_1$
R^3: If x_2 is big$_2$ then $y = 3 \times x_2$

Table 1 shows the reasoning process by each implication when we are given $x_1 = 12, x_2 = 5$. The column "Premise" in Table 1 shows the membership functions of the fuzzy sets "small" and "big" in the premises. The column "Consequence" shows the value of y^i calculated by the function g^i of each consequence and "Tv" shows the truth value of $|y - y_i|$. For example, we have

$$|y = y^1| = |x_1^0 = \text{small}_1| \wedge |x_2^0 = \text{small}_2|$$
$$\text{small}_1(x_1^0) \wedge \text{small}_2(x_2^0) \tag{8}$$
$$= 0.25.$$

The value inferred by the implications is obtained by referring to Table 1:
$$y = \frac{0.25 \times 17 + 0.2 \times 24 + 0.375 \times 15}{0.25 + 0.2 + 0.375} \tag{9}$$
$$= 17.8.$$

Table 1. Reasoning process for Example 2

Implication	Premise		Consequence	Tv
R1	small$_1$, .25 at 0, 16	small$_1$, .375 at 0, 8	$y = 12 + 5 = 17$	.25 ∧ .375 = .25
R2	big$_1$, .2 at 10, 20		$y = 2 \times 12 = 24$	.2
R3		big$_2$, .375 at 2, 10	$y = 3 \times 5 = 15$	.375
	$x_1 = 12$	$x_2 = 5$		

C. Properties of Reasoning

We show two illustrative examples to find the performance of the presented reasoning algorithm.

Example 3: Suppose we have two implications.

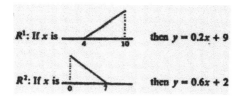

Then Fig. 1 shows the relation of x and y, which is marked by the symbol +. The line R^1 shows the function in the consequence of R^1. The equation in a consequence can be interpreted to represent a law that holds in the fuzzy subspace defined in a premise.

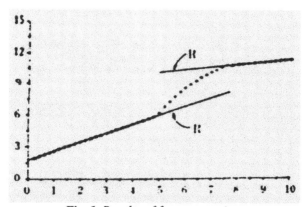

Fig. 1. Results of fuzzy reasoning.

Let us consider the difference between the ordinary piecewise linear approximation method and the presented method. If we take piecewise linear approximation, we first divide input space into crisp subspaces and next build a linear relation in each subspace. For example, in the case shown in Fig. 1, we need another linear relation connecting R^1 and R^2. It is easily seen that those three straight lines are not smoothly connected. On the other hand, the presented method enables us to reduce the number of piecewise linear relations and also to connect them smoothly. It is of crucial importance to reduce the number of linear relations in a multidimensional case.

Further, with the fuzzy partition of input space, we can put linguistic conditions to linear relations such as "x_1 is small and x_2 is big." Thus, for example, we can use the variable that is observed only by humans (see Section 4-B).

Example 4: Fig. 2 shows the input-output relation expressed by the implications of Example 2. In this case the premises are two-dimensional. In the figure the curved surface shows a highly nonlinear input-output relation whose shape reflects the dominance of each implication in its essentially applicable area and also the conflict of implications in an overlapped area.

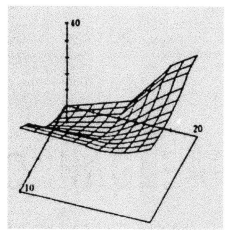

Fig. 2. *Results of fuzzy reasoning in Example 2.*

3. Algorithm of Identification

As has been stated, we consider a fuzzy model consisting of some number of implications that are of the format:

If x_1 is A_1 and $\cdots$ and x_k is A_k,

then $y = p_0 + p_1 \cdot x_1 + \cdots + p_k \cdot x_k$

characterized by "and" connective and a linear equation.

For identification we have to determine the following three items by using the input-output data of an objective system.

1) $x_1, \ldots, x_k$ Variables composing the premises of implication
2) $A_1, \ldots, A_k$ Membership functions of the fuzzy sets in the premises abbreviated as premise parameters.
3) $p_0, \ldots, p_k$ Parameters in the consequences.

Notice that all the variables in a premise may not always appear. Items 1) and 2) are related to the partition of space of input variables into some fuzzy subspaces. Item 3) is related to describing an input-output relation in each fuzzy subspace.

We can consider the relation among three items hierarchically from 1) down to 3). The algorithm of the identification of implications is divided into three steps corresponding to the above three items. We first give a brief explanation of the algorithm at each step.

1) *Choice of Premise Variables:* First a combination of premise variables is chosen out of possible input variables we can consider. Next the optimum premise and consequence parameters are identified according to Steps 2) and 3), and also the efforts between the output values of the model and the output data of the objective system are calculated. We then improve the choice of the premise variables so that the performance index is decreased, which is defined as the root mean square of the output efforts.

2) *Premise Parameters Identification:* In this step the optimum premise parameters are searched for the premise variables chosen at Step 1). Assuming the values of premise parameters, we can obtain the optimum consequence parameters together with the performance index according to Step 3). So the problem of finding the optimum premise parameters is reduced to a nonlinear programming problem minimizing the performance index.

3) *Consequence Parameters Identification:* The consequence parameters that give the least performance index are searched by the least squares method for the given premise variables in Step 1) and parameters in Step 2).

The outline of the algorithm is shown in Fig. 3. From the next we shall discuss the method in detail with an illustrative example at each step.

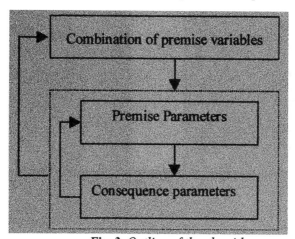

Fig. 3. Outline of the algorithm.

A. Consequence Parameter Identification

In this section we show how to determine the optimum consequence parameter to minimize the performance index, provided that both the premise variables and parameters are given. The performance index has been defined above as a root mean square of the output errors, which means the differences between the output data of an original system and those of a model.

Let a system be represented by the following implications:

R^1 If x_1 is $A_1^1, \cdots,$ and x_k is A_k^1,

then $y = p_0^1 + p_1^1 \cdot x_1 + \cdots + p_k^1 \cdot x_k$

$\vdots \quad \vdots$

R^n If x_1 is $A_1^n, \cdots,$ and x_k is A_k^n,

then $y = p_0^n + p_1^n \cdot x_1 + \cdots + p_k^n \cdot x_k.$

Then the output y for the input $(x_1, \ldots, x_k)$ is obtained as:

$$y = \frac{\sum_{i=1}^{n}(A_1^i(x_1) \wedge \cdots \wedge A_n^i(x_n)) \cdot (p_0^i + p_1^i \cdot x_1 + \cdots + p_k^i \cdot x_k)}{\sum_{i=1}^{n}(A_1^i(x_1) \wedge \cdots \wedge A_n^i(x_n))} \qquad (10)$$

Let β_i, be

$$\beta = \frac{A_1^i(x_1) \wedge \cdots \wedge A_n^i(x_n)}{\sum_{i=1}^{n}(A_1^i(x_1) \wedge \cdots \wedge A_n^i(x_n))} \qquad (11)$$

then

$$\begin{aligned}y &= \sum_{i=1}^{n} \beta_i (p_0^i + p_1^i \cdot x_1 + \cdots + p_k^i \cdot x_k) \\ &= \sum_{i=1}^{n}(p_0^i \cdot \beta_i + p_1^i \cdot x_1 \cdot \beta_i + \cdots + p_k^i \cdot x_k \cdot \beta_i).\end{aligned} \qquad (12)$$

When a set of input-output data $x_{1j}, x_{2j}, \ldots, x_{kj} \to y_j$ ($j = 1 \ldots m$) is given, we can obtain the consequence parameters $p^i{}_0, p^i{}_1, \ldots, p^i{}_k$ ($i = 1 \ldots n$) by the least squares method using (12).

Let X ($m \times n(k+1)$ matrix), Y (m vector) and P ($n(k+1)$ vector) be

$$X = \begin{bmatrix} \beta_{11}, \cdots, & \beta_{n1}, x_{11} \cdot \beta_{11}, \cdots, & x_{11} \cdot \beta_{n1}, \cdots, \\ & \cdots & x_{k1} \cdot \beta_{11}, \cdots, \quad x_{k1} \cdot \beta_n \\ \vdots & & \vdots \\ \beta_{1m}, \cdots, & \beta_{nm}, x_{1m} \cdot \beta_{1m}, \cdots, & x_{1m} \cdot \beta_{1m}, \cdots, \\ & & x_{k1} \cdot \beta_{1m}, \cdots, \quad x_{k1} \cdot \beta_{nm} \end{bmatrix} \qquad (13)$$

where

$$\beta_{ij} = \frac{A_{i1}(x_{1j}) \wedge \cdots \wedge A_{ik}(x_{kj})}{\sum_j A_{i1}(x_{ij}) \wedge \cdots \wedge A_{ik}(x_{kj})} \qquad (14)$$

$$Y = [y_1, \cdots, y_m]^T \qquad (15)$$
$$P = \left[p_0^1, \cdots, p_0^n, p_1^1, \cdots, p_1^n, \cdots, p_k^1, \cdots, p_k^n \right]^T \qquad (16)$$

Then the parameter vector P is calculated by

$$P = (X^T X)^{-1} X^T Y. \qquad (17)$$

It is noted that the proposed method is consistent with the reasoning method. In other words, this method of identification enables us to obtain just the same parameters as the original system, if we have a sufficient number of noiseless output data for the identification.

In this paper the parameter vector P is calculated by a stable-state Kalman filter. The so-called stable-state Kalman filter is an algorithm to calculate the parameters of a linear algebraic equation that gives the least squares of errors. Here we apply it to calculate the parameter vector P in (17).

Let the ith row vector of matrix X defined in (13) be x_i and the ith element of Y be y_i. Then P is recursively calculated by (18) and (19) where S_i is $(n \cdot (k+1)) \times (n \cdot (k+1))$ matrix.

$$P_{i+1} = P_i + S_{i+1} \cdot x_{i+1} \cdot (y_{i+1} - x_{i+1} \cdot P_i) \tag{18}$$

$$S_{i+1} = P_i - \frac{S_i \cdot x_i + x_{i+1} \cdot S_i}{1 + x_{i+1} \cdot S \cdot x_{i+1}^T}, \quad i = 0, 1, \cdots, m-1 \tag{19}$$

$$P = P_m \tag{20}$$

where the initial values of P_0 and S_0 are set as follows.

$$P_0 = 0 \tag{21}$$

$$S_0 = \alpha \cdot I \ (\alpha = \text{big number}) \tag{22}$$

where I is the identity matrix.

Example 5: Suppose in a system

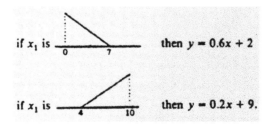

Under the condition that the premise of the model is fixed to that of the original system, the consequences are identified from input-output data as follows, where the noises are added to the data.

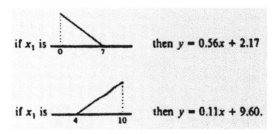

Fig. 4 shows the noised input-output data, the original consequences, and the identified consequences.

B. Premise Parameters Identification

In this section we show how to identify the fuzzy sets in the premises, that is, how to partition the space of premise variables into fuzzy subspaces, provided that the premise variables are chosen.

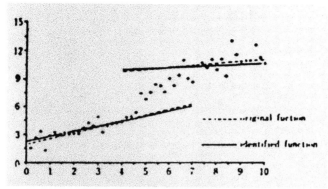

Fig. 4. Results of identification.

For example, looking at the input-output data shown in Fig. 5, we can see that the input-output characteristics change as the input x increases. So dividing the space of x into two fuzzy subspaces that x is *small* or x is *big*, we have a model with the following two implications:

if x is *small*, then $y = a_1 x + b_1$

if x is *big*, then $y = a_2 x + b_2$.

We next have to determine the membership functions of "*small*" and "*big*" as well as the parameters a_1, b_1, a_2 and b_2 in the consequences.

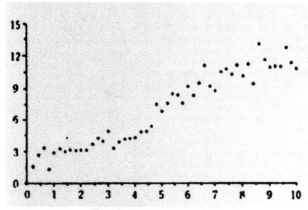

Fig. 5. Input-output data.

As it is easily seen, to divide the spaces into some fuzzy subspaces is to determine the membership functions of the fuzzy sets in the premises. The problem is thus to find the optimum parameters of their membership functions by which the performance index is minimized.

We call this procedure "premise parameter identification". The algorithm is as follows.

1) Assuming the parameters of the fuzzy sets in the premises, we can obtain the optimum parameters in the consequences that minimize the performance index as discussed in the previous section.

2) The problem of finding the optimum premise parameters minimizing the performance index is reduced to a nonlinear programming problem. In this study we use the well-known complex method for the minimization. Each fuzzy set in the premises is determined by two parameters that give the greatest grade 1 and the least grade 0, since a fuzzy set is assumed to have a linear membership function.

Example 6: This example shows the identification using the input-output data gathered from a preassumed system with noises. The standard deviation of the noises is 5% of that of the outputs. It also has to be noted that we can identify just the same parameters of the premises as the original system if noises do not exist.

It is of great importance to point out the above fact. If it is not the case, we cannot claim the validity of an identification algorithm together with a fuzzy system description language.

Suppose the original system exists with the following two implications:

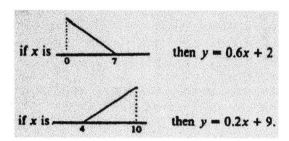

The functions in the consequences of the implications and the noised input-output data are shown in Fig. 6.

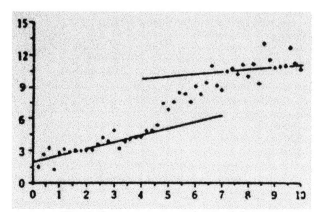

Fig. 6. Consequences and noised data.

The identified premise parameters are as follows. We can see that the parameters have been derived.

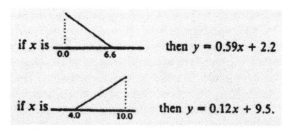

C. Choice of Premise Variables

In this section we suggest an algorithm to choose premise variables from the considerable input variables. As has been stated previously, all the variables of the consequences do not always appear in the premises. There are two problems concerned with the algorithm. One is the choice of variables: to choose a variable in the premise implies that its space is divided. The other is the number of divisions. The whole problem is a combinatorial one. So, in general, there seems to be no theoretical approach available. Here we just take a heuristic search method described in the following steps.

Suppose we build a fuzzy model of a k-input $x_1, \ldots, x_k$ and single output system.

Step 1: The range of x_1 is divided into two fuzzy subspaces "big" and "small" and the ranges of the other variables $x_2, \ldots, x_k$ are not divided, which means that only x_1 appears in the premises of the implications. This model consisting of two implications is thus:

if x_1 is big_1 then $\cdots$

if x_1 is small_1 then $\cdots$

It is called model 1-1. Similarly, a model in which the range of x_2 is divided and the ranges of the other variables $x_1, x_3, \ldots x_k$ are undivided is called model 1-2. In this way we have k-models, each of which is composed of two implications. In general, the model $1 - i$ is of the form.

Step 2: For each model the optimum premise parameters and consequence parameters are found by the algorithm described in the previous sections. The optimum model with the least performance index is adopted out of the k-models. It is called a stable state.

Step 3: Starting form a stable state at step 1.

Step 4: Repeat step 3 in a similar way by putting another variable into the premise.

Step 5: The search is stopped if either of the following criteria is satisfied:
1) The performance index of a stable state becomes less than the predetermined value.
2) The number of implications of a stable state exceeds the predetermined number.

The choice of the variables in the premises proceeds as is shown in Fig. 7.

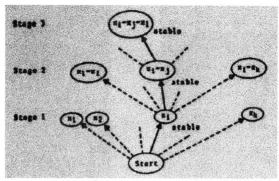

Fig. 7. Choice of premise variables.

Example 7: We show an example of identification. The original system is also a fuzzy system with two inputs and single output expressed by the implications as are shown below.

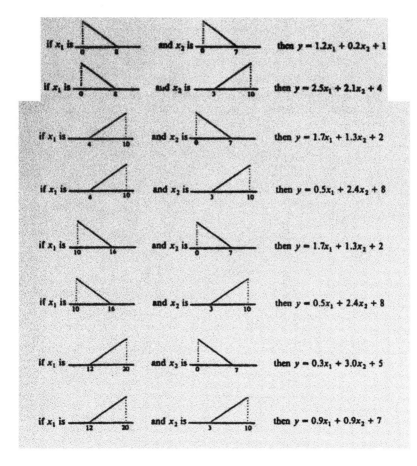

In this system, the range of x_1 is divided into four fuzzy subspaces and that of x_2 into two fuzzy subspaces. So the number of implications is 4 × 2 altogether.

Fig. 8 shows a fuzzy partition of the input space in the original system, i.e., the membership functions of the fuzzy relation expressed in the premises.

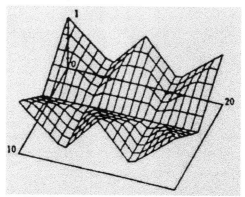

Fig. 8. Fuzzy partition of input space.

Fig. 9 shows the input-output relation of the above system. Now 441 input-output data of this system are taken for the identification and noises are added to the outputs, where the standard deviation of the noises is 2% of that of the outputs. Fig. 10 shows the noise data. Now the system is identified using these data.

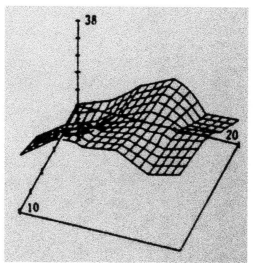

Fig. 9. Input-output relation of the original system.

Stage 1: Let us start from two models, each of which consists of two implications. They are shown together with their performance indices. Figs. 11 and 12 show the input-output relation of the models.

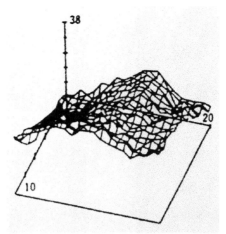

Fig. 10. *Identification data with noises.*

Model 1-1: (*the Range of* x_1, *is Divided*)
   Implications

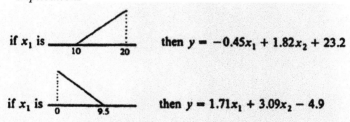

Performance Index =2.55.

Model 1-2: (*the Range of* x_2 *is Divided*)
   Implications

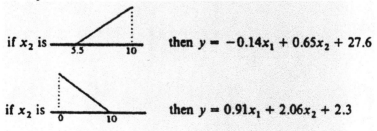

Performance Index =2.55.

The model 1-1 is found to be a stable state at the stage 1 since its performance index is the minimum of the two. In fact, Fig. 9 seems more similar to Fig. 11 than to Fig. 12. At the next stage we fix the variable x_1 in the premises.

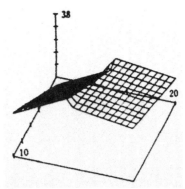

Fig. 11. Input-output relation of the model 1-1.

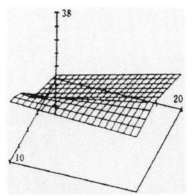

Fig. 12. Input-output relation of the model 1-1.

Stage 2: In this stage the space of inputs is further divided. In the model 2-1 (Fig. 13), which is the extended one of the model 1-1, the range of x_1 is divided into four subspaces, leaving that of x_2 undivided.

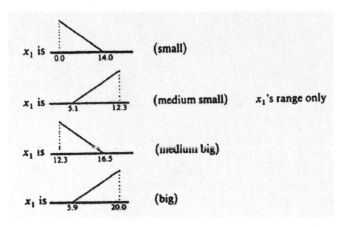

In the model 2-2 (Fig. 14), the ranges of x_1 and x_2 are newly divided in two fuzzy subspaces, respectively.

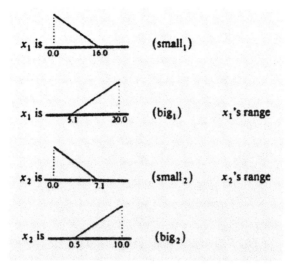

Notice that this partition of the range of x_1 is different from that of model 1-1. This is because the optimization of fuzzy sets concerned with x_1 is performed together with those concerned with x_2.

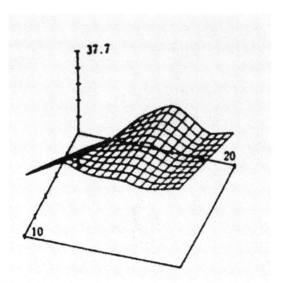

Fig. 13. Input-output relation of the model 2-1.

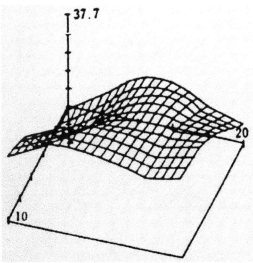

Fig. 14. Input-output relation of the model 2-2.

The implications of the two models and their performance indices are obtained this time as follows.

Model 2-1:
  Implications

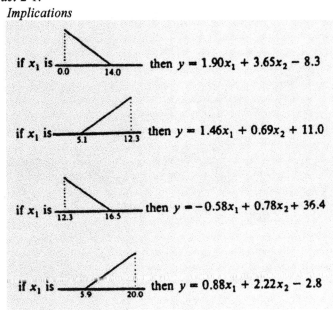

Performance Index = 1.91

Model 2-2:
Implications

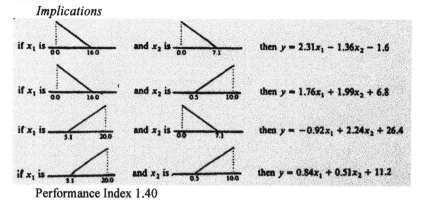

Performance Index 1.40

The model 2-2 is now found to be a stable state at stage 2.

Stage 3. We show the implications and the performance indices of two models (Figs. 15 and 16) extended from the model 2-2.

Model 3-1:
Implications

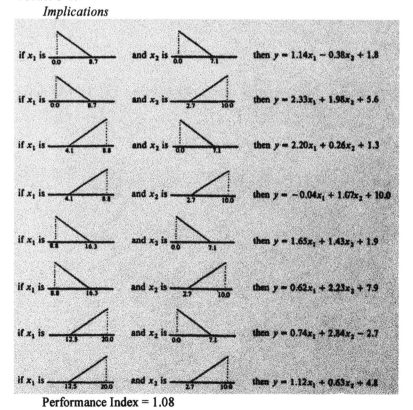

Performance Index = 1.08

where the partition of the range of x_1 is four and that of x_1 is two, as is seen.

Model 3-2:
  Implications

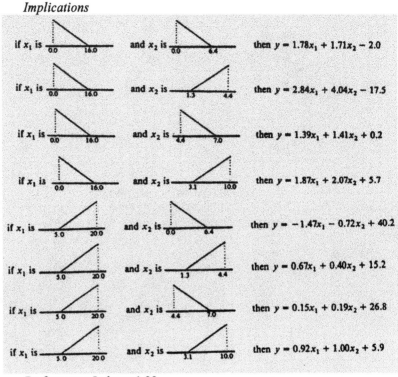

Performance Index = 1.33

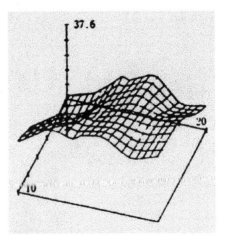

Fig. 15. *Input-output relation of the model 3-1.*

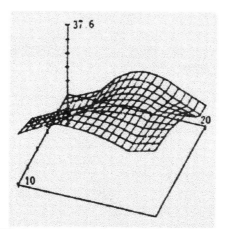

Fig. 16. Input-output relation of the model 3-2.

At stage 3, the model 3-1 is found to be a model with the same structure as the original system, which is of course a stable state.

We can say that the presented method enables us to derive almost the same premise parameters as those of the original system. We can also recognize that Fig. 9 is almost the same as Fig. 15.

The choice of the premise variables has proceeded in this example as is shown in Fig. 17.

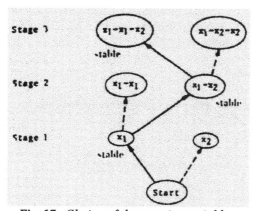

Fig. 17. Choice of the premise variables.

4. Application to Fuzzy Modeling

This chapter shows two practical applications of the proposed method to real industrial processes. The first one is the fuzzy modeling of a human operator's control actions in a water cleaning process. The obtained model may be directly used in place of an operator to control the process.

The other one is the fuzzy modeling of a converter in a steel-making process. The relation between the input-output of the converter is so complex that an appropriate algebraic model has not been developed. The obtained fuzzy model is applied to the control of the converter, and the results are compared with the case when an operator controls it without a model.

A. *Fuzzy modeling of human operator's control actions*

Water Cleaning Process: We shall now show an example where an operator's control actions are fuzzily modeled.

The control process is a water cleaning process for civil water supply as is illustrated in Fig. 18. In the process, turbid river water first comes into a mixing tank where chemical products called PAC and also chlorine are put and mixed in the water. Then the mixed water flows into a sedimentation tank where the turbid part of water is cohered with the aid of PAC and settled to the bottom.

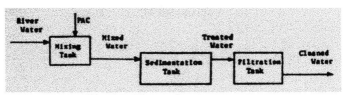

Fig. 18. Water cleaning process.

After sedimentation, which takes about 3 to 5 hours depending on the capacity of the tank, the treated water finally flows into a filtration tank producing clean water. Chlorine is added only for the sterilization of the water.

The main control problem of a human operator in this process is to determine the amount of PAC to be added so that the turbidity of the treated water is kept below a certain level. The optimal amount, not too little, nor too much, depends on the properties of the turbid water. The amount of PAC must be controlled also from an economical point of view.

The process is characterized by a lack of any physical model, significant variation of the turbidity of river water and the fact that turbidity itself is not clearly defined nor accurately measured. Therefore an operator's experience is a key factor in this control process.

However, a number of variables influencing sedimentation process have been found so far that can be measured. Let us first list all the variables concerned.

TB1 = Turbidity of the original water (ppm)   PH = PH
TB2 = Turbidity of the treated water (ppm)   AL = Alkalinity
PAC = Amount of PAC (ppm)   CL = Amount of chlorine (ppm)
TE = Temperature of water (C)

For example, if TE is lower, then more PAC is necessary. Both PH and AL nonlinearly affect the necessary amount of PAC. The optimal PAC depends on these variables; the relation among them is not clear. There are some other variables influencing the process, e.g., plankton in the river water, which increases in springtime but cannot be measured at present.

In most water cleaning processes a statistical model has been built. However the models are not accurate. These cover only steady state, i.e., a small range of TB1. TB1 increases, for example, 100 times more when it rains. So an operator controls PAC taking into account TB1, TE, PH, AL, and TB2. Now our process can be illustrated as in Fig. 19.

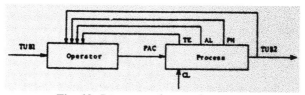

Fig. 19. Diagram of control process.

Derivation of Control Rules

We have a lot of operation data where all the variables are measured every hour for four months. That is, the number of data is 24 hours x 30 days x 4 months = 2880. Table 2 shows a part of these.

Among the data we have used about 600 for the identification taken in June and July. June in Japan is a rainy season and July is summer.

According to the identification algorithm discussed previously, eight control rules are derived that can be called a fuzzy model of operator's control as is stated, where a control rule is of the form:

If (*PH* is *), (*AL* is *) and (*TE* is *)

then $PAC = p_0 + p_1 \cdot TB1 + p_2 \cdot TB2 + p_3 \cdot PH + p_4 \cdot AL + p_5 \cdot TE$

PH, AL, and TE are picked up as premise variables and their ranges divided into small and big, as shown in Fig. 20. The number of the control rules is thus $2^3 = 8$. Those are shown in Fig. 21.

Table 2. Data for water cleaning process

TUB1	PH	TE	AL	PAC	TUB2
10.0	7.1	18.8	53.0	1300	1.0
17.0	7.0	18.6	50.0	1300	1.0
22.0	7.3	19.4	46.0	1400	2.0
50.0	7.1	19.5	40.0	1400	1.0
9.0	7.3	23.3	48.0	900	4.0
11.0	7.1	20.7	50.0	900	1.0
12.0	7.2	21.3	50.0	900	3.0
14.0	7.2	23.6	53.0	900	4.0
35.0	7.0	17.8	35.0	1200	1.0
20.0	7.0	16.6	40.0	1100	1.0
20.0	6.9	17.8	42.0	1100	1.0
18.0	7.1	17.3	40.0	1100	1.0
12.0	7.2	18.8	55.0	900	3.0
8.0	7.2	18.0	50.0	1000	1.5
11.0	7.1	19.2	49.0	1000	2.0
50.0	7.0	18.0	37.0	1200	1.5
35.0	7.0	17.7	42.0	1200	1.5
30.0	7.0	17.3	41.0	1100	1.5
16.0	7.1	19.3	42.0	1100	3.0

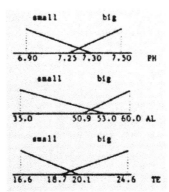

Fig. 20. Partition of the ranges of premise variables.

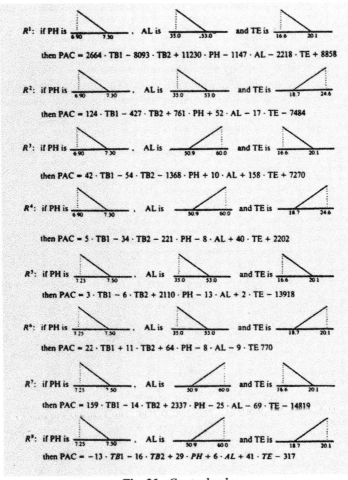

Fig. 21. Control rules.

Results of Fuzzy Control

The performance of the derived control rules is tested by using testing data. The results are shown in Table 3 as well as operators control input and the results of a statistical model where we used 38 testing data. The statistical model is represented in (23) that is usually used in a water cleaning process

$$PAC = 9.11\sqrt{TB1} - 79.8PH + 12.7CL + 1255.6. \qquad (23)$$

It is seen that operator's control actions are well modeled in the form of fuzzy control rules. The average of the absolute differences between the results of the fuzzy model and the operator, and those between the results of a statistical model and the operator are, respectively,

$$\begin{array}{ll}\text{fuzzy model} & 48.5 \\ \text{statistical model} & 128.0.\end{array}$$

These results show the excellence of the fuzzy model.

Table 3. Comparison of results

Operator	Statistical Model	Fuzzy Model
1300	994.7	1308.6
1300	995.9	1027.4
1300	1119.6	1063.0
1400	1151.1	1386.2
1400	1409.4	1551.1
900	1066.4	923.9
900	1068.9	965.5
900	1012.3	875.4
1200	1286.8	1236.7
1200	1246.8	1172.6
1100	1151.4	1075.9
1100	1199.5	1115.1
1100	1159.4	1130.5
1000	985.7	934.1
1000	1009.3	973.8
1000	1038.2	984.6
1200	1398.3	1285.3
1200	1290.6	1160.8

B. Fuzzy Modeling of Converter in a Steel-Making Process and its Control

The Problem

The steel-making process consists of the following four steps.
1) Iron ore is melted in a blast furnace. The obtained molten iron called hot pig is removed by a torpedo car into a converter after desulfurization.
2) In a converter, scrap, iron ore, and burnt lime are first added to hot pig, and then decarbonization and dephosphor are performed by oxygen below. After that various alloys are added for adjusting the ingredients of produced steel.
3) Floating slag is taken away and the amount of ingredients is readjusted in a ladle refining process.
4) It is then cast and finally cut into appropriate figures.

Each step except the final half of the fourth step depends on a human operator's trained control because it is very difficult to build a process model.

A steel-making plant produces various kinds of steel according to its ingredients. Especially the manganese ratio in the products is required to be variously adjusted.

In this section we deal with the problem of determining the amount of manganese alloy to adjust the manganese ingredient of produced steel in step (2). This process is the most difficult to be controlled among the ingredients adjustments.

We now list all the variables possibly concerned with the process.

Mn1   Original ratio of manganese in input iron.
Mn2   Final ratio of manganese in produced steel.
Mn2* Required ratio of manganese in produced steel.
MA   Ratio of manganese alloy put into input iron.
HP   Hot pig ratio of input iron.
[O]   Ratio of oxygen in input iron after oxygen blow.
[Si] Ratio of silicon of hot pig.
SG   State of floating slag.

Here input iron consists of hot pig and scrap. Among those, SG is physically measured only after the conversion process is finished. But it can be evaluated by operator's observation before that. Fig. 22 shows the conversion process.

Manganese alloy is added into hot pig after the oxygen blow process to produce steel such that the percentage of manganese ingredient, Mn2, becomes a required value Mn2* and its ideal amount can be calculated by (24), based on physical analysis.

$$MA = (Mn2^* - Mn1). \tag{24}$$

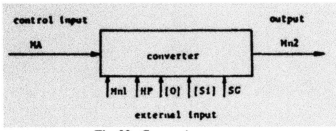

Fig. 22. Conversion process.

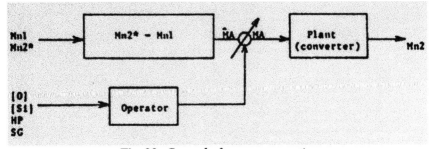

Fig. 23. Control of manganese ratio.

However, the actual ratio of manganese in the products is usually less than the physically estimated value from Mn1 and MA because of absorption by slag or other reasons due to, for example, various other ingredients in steel. So a human operator predicts Mn2 by referring to [O], [Si], HP, and observing the state of slag SG, etc. and controls MA by correcting the guided value by (24) so that Mn2* is attained. Fig. 23 shows the present situation of manganese control.

The past trials to derive a mathematical model of the converter with respect to manganese control have not been successful where the inputs are the observable variables and the output is the manganese ratio of the produced steel.

For the purpose of control, we try to build a fuzzy model of the converter by finding some clues from an experienced operator's way of control. This approach has the following advantages.

1) His manner tells us how he recognizes the characteristics of the conversion process based on his experience. For example, we can know important variables that should be put into promises of fuzzy implications.

2) We can even use input variables that only he can measure, for example, by just watching. Those variables are easily used as premise variables of our model. Since a model is of the form "if . . . then. . . .", the obtained model may be refined by his knowledge.

3) We can derive fuzzy control rules from the model, rather than from his control actions which may not be the best from a quantitative point of view. Needless to say, fuzzy control rules are easily understood qualitatively by him and we can adjust control rules also by his way of control. This is a very important point if an operator remains as a key essence in process control.

Modeling: We have taken 61 operation data from among the ones obtained in one month, and have used them for the identification of the conversion process. Further, we prepared 20 testing data different from the above identification data to check the validity of the obtained model. Table 4 shows some of the data. The input and output variables of the converter model are as follows:

[input]
$$HP = \frac{\text{hot pig}}{\text{input iron}} = \frac{\text{hot pig}}{\text{hot pig} + \text{scrap}} \quad (\text{percent})$$

$$MA = \frac{\text{manganese alloy}}{\text{input alloy}} \times 10^{-1} \quad (\text{percent})$$

SG - indication about softness of slag
[output]
$$\Delta Mn = Mn2 - Mn1$$

In this study, other variables [O] and [Si] are found not to seriously affect the process and so are deleted. For SG we conventionally use the measured values after the conversion is finished, which can be replaced by operator's observation. An experienced operator can measure SG rather qualitatively such as soft, medium, or hard. For this reason we put SG only into the premises for conditioning the input-output relation and do not use it in the consequences.

Table 4. Converter input-output data

HP	ΔM	SG	MA
93.90	1.11	0.30	14.00
94.00	14.52	-0.06	135.00
93.10	5.22	-0.05	54.00
93.60	5.36	-0.43	53.00
85.60	13.56	0.11	129.00
85.80	11.55	0.12	106.00
86.40	11.88	-0.06	113.00
93.10	13.85	0.16	119.00
88.50	12.27	-0.02	129.00
93.00	11.69	-0.03	127.00
93.00	9.86	-0.01	98.00
95.00	4.96	-0.13	46.00
94.70	2.19	0.01	25.00
85.00	1.89	-0.10	23.00
87.50	14.26	-0.09	121.00
90.50	11.52	-0.10	112.00
90.10	12.59	-0.06	116.00
90.20	5.67	-0.28	60.00

According to the proposed identification algorithm, each range of MA and SG is divided. The range of HP has remained undivided. Finally we have obtained four implications as are shown in Fig. 24.

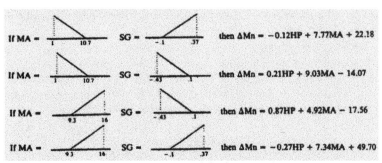

Fig. 24. Fuzzy model of converter.

The premise variables, the performance indices of the models, the correlation coefficients of the original output and model's output through the identification process are shown in Table 5.

As is seen in the model the space of each premise variable is divided only into two fuzzy subspaces. This is mainly because of the shortage of data. Notice that there are five parameters in one implication: 4 x 5 = 20 altogether. On the other hand, the number of data is only 61.

Results of Fuzzy Model. Table 6 shows the results of the fuzzy model, those of a statistical model and converter outputs, when the models are applied for the testing data. The statistical model is represented in (25), whose parameters were obtained by linear regression using the identification data.

Table 5. Identification parameters

choice of premise variable		performance index	correlation coefficient
[Stage 1]	HP	5.20	0.98975
	MA*	4.87	0.99105
	SG	5.54	0.98840
[Stage 2]	MA - HP	4.17	0.99344
	MA - MA	4.23	0.99234
	MA - SG*	4.06	0.99378

* indicates a stable state at each stage

$$\Delta Mn = -0.24 HP + 8.64 MA + 45.60 \tag{25}$$

The performance indicies of the results of the fuzzy model and the statistical model for 20 testing data are as follows:
fuzzy model   7.15.
statistical model   7.77.

The results are better than those obtained by a statistical model.

It should be noted that the fuzzy partition of the state of slag, SG, derived from the data shows a good agreement with that by an operator: he usually recognizes SG according to a similar partition and uses this information in his control.

Table 6. Results of fuzzy model

Actual output	Fuzzy model	Statistical model
143.00	135.18	136.48
128.00	124.06	125.09
29.00	27.29	29.40
129.00	128.65	126.57
101.00	99.31	97.07
112.00	114.27	111.57
107.00	111.74	111.63
143.00	117.70	115.08
13.00	13.72	15.85
121.00	120.66	120.73
58.00	52.59	53.89
112.00	109.61	105.14
126.00	126.44	127.22
58.00	57.32	58.33
57.00	55.44	56.34
72.00	75.03	75.44
40.00	43.78	43.94
104.00	110.12	102.14
113.00	114.60	114.50
110.00	96.93	94.99

Control of Converter: We now try to control the converter by using its model. Given a desired output ΔMn^*, we can calculate a necessary input MA from a model. Here for simplicity we use this MA instead of designing a fuzzy controller.

The problem is how to compare the results of the model-based control with those of an operator, because we cannot make an experiment at present. Let us take as an index of control performance

$$AC = \text{average of } \frac{|\Delta Mn^* - \Delta Mn|}{\Delta Mn^*}$$

where
ΔMn^* desired output
ΔMn actual output

As for an operator's control, this index is obtained from input-output data since, given ΔMn^*, he controls a converter. Let us denote it AC_{ope}. In case of a model-based control we assume that the output ΔMn of a converter is a desired output. Then we get the optimal input MA^0 from a model, input MA^0 to a process and see its output $\Delta \tilde{Mn}$. This output $\Delta \tilde{Mn}$ can be estimated by taking into account the accuracy of the model without experiments. So we can set $\Delta \tilde{Mn} = \Delta Mn \pm \varepsilon$ where ε is the error of the model. We now have:

$$AC_{model} = \text{average of } \frac{|\Delta Mn - \Delta \tilde{Mn}|}{\Delta Mn}$$
$$= \text{average of error of the model}.$$

We then obtain the following results:
$AC_{model} = 4.7\,\%$
$AC_{ope} = 6.7\,\%$

From the results we can expect that the control based on the obtained fuzzy model gives us better results than the present control by the operator.

Apart from the above method to calculate the input MA, we can directly derive fuzzy control rules in this case from the data $(\Delta Mn, HP, SG) \rightarrow MA$. Those are shown in Fig. 25.

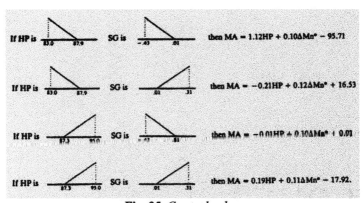

Fig. 25. Control rules.

5. Conclusion

We have suggested a mathematical tool to fuzzily describe a system. It has a quite simple form, but it can represent highly nonlinear relations as has been shown in examples. An algorithm of identification has also been shown and two applications to industrial processes have been discussed. The applications as well as illustrative examples show that the proposed method is general and thus very useful. We can put the results of fuzzy measurements by people such as "temperature is high" into the premises of implications. Linear relations in the consequences enable us to easily deal with this mathematical tool, as we well know in linear systems theory.

However, to claim the validity of the method, more studies have to be performed. The system theoretic approach is especially important. For example, we have minimal realization problems, decomposition problems, design problems of controller, etc. To solve these problems, it is required for us to deal with fuzzy system representation just as we do with a linear system.

In this paper, modeling of a human operator's control actions, rather than that of a process, has been mainly discussed. It is, however, possible to apply the method presented for the identification of a dynamic and distributed system. Modeling of a multilayer incineration furnace, a dynamic and distributed parameter system, is now under study.

References

[1] E. H. Mamdani, "Application of fuzzy algorithms for control of simple dynamic plant," *Proc, IEEE*, vol. 121, no. 12, pp. 1585-1588, 1976.
[2] R. M Tong, M.B. Beck and A. Latten, "Fuzzy control of the activated sludge wastewater treatment process," *Automatica*, vol 16, pp 659-701, 1980.
[3] P. M. Larsen, " Industrial application of fuzzy logic control," *Int. J. Man - Machine Studies*, Vol. 12, pp. 3 - 10, 1980.
[4] R. M. Tong, "The construction and evaluation of fuzzy models," in *Advances to Fuzzy Set Theory and Applications*, M. M. Gupta Ed. New York: Plenum, 1979.
[5] M. B. Gorzakzany, J. B. Kiszka, and M. S. Stachowicz, "Some problems of studying adequacy of fuzzy models," in *Fuzzy Set and Possibility Theory*, R. R. Yager, Ed. New York: Pergamon, 1982.
[6] M. Sugeno and T. Takagi, "A Multi-Dimensional fuzzy reasoning," *Fuzzy Sets and Systems*, Vol. 9, no. 2, 1983.

Structure Identification of Fuzzy Model[*]

M. Sugeno and G.T. Kang
Department of Systems Science.
Tokyo Institute of Technology,
4259 Nagatsuda, Midori-ku,
Yokoharna 227, Japan

Received May 1986
Revised February 1987

Abstract

The problems of structure identification of a fuzzy model are formulated. A criterion for the verification of a structure is discussed. Using the criterion, an algorithm for identifying a structure is suggested. Further, a successive identification algorithm of the parameters is suggested. The proposed methods are applied to an example.

Keywords: fuzzy model, identification of fuzzy model, structure identification.

1. Introduction

In most studies of identification of a process by using its input-output data, it is assumed that there exists a global functional structure between the input and the output such as a linear relation. Statistical methods are used to identify the parameters in a mathematical model.

It is, however, very difficult to find a global functional structure for a nonlinear process. On the other hand, fuzzy modeling is based on the idea to find a set of local input-output relations describing a process. So it is expected that the method of fuzzy modeling can express a nonlinear process better than an ordinary method. There are, however, many problems concerned with the structure identification of a fuzzy model when there is no *a priori* information about its structure.

[*] © 1988 North-Holland. Retyped with written permission from *Fuzzy Sets and Systems*, **28** (1988) 15-33.

This paper discusses the problems of structure identification of a fuzzy model. Section 2 describes a fuzzy model. Section 3 formulates the problems and shows their solutions. Section 4 shows an example.

2. Fuzzy model

As the expression of a fuzzy model, we use the fuzzy implications and the fuzzy reasoning method suggested by Takagi and Sugeno [1]. A fuzzy implication is of the following form:

$$L^i : \text{If } x_1 \text{ is } A_1^i, x_2 \text{ is } A_2^i, \ldots, x_p \text{ is } A_p^i,$$
$$\text{then } y^i = c_0^i + c_1^i x_1 + c_2^i x_2 + \ldots + c_p^i x_p \qquad (1)$$

where L^i means the i-th implication and A^i_j is a fuzzy variable, x_j an input variable, y^i the output from the i-th implication, and c^i_j a consequent parameter. For simplicity, the grade of membership of a fuzzy set A is written as $A(x)$. In the sequel, the membership function of a fuzzy variable will be of a convex type and formed by straight lines.

The fuzzy implications are formed by fuzzily partitioning the inputs space. Therefore, the premise of a fuzzy implication indicates a fuzzy subspace of the inputs space. Each implicational relation expresses a local input-output relation. Here it should be noted that not all the input variables appear in a premise. For example, at the fuzzy model shown in Figure 1, the inputs space is divided into four fuzzy subspaces as shown in Figure 2, where the first fuzzy implication L^1 has only one variable x_1 in its premise.

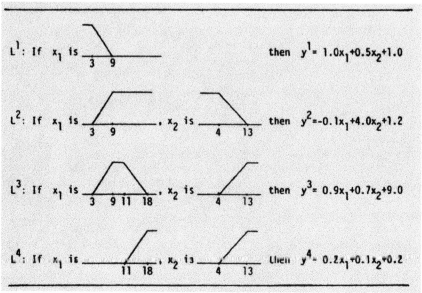

Fig. 1. Fuzzy implications.

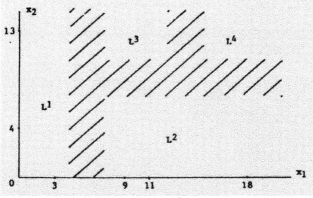

Fig. 2. Fuzzy subspaces.

Given an input $(x^0_1, x^0_2, \ldots, x^0_p)$, the output y^* is inferred by taking the weighted average of the y^i's

$$y^* = \sum_{i=1}^{q} w^i y^i \bigg/ \sum_{i=1}^{q} w^i : \qquad (2)$$

where q is the number of fuzzy implications, y^i is calculated for the input by the consequent equation of the i-th implication, and the weight w^i implies the overall truth value of the premise of the i-th implication for the input calculated as

$$w^i = \prod_{i=1}^{q} A^i_j(x^0_j). \qquad (3)$$

Figure 3 shows the input-output relation expressed by the four fuzzy implications in Figure 1. We can see that those implicational relations express a highly nonlinear functional relation.

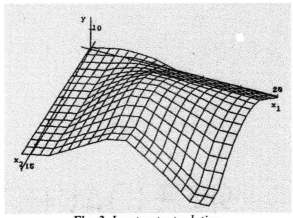

Fig. 3. Input-output relation.

The identification of a fuzzy model using input-output data consists of two parts: structure identification and parameters identification. The structure identification consists of premise structure identification and consequent structure identification. The parameters identification also consists of premise parameters identification and consequent parameters identification. The consequent parameters are the coefficients of linear equations. An algorithm of identification of those parameters has been suggested by Takagi [1] as follows. In Eq. (2), the inferred output y^* can be represented by a linear equation of which the coefficients are the consequent parameters:

$$y^* = c_0^1(g^1) + c_1^1(g^1 x_1^0) + \cdots + c_p^1(g^1 x_p^0)$$
$$+ c_0^2(g^2) + c_1^2(g^2 x_1^0) + \cdots + c_p^2(g^2 x_p^0) \qquad (4)$$
$$+ \cdots$$
$$+ c_0^q(g^q) + c_1^q(g^q x_1^0) + \cdots + c_p^q(g^q x_p^0)$$

where

$$g^i = w^i \bigg/ \sum_{k=1}^{q} w^k \quad (i = 1, 2, \ldots, q). \qquad (5)$$

Therefore, when a set of input-output data is given, the consequent parameters can be identified by, for example, the least squares method.

Figure 4 shows the steps in the identification of a fuzzy model. In this paper, we mainly deal with the structure identification and the premise parameters identification.

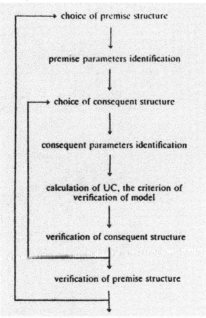

Fig. 4. *The steps in the identification of a fuzzy model.*

3. Structure identification

As stated in the previous section, the structure identification of a fuzzy model consists of two parts. The first part is concerned with the premise structure. This has two problems. One is that we have to find which variables are necessary in the premises. The other is that we have to find an optimal fuzzy partition of the inputs space, which is a problem peculiar to fuzzy modeling. For example, suppose that a fuzzy model of a process with three inputs and one output is represented by three fuzzy implications such as:

L^1 : If x_1 is Small$_1$ and x_3 is Small$_2$ then $y^1 = 0.5 + 1.2\, x_2 - 0.6\, x_3$,

L^2 : If x_1 is Small$_1$ and x_3 is Large$_2$ then $y^2 = 1.2 + 0.2\, x_1 + x_2 - 2.3\, x_3$, (6)

L^3 : If x_1 is Large$_1$ then $y^3 = 1.5 + 2.3\, x_1 - 0.9\, x_3$.

The variable x_2 does not appear in all the premises and L^3 has the only one variable x_1 in its premise. A fuzzy implication represents an input-output relation holding in a fuzzy subspace of the inputs space represented by its premise. For example, L^2 represents the input-output relation in the fuzzy subspace defined by x_1 = Small$_1$, and x_3 = Large$_2$. Therefore, the premise structure identification is the same as the problem to find out the fuzzy partition of the inputs space, where the number of the fuzzy subspaces corresponds to that of the necessary implications. The premise parameters are those of the fuzzy variables of premises.

The second part is concerned with the consequent structure identification. We have to find which variables are necessary in the consequent of an implication. For example, L^1 has two variables x_2 and x_3 in its consequent, while L^3 has x_1 and x_3.

In addition, we need to find a criterion for the verification.

3.1. Criterion for verification of structure

When a model is identified by using the observed data, there are generally a number of structures which seem to be adequate. Hence a criterion for the verification of a structure is of crucial importance.

Takagi [1] has suggested choosing a structure minimizing the mean square error. The structure identification of a fuzzy model, however, has the same characteristics as the input variable identification of a linear model, and it is known that the mean square error is not adequate as a criterion for the structure verification of a linear model.

As a criterion of fitting a statistical model, Akaike [5] has suggested the information theoretic criterion AIC on the assumption that residuals obey a normal distribution. There are, however, many problems not satisfying the assumption in practice.

Here we take the unbiasedness criterion which is used as a proper criterion for the verification of structure in GMDH [2]: a method for identifying nonlinear models. The basic idea embodied in this criterion is the following: in the presence of moderate noise, the parameters of the model with the true structure are the least sensitive to the changes of the observed data which are used for identifying the parameters. The unbiasedness criterion of solutions for a structure is obtained as follows. We first divide the observed data into two sets N_A and N_B, and identify the consequent parameters for each set of data separately. Then the unbiasedness criterion UC is calculated as:

$$\text{UC} = \left[\sum_{i=1}^{n_A} (y_i^{AB} - y_i^{AA})^2 + \sum_{i=1}^{n_B} (y_i^{BA} - y_i^{BB})^2 \right]^{1/2}, \quad (7)$$

where n_A is the number of the data set n_A, y_i^{AA} the estimated output for the data set N_A from the model identified by using the data set N_A, and y_i^{AB} the estimated output for the data set N_A from the model identified by using the data set N_B.

As a fuzzy model is constructed by partitioning the inputs space, the data for the identification should be distributed uniformly over the space where the fuzzy model is applied. Therefore, the data of N_A and N_B should be also be distributed uniformly over the space. When the number of the observed data is small, we make N_A and N_B have some data in common.

3.2. Premise structure identification

The identification of the premise structure can be regarded as the problem to partition the inputs space into the fewest fuzzy subspaces so that the identified fuzzy model represents the object system adequately.

When identifying the premise structure, we use the following idea. As the number of fuzzy subspaces, i.e., the number of fuzzy implications, increases, the UC of the fuzzy model decreases. But, if the number of fuzzy subspaces exceeds that of the optimal premise structure, the parameters of the model become sensitive to the changes of the data used for identifying the parameters, and the UC of the model increases.

Here we choose the premise structure which minimizes the unbiasedness criterion UC, and use the following algorithm resembling the forward selection of variables which is a method for finding variables in a linear model. That is, we start the process from the identification of a model with one implication, i.e., a linear model, and increase the number of fuzzy implications until the UC of the fuzzy model begins to increase. In the process of the premise structure identification, the premise parameters and the consequence of the model are also identified as shown in Figure 4.

The description of the algorithm

Define stage i in the identification process such that a fuzzy model at this stage consists of i fuzzy implications. Note that a fuzzy model has only one implication at stage 1, that is, it is just an ordinary linear model. The identification process starts from stage 1. Suppose that k variables $x_1, x_2, ..., x_k$ necessarily appear in the premises.

Stage 1: An ordinary linear model is identified. Its UC is calculated and written as $\text{UC}_{[1]}$.

Stage 2: A fuzzy model consisting of two fuzzy implications is constructed by first dividing the range of x_1 into two fuzzy subspaces 'Small' and 'Large' as follows:

L^1 : If x_1 is 'Small' then ...,

L^2 : If x_1 is 'Large' then ..., (8)

Then the premise parameters, the consequent structure and parameters are identified. The UC of the model is calculated. Similarly, a fuzzy model dividing the range of x_2 is identified and its UC is calculated. In this way, k fuzzy models are identified and their UC's are calculated. Among the k models, one with the least *UC* is picked up. Its premise structure and *UC* are writtten as $ST_{[2]}$ and $UC_{[2]}$, respectively.

Stage 3: Suppose that the input variable x_j is found to be put into the premises at stage 2. Then $ST_{[2]}$ is as shown in Figure 5(a). At this stage, each premise structure consists of three fuzzy subspaces and there are two ways in constructing a premise structure. One is that the range of x_j itself is divided into three fuzzy subspaces as shown in Figure 5(b). The other is that another variable, say x_j, is put into $ST_{[2]}$ and its range is divided into two fuzzy subspaces as shown in Figures 5(c) and (d). The fuzzy model with the premise structure of Figure 5(b) is

L^1 : If x_j is 'Small' then ...,

L^2 : If x_j is 'Medium' then ..., (9)

L^3 : If x_j is 'Large' then ...,

and the fuzzy model with the premise structure of Figure 5(d) is

L^1 : If x_j is 'Small$_1$' then ...,

L^2 : If x_j is 'Large$_1$' and x_i is 'Small$_2$' then ..., (10)

L^3 : If x_j is 'Large$_1$' and x_i is 'Large$_2$' then ....

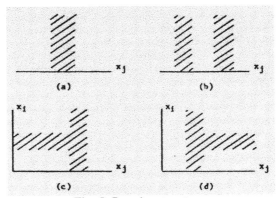

Fig. 5. Premise structures.

The variables which are put into the premises at stage i ($i \geq 3$) are, however, not all the k input variables, which will be explained later.

For each premise structure which can be constructed at this stage, a fuzzy model is identified and its UC is calculated. Among those fuzzy models, one with the least UC is picked up. Its premise structure and UC are similarly written as $ST_{[3]}$ and $UC_{[3]}$, respectively. If $UC_{[3]}$ is larger than $UC_{[2]}$, the process is terminated and the optimal premise structure is found to be $ST_{[2]}$. Otherwise the process is proceeded to the next stage 4.

The stages after stage 3 are similar to stage 3. That is, at stage i ($i \geq 3$) fuzzy models consisting of i fuzzy implications are constructed in two ways. One is that one of the fuzzy subspaces of $ST_{[i-1]}$, is divided with respect to one of the premise variables of $ST_{[i-1]}$ as illustrated in Figure 5(b). The other is that another new variable is put into $ST_{[i-1]}$, and one of the fuzzy subspaces of $ST_{[i-1]}$ is divided with respect to the new variable just as in Figures 5(c) and (d). $ST_{[i]}$ and $UC_{[i]}$ are determined in the sense of the least UC. Whether the process is terminated or not is decided by comparing $UC_{[i]}$ with $UC_{[i-1]}$.

When the number of input variables is large, the number of the possible structures at each stage becomes combinatorialy large. Therefore, it is necessary to heuristicaly find which variables are possibly put into the premises. At stage 2 we divide the one-dimensional space of each premise variable into two fuzzy subspaces: a fuzzy model consists of two implications. If a model at this stage has a larger UC than the $UC_{[i]}$ (UC of a linear model), its structure is not adequate and this implies that its premise variable should not appear in the premises at the stages after stage 2.

Example: In order to verify the algorithm, we use the following nonlinear system.

$$y = (1 + x_1^{0.5} + x_2^{-1} + x_3^{-1.5})^2. \tag{11}$$

For this system, Kondo [4] identified a model by using the improved GMDH method which can construct polynomial models with non-integer degrees. He generated 20 input-output data for the identification and 20 data for the verification of the model. The identified model is

$$y = -3.1 + 5.2 x_1^{0.5118} x_2^{-0.3044} + 3.3 x_1^{0.4456} x_3^{-0.3371} \\ + 10.2 x_2^{-0.3174} x_3^{-0.5859}. \tag{12}$$

We identified a fuzzy model of the system of Eq. (11) by using the same identification data. The process of the premise structure identification is as follows.

Stage 1: A linear model is identified and the $UC_{[1]}$ is 3.8:

$$y = 15.3 + 19.7 x_1 - 1.35 x_2 - 1.57 x_3. \tag{13}$$

Stage 2: As there are four input variables in the data, four premise structures can be formed at this stage. Table 1 shows the UC and premise variable of each structure. Since the structure (1) with x_3 in the premise gives the least UC, it is adopted as $ST_{[2]}$, and $UC_{[2]}$ is 3.3. The fuzzy model with $ST_{[2]}$ is shown in Figure 6.

Table 1. The UC and the premise variable of each structure

Structure number	1	2	3	4
Premise variable	x_1	x_2	x_3	x_4
UC	5.4	3.5	3.3	4.6

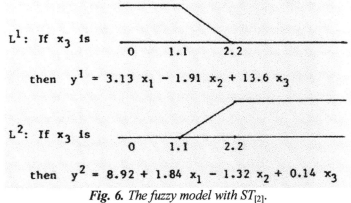

Fig. 6. The fuzzy model with $ST_{[2]}$.

The variables x_1 and x_4 should not be put into the premises at the stages after stage 2 because the structures with those variables in the premises give the larger UCs than $UC_{[1]}$.

Stage 3: Three premise structures having three fuzzy subspaces are formed, as shown in Figure 7, by dividing a fuzzy subspace of the $ST_{[2]}$. Since the UC of the structure (3) in Figure 7 is minimal, this structure is adopted as $ST_{[3]}$, and $UC_{[3]}$ is 2.8. The fuzzy model with $ST_{[3]}$ is shown in Figure 8. We proceed to the next stage since $UC_{[3]}$ is smaller than $UC_{[2]}$.

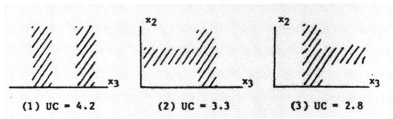

Fig. 7. Premise structures in stage 3.

Stage 4: The structures of this stage are shown in Figure 9. $ST_{[4]}$ is the structure (3) in Figure 9, and $UC_{[4]}$ is 3.4. The fuzzy model with $ST_{[4]}$ is shown in Figure 10. Since $UC_{[4]}$ is larger than $UC_{[3]}$, the optimal premise structure is found to be $ST_{[3]}$.

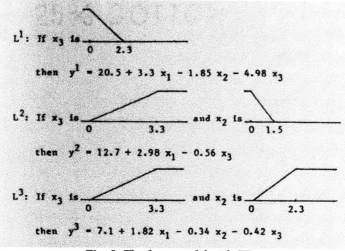

Fig. 8. The fuzzy model with $ST_{[3]}$.

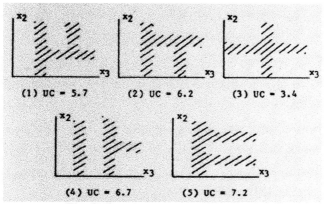

Fig. 9. Premise structures in stage 4.

Kondo [5] used the following function to evaluate a model:

$$J = \left(\sum_{i=1}^{n} |y_i - y_i^*|/y_i\right) \times (1/n) \times 100(\%), \tag{14}$$

where n is the number of data and y_i^* is the estimated value of y_i. Let J_1 be the value of J for the identification data, and J_2 that for the verification. At the improved GMDH model of Eq. (12), J_1 is 4.7% and J_2 is 5.7%. At the fuzzy model of Figure 8, J_1 is 1.5% and J_2 is 2.1%. J_1 of the fuzzy model of Figure 10 is 0.59% and better than that of the model of Figure 8 since the former model has one more fuzzy implication than the latter. J_2 of the model of Figure 10 is, however, 3.4% and worse than that of the model of Figure 8. Therefore, we can see that the model of Figure 10 does not have the true structure. The criterion for verification of structure, UC, also shows that the model of Figure 8 is better than that of Figure 10. Table 2 shows the results.

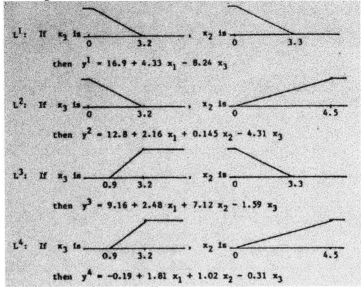

Fig. 10. The fuzzy model with $ST_{[4]}$.

Table 2. The error and UC of each model

Model	$J_1(\%)$	$J_2(\%)$	UC
Linear Model (Eq. (13))	12.7	11.1	3.8
Improved GMDH Model (Eq. (12))	4.7	5.7	-
Fuzzy Model (Figure 8)	1.5	2.1	2.1
Fuzzy Model (Figure 10)	0.59	3.4	3.4

3.3. Premise parameters identification

In this paper, as stated before, the membership function of a fuzzy variable is of a convex type and formed by straight lines as shown in Figure 11. The parameters of a premise variable are the coordinates such as $p_1, p_2, ..., p_8$ corresponding to its vertices in Figure 11: the grade of membership of a parameter is either zero or one.

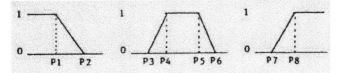

Fig. 11. The types of fuzzy variables and premise parameters.

When input-output data are given, for the premise parameters identification we have used a complex method: an optimization technique of nonlinear programming problems. The complex method is, however, not adequate when we are to successively adjust the parameters by using the obtained data. Here we suggest a successive identification algorithm: as a piece of input-output data is processed, the premise parameters are adjusted.

The algorithm of the adjustment of the premise parameters

When a piece of input-output data is processed, we adjust the parameters so that the error of the inferred output decreases, where the parameters are those that have been estimated by using the past data. For the data $(y^0, x^0_1, x^0_2, \ldots x^0_p)$, the error of the output y^* inferred from Eqs. (2) and (3) becomes

$$\text{error} = y^0 - y^*$$

$$= y^0 - \sum_{i=1}^{q} \prod_{j=1}^{p} A_j^i(x_j^0) y^j \bigg/ \sum_{i=1}^{q} \prod_{j=1}^{p} A_j^i(x_j^0). \tag{15}$$

The algorithm for adjusting the premise parameters by using the data $(y^0, x_1^0, x_2^0, \ldots, x_p^0)$ is as follows.

(1) For the input $(x_1^0, x_2^0, \ldots, x_p^0)$, calculate the grade of membership of each premise fuzzy variable and the output y^i ($i = 1, 2, \ldots, q$) inferred from each implication.

(2) Select the fuzzy variables to be adjusted, as follows.

For each premise variable, two fuzzy variables are selected. When there are p variables in the premise, the number of variables to be selected is $2 \times$ accordingly.

Suppose that we are to select the fuzzy variables concerned with the premise variable x_k. There are two cases according to the value of x_k^0. One is that for x_k^0, two fuzzy variables have grades of membership greater than zero as seen in Figure 12(a). In this case, the two fuzzy variables (A_k^m and A_k^n of Figure 12(a)) are selected. The other is that for x_k^0, one fuzzy variable has grade of membership greater than zero as seen in Figure 12(b). Here the fuzzy variable (A_k^n of Figure 12(b)) and a fuzzy variable which has parameter nearest to x_k^0 (A_k^m of Figure 12(b)) are selected.

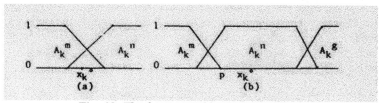

Fig. 12. *The fuzzy variables to be adjusted.*

(3) Adjust the grades of membership of the selected fuzzy variables as follows.

Suppose that we are to adjust the grades of membership of A_k^m and A_k^n for x_k^0

For simplicity, the present values of $A_k^m(x_k^0)$ and $A_k^m(x_k^0)$ will be written as W_m^0 and W_n^0, respectively.

By setting the error at zero, we can obtain the next equation from Eq. (15):

$$\left[\sum_{i=1}^{q}\prod_{j=1}^{p}A_j^i(x_j^0)\right]y^0 - \sum_{i=1}^{q}\prod_{j=1}^{p}A_j^i(x_j^0)y^i = 0. \tag{16}$$

In Eq. (16), substitute the values obtained at (1) for $A^i(x_j^0)$ and y_i ($i = 1, 2, \ldots,$ q and $j = 1, 2, \ldots, p$) except $A^m{}_k(x^0{}_k)$ and $A^n{}_k(x^0{}_k)$. Rewriting $A^m{}_k(x^0{}_k)$ and $A^n{}_k(x^0{}_k)$ in Eq. (16) as w_m and w_n, respectively, we obtain the next linear equation of w_m and w_n:

$$a_1 w_n + a_2 w_m + a_3 = 0 \tag{17}$$

where, $a1$, $a2$ and $a3$ are constants. Find the nearest value (w^*_m, w^*_n) to (w^0_m, w^0_n) on the line expressed by Eq. (17) under the constraint Eq. (18), as shown in Figure 13:

$$\omega_1 < w_m + w_n < \omega_2 \tag{18}$$

where, ω_1 and ω_2 are constants: we set $\omega_1 = 0.7$ and $\omega_2 = 1$ in this algorithm.

When there is none satisfying the constraint, $A^m{}_k$ and $A^n{}_k$ are not adjusted. The new grades of membership ($(\tilde{w}_m, \tilde{w}_n)$) are calculated, using the adaptive gain α, from (w_n^0, w_n^0) and (w_n^*, w_n^*)

$$\tilde{w}_n = w_m^0 + \alpha(w_n^* - w_m^0) \tag{19}$$

where $\alpha = \alpha_1 \times \alpha_2$, and

$$\alpha_1 = (1/h)^{1/2}, \tag{20}$$

$$\alpha_2 = 1/(1+ | w_m^* - w_m^0 |), \tag{21}$$

with h the number of all the fuzzy variables to be adjusted. The adaptive gain α has a significant effect on the convergence properties. The gain α_2 is set to avoid an excessive adjustment.

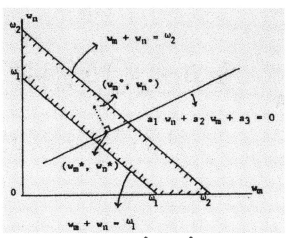

Fig. 13. w_m^* and w_n^*.

(4) Adjust the premise parameters by using the new grade of membership as follows.

Suppose we are to adjust the parameters of A_k^m and A_k^n by using $\tilde{w}_m$ and $\tilde{w}_n$.

(a) When $w_m^0 > 0$ and $w_n^0 > 0$ (see Figure 12(a)): adjust the parameters of A_k^m and A_k^n so that their grades of membership for x_k^0 become $\tilde{w}_m$ and $\tilde{w}_n$, respectively.

(i) When $\tilde{w} \geq 0.5$: adjust the parameter of which the grade of membership is one, as shown in Figure 14.
(ii) When $w < 0.5$: adjust the parameter of which the grade of membership is zero, as shown in Figure 15.

(b) When $w_m^0 = 0$ (see Figure 12(b)): If the parameter of A_k^m is adjusted so that its grade of membership for x_k^0 is $\tilde{w}_m$, the amount of adjustment becomes excessive. Therefore, adjust the parameter of which the grade of membership is zero so that it gets nearer to x_k^0 (see Figure 16).

For A_k^n, adjust the parameter of which grade of membership is one, as shown in Figure 16.

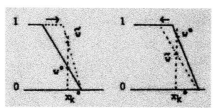

Fig. 14. Adjustment when $\tilde{w} \geq 0.5$.

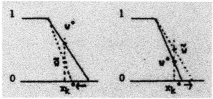

Fig. 15. Adjustment when $\tilde{w} < 0.5$. **Fig. 16.** Adjustment when $w_m^0 = 0$.

The amounts of the adjustments in Figure 16 are as follows:

$$\Delta p_m = \tilde{w}_m \beta \,|\, x_k^0 - p_m\,|, \qquad (22)$$

$$\Delta p_m = \tilde{w}_m \beta \,|\, x_k^0 - p_m\,|, \qquad (23)$$

The gain β decreases the amount of adjustment when the parameter p_m of A_k^m is far from x_k^0, as shown in Figure 17:

$$\beta = (1 - |\, x_k^0 - p_m\,| / L)^3 \qquad (24)$$

where L is a large constant.

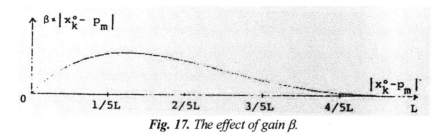

Fig. 17. *The effect of gain β.*

Example: In order to verify the performance of the successive identification, we use a fuzzy system shown in Figure 18. Now suppose that we have a fuzzy model which is the same as Figure 18, and the values of parameters of the system, $p_1, p_2, \ldots, p_8$ are changed from (4.5, 5.5, 4.5, 5.5, 4.5, 5.5, 4.5, 5.5) to (2.0, 5.5, 3.0, 6.0, 5.5, 8.0, 6.0, 9.0). By using data obtained from the new fuzzy system, we adjust the premise parameters of the model. Table 3 shows the results of the adjustment. We can see that after about 100 adjustments the parameters nearly converge to the new values. In order to evaluate the precision of the model, we generate 200 data from the new fuzzy system, and calculate the correlation of the data and the estimated values of the data. At Table 3, CR is the correlation.

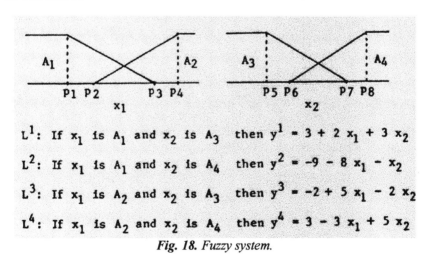

Fig. 18. *Fuzzy system.*

3.4. Consequent structure identification

All the input variables do not always appear in the consequent of an implication. So we have to find which variables are necessary in the consequent of an implication. This is the same kind of problem as finding input variables in a linear model. If a fuzzy model has q fuzzy implications and p input variables, and if each consequent includes all the p input variables, the output is represented by a linear equation with $q \times (p+1)$ terms as shown in Eq. (4). The $q \times (p+1)$ terms are as follows:

$$(g^1), (g^1 x_1), (g^1 x_2), \ldots, (g^1 x_p),$$
$$(g^2), (g^2 x_1), (g^2 x_2), \ldots, (g^2 x_p),$$
$$\vdots$$
$$(g^q), (g^q x_1), (g^q x_2), \ldots, (g^q x_p).$$

Actually we have to choose some necessary ones from the above $q \times (p + 1)$ terms. For example, suppose that the number of implications is 2, i.e. $q = 2$, and the number of input variables is 3, i.e. $p = 3$. When the chosen terms are g^1, $g^1 x_2$, $g^1 x_3$, $g^2 x_1$ and $g^2 x_2$, the fuzzy model is as follows:

$$\begin{aligned} L^1 &: \text{If} \ldots \text{then } y^1 = c_1^1 + c_2^1 x_2 + c_3^1 x_3, \\ L^2 &: \text{If} \ldots \text{then } y^2 = c_1^2 x_1 + c_2^2 x_2, \end{aligned} \quad (25)$$

Here we also use the unbiasedness criterion UC as the criterion of choice: We choose the terms which give the minimal UC.

As the method for sequentially choosing the terms, we use the backward elimination method or the stepwise backward regression method which is the method for selecting variables in a linear model.

4. Application to prediction of water flow rate in the river Dniepr

4.1. *The problem*

The problem is to predict the average annual rate of water flow in the river Dniepr using those values of the river Niemen, a main upper stream of the Dniepr, and Wolf's number concerned with sunspot activity. The prediction duration is 7 years and the data from 1812 to 1971 are given. Ivakhnenko [3] has constructed a prediction model by using the GMDH method, where the data from 1812 to 1964 are used for constructing the model and those from 1965 to 1971 for checking the accuracy of predictions. Ivakhnenko's model consists of two parts, $f_{tr}(t)$, the harmonic trend of time, and $f(x_1, x_2, \ldots)$, the prediction of the remainder on the basis of input variables. It is expressed as follows:

$$y_{t+7} = f_{tr}(t) + f(y_{t-1}, y_{t-7}, y_{t-12}, r_{t-1}, r_{t-2}, r_{t-8}, r_{t-9}, w_t, w_{t-10}, w_{t-12}) \quad (26)$$

where y_{t-i} is the water flow rate of Dniepr ($\times$ 100 m^3/sec), r_{t-i} that of Niemen ($\times$ 100 m^3/sec), and w_{t-i} Wolfs number ($\times$ 10 mm^{-2}). First, the harmonic trend of time is constructed. Then, for the remainder, $\Delta y = y_{t+7} - f_{tr}$, a prediction function of input variables is constructed. In Ivakhnenko's model, this prediction function has 76 parameters.

Here we construct a fuzzy prediction model for the remainder.

4.2. The results

Figure 19 shows the identified fuzzy model with three implications. In order to check the prediction accuracy, the mean square error Δ of the checking data is calculated.

$$\Delta = \sum_{t=1965}^{1971} (y_t - y_t^*)^2 \bigg/ \sum_{t=1965}^{1971} y_t^* \quad (27)$$

Where y_t^* is the predicted value of y_t.

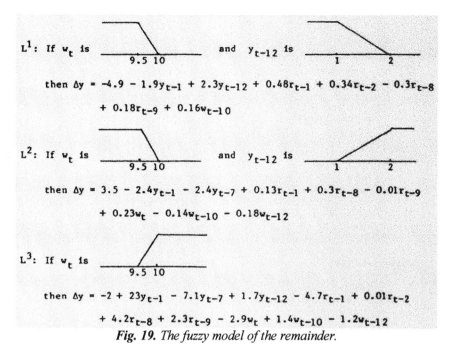

Fig. 19. The fuzzy model of the remainder.

While Δ is 1.3% at the GMDH model, it is 0.93% at the fuzzy model and the number of parameters is much smaller. Figure 20 shows the results of the prediction by the fuzzy model.

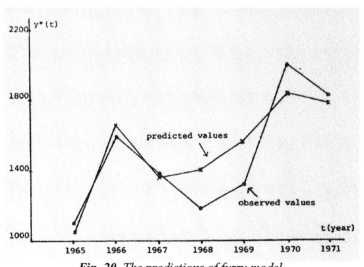

Fig. 20. The predictions of fuzzy model.

5. Conclusion

We have formulated the problems of structure identification of a fuzzy model. A criterion for the verification of a structure has been found. Using the criterion, an algorithm for choosing a premise structure has been suggested.

Further, a successive identification algorithm of the premise parameters and a method to find the consequent variables have been suggested.

Finally, the proposed methods have been applied to an example and their feasibility has been confirmed.

References

[1]. T. Takagi and M. Sugeno, Fuzzy identification of systems and its applications to modeling and control, *IEEE Trans. Systems Man Cybernet.* **15** (1985) 116-132.
[2]. A.G. Ivakhnenko, V.N. Vysotskiy and N.A. Ivakhnenko, Principal versions of the minimum bias criterion for a model and an investigation of their noise immunity, *Soviet Automat. Control* **11** (1978) 27-45.
[3]. A.G. Ivakhnenko and M.M. Todua, Prediction of random processes using self-organization of the prediction equation - part 1. Problems of simple medium-term prediction, *Soviet Automat. Control* **5** (1972) 35-51.
[4]. T. Kondo, Revised GMDH algorithm estimating degree of the complete polynomial, *Trans. Soc. Instrument and Control Engrs.* **22** (1986) 928-934 (in Japanese).
[5]. H. Akaike, A new look at the statistical model identification, *IEEE Trans. Automat. Control* **19** (1974) 716-723.

Successive Identification of a Fuzzy Model and its Applications to Prediction of a Complex System[*]

Michio Sugeno and Kazuo Tanaka
Department of Systems Science,
Tokyo Institute of Technology,
4259 Nagatsula, Midori-ku.
Yokohama 227, Japan

Received March 1989
Revised June 1989

Abstract

A successive identification method of a fuzzy model is suggested. The identification mechanism consists of two levels. One is the supervisor level and the other is the adjustment level. The supervisor level determines a policy of parameter adjustment using a set of fuzzy adjustment rules. The adjustment rules are derived from the fuzzy implications of a fuzzy model and are extended to fuzzy adjustment rules by using an extended concept of Zadeh's contrast intensification. The adjustment level executes the policy of parameter adjustment determined with the fuzzy adjustment rules. The parameter adjustment consists of premise parameter adjustment and consequent parameter adjustment. Both of them are realized by the weighted recursive least square algorithm. Finally, it is shown from two examples that the method is very useful for modeling complex systems

---

[*] © 1991 North-Holland. Retyped with written permission from *Fuzzy Sets and Systems, 42 (1991) 315-334.*

Keywords: fuzzy model, successive fuzzy modeling, contrast intensification, fuzzy adjustment rule.

1. Introduction

Many kinds of fuzzy models in control processes have been developed since Mamdani's paper [3] was published. Most of these models are expressed by a set of fuzzy linguistic propositions which are derived from the experience of skilled operators and knowledge of manual controls. However, a multi-input/multi-output fuzzy model has the following difficult problems to describe the behavior of a complex system:
(1) many fuzzy linguistic propositions, and
(2) a multidimensional fuzzy relation.

Identification and parameter adjustment for such a fuzzy model are very difficult. One of the possible approaches to overcome the difficulties is to develop a new type of fuzzy model.

One of the authors has reported in previous papers [5, 6, 7] that the modeling method using a new type of fuzzy model is very useful in the field of system modeling. The new type of fuzzy model is described by fuzzy implications which locally represent linear input-output relations of a system. The modeling method is called 'fuzzy modeling'. This is a method of off-line identification. This paper presents a successive identification algorithm, of the new type of fuzzy model. The method is named 'successive fuzzy modeling'. It is an extended version of off-line fuzzy modeling. It is possible to realize fuzzy adaptive control and fuzzy learning control if we have a successive identification method of a fuzzy model.

This paper is organized as follows. Section 2 shows two types of fuzzy models. Section 3 describes the successive identification algorithm Section 4 gives two numerical examples.

2. Fuzzy models

We have two types of fuzzy implications. One is a type of linguistic proposition such as 'IF x is A THEN y is B', where A and B are fuzzy sets. The other is a new type of fuzzy implication such as 'IF x is A THEN $Y = f(x)$', where f is a linear function. This paper deals with the new type of fuzzy implication

2.1. *Ordinary fuzzy model*

The fuzzy linguistic propositions of ordinary fuzzy models are of the following form:

$$L^i : \text{IF } x_1 \text{ is } A_1^i \text{ and} \cdots x_1 \text{ is } A_1^i \text{ THEN } y \text{ is } C^i. \tag{1}$$

where L^i ($i = 1, 2, \ldots, l$) denotes the i-th implication, l is the number of fuzzy implications, A_j^i and C^i are fuzzy sets, x_j is the input variable, and y is the output variable. For simplicity, the membership function in the fuzzy set A is written as $A(x)$ from now on.

From (1), we can obtain the following fuzzy relational equations:

$$C^i : R^i \circ (A^i_1 \times ... \times A^i_n), \quad i = 1, 2, ..., l, \tag{2}$$

where $\circ$ denotes the max-min composition and $\times$ denotes the Cartesian product. If we use Mamdani's method [3], the fuzzy relations R^i are represented as follows:

$$R_i = \int_{X_1 \times X_2 \times \cdots \times X_n, Y} \min(A^i_1(x_1), A^i_1(x_2), ..., A^i_n(x_n), C^i(y) / (x_1, x_2, ... x_n, n)). \tag{3}$$

The final relation matrix R is calculated as:

$$R = \bigcup_{i=1}^{l} R^i, \tag{4}$$

where $\cup$ denotes the union operator.

Some studies [2, 8] have reported on fuzzy model identification algorithms. It is well known that a fuzzy model identification algorithm using referential fuzzy sets have been proposed by Pedryez [4].

2.2. Takagi and Sugeno's fuzzy model

A new type of fuzzy implication, suggested by Takagi and Sugeno [7], is of the following form:

$$\begin{aligned} L_i : \text{IF } x_i \text{ is } A^i_1 \text{ and} ... \text{and } x_n \text{ is } A^i_n \\ \text{THEN } y^i = c^i_0 + c^i_1 x_1 + ... + c^i_n x_n, \end{aligned} \tag{5}$$

where y^i is the output from the i-th implication, and A^i_j is a fuzzy set whose membership function is of convex type and formed by straight lines. c^i_j is a consequent parameter.

Given an input $(x_1, x_2, ..., x_n)$, the final output of the fuzzy model is inferred by taking the weighted average of the y^i=s:

$$y = \sum_{i=1}^{l} w^i y^i / \sum_{i=1}^{l} w^i, \tag{6}$$

where y^i is calculated for the input by the consequent equation of the i-th implication, and the weight w_i implies the overall truth value of the premise of the i-th implication for the input calculated as

$$w^i = \prod_{k=1}^{n} A^i_k(x_k). \tag{7}$$

3. Algorithm of successive fuzzy modeling

In order to successively identify a fuzzy model, we must determine the structure and the initial parameters of a fuzzy model. This model is called the 'initial model'.

3.1. *Identification of the initial model* [6]

The initial model is identified by the off-line fuzzy modeling method using some pairs of input-output data. The identification of the initial model is divided into two parts: structure identification and parameter identification. The former consists of premise structure identification and consequent structure identification. The latter also consists of premise parameter identification and consequent parameter identification. The most important thing for identifying the initial model is to determine the structure of the fuzzy model. In the structure identification, we must find the input variables which affect the outputs. Here we utilize the unbiasedness criterion which is used as a proper criterion for the verification of structure in GMDH [6]. The outline of the identification of the initial model is discussed in Appendix A.

3.2. *Successive identification of the fuzzy model*

The parameters of the initial model are successively identified by successive fuzzy modeling. Figure 1 shows the outline of the successive identification algorithm. It consists of two major levels, the supervisor level and the adjustment level.

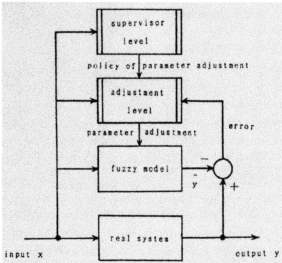

Fig. 1. Outline of the successive identification algorithm.

The supervisor level determines a policy of parameter adjustments using a set of fuzzy adjustment rules (FAR). In the adjustment level, the parameters of a fuzzy model are adjusted according to the policy determined by the FAR. The parameter adjustment consists of premise parameter adjustment and consequent parameter adjustment. Both are performed using the weighted recursive least square algorithm (WRLSA). The details of WRLSA are discussed in Appendix B. The consequent parameters c^i_js can be easily adjusted in the same manner as the successive identification of linear equations. However, it is difficult to identify successively the parameters of the membership functions A^i_js. It is convenient to use the following method to identify the premise parameters since the membership functions of a fuzzy model are of convex type and formed by straight lines. For example, the membership functions can be expressed as follows:

$$A(x) = \begin{cases} 0, & x < q_1 \\ ax+b, & q_1 \leq x \leq q_2 \\ 1, & q_2 < x, \end{cases} \quad (8)$$

$$B(x) = \begin{cases} 0, & x < q_3, \\ cx+d, & q_3 \leq x \leq q_4, \\ 1, & q_4 < x < q_5, \\ ex+f, & q_5 \leq x \leq q_6, \\ 0, & q_6 < x, \end{cases} \quad (9)$$

where

$$q_1 = -b/a, \quad q_2 = (1-b)/a, \quad q_3 = -d/c,$$
$$q_4 = (1-d)/c, \quad q_5 = (1-f)/e, \quad q_6 = -f/e,$$
$$q_1 \leq q_2, \quad q_3 \leq q_4 \leq q_5 \leq q_6,$$

q_1 - q_6 are break points of the membership functions. In (8) and (9), a - f are the parameters of oblique lines of the membership functions. a - f are named premise parameters. The premise parameter adjustment is realized by identifying successively the premise parameters a - f of the oblique lines.

Since the fuzzy model usually has many parameters, it is not easy to adjust these parameters. It is necessary to impose a restriction on the premise parameters of fuzzy models. Figure 2(b) shows the restriction. We can see from Figure 2(b) that the break points of two membership functions, which overlap each other, are equalized, that is, $q_1 = q_3$ and $q_2 = q_4$. This restriction reduces the number of the premise parameters by half.

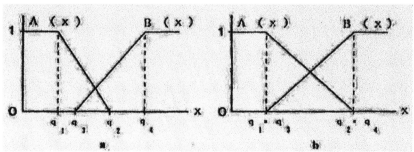

Fig. 2. Restriction of membership functions. (a) Unrestricted membership functions, (b) restricted membership functions.

Let us explain the adjustment algorithms of two levels through the fuzzy model shown in Figure 3.

3.2.1. Supervisor level

(A) Derivation of FAR: The FAR is derived from the fuzzy implications of the fuzzy model shown in Figure 3. The derivation process of FAR consists of two steps:
(1) derivation of adjustment rules,
(2) extension to fuzzy adjustment rules.

Step 1: *Derivation of adjustment rule.* It is possible to partition the input space of the fuzzy model into some areas according to adjustment policy. Figure 4(a) shows five areas of the input space for the fuzzy model. The areas U^1 - U^5 are partitioned by the break points (s_1 - s_4) of the membership functions. The areas U^1 - U^5 are crisp subsets on $X_1 \times X_2$. U^1 - U^5 can be classified into the following types.

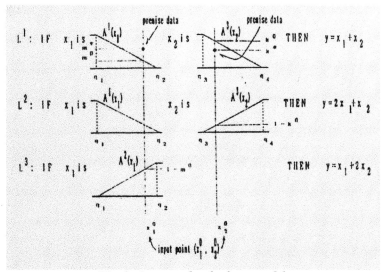

Fig. 3. An example of a fuzzy model.

(**Type A1**: areas U^1 - U^3) U^1, U^2 and U^3 are the areas which are expressed only by the linear equation of the consequent parts of L^1, L^2 and L^3, respectively.

(**Type A2**: area U^4) U^4 is the area mainly characterized by the grades of membership of x_2 in A^3 and A^4.

(**Type A3**: area U^5) U^5 is the area mainly characterized by the grades of membership of x_1 in A^1 and A^2.

From these viewpoints, the following adjustment rules can be derived
R^1: IF (x_1, x_2) is U^1 THEN adjust the consequent parameters of L^1,
R^2: IF (x_1, x_2) is U^2 THEN adjust the consequent parameters of L^2,
R^3: IF (x_1, x_2) is U^3 THEN adjust the consequent parameters of L^3,
R^4: IF (x_1, x_2) is U^4 THEN adjust the premise parameters of $A^3(x_2)$, and $A^4(x_2)$,
R^5: IF (x_1, x_2) is U^5 THEN adjust the premise parameters of $A^1(x_1)$, and $A^2(x_1)$,

where R^1, R^2 and R^3 are rules for the consequent parameter adjustment, and R^4 and R^5 are rules for the premise parameter adjustment. R^1 means that if a given input (x_1, x_2) belongs to U^1, then adjust the consequent parameters of L^1, and R^4 means that if a given input (x_1, x_2) belongs to U^4, then adjust the premise parameters of $A^3(x_2)$, and $A^4(x_2)$,. These rules are based on types $(A1)$, $(A2)$ and $(A3)$.

Here, assume that four input points $(p_1 - p_4)$ on the input space are given as shown in Figure 4(a). Moreover, assume that $(p_1 - p_4)$ have the following coordinates on $X_1 \times X_4$, respectively:

$$p_1: (x^0{}_{11}, x^0{}_{21}),$$
$$p_2: (x^0{}_{12}, x^0{}_{22}),$$
$$p_3: (x^0{}_{13}, x^0{}_{31}),$$
$$p_4: (x^0{}_{14}, x^0{}_{41}).$$

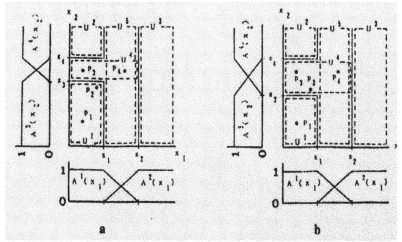

Fig. 4. *Fuzzy partition on input space.* (a) *Time instant t-1,* (b) *time instant t.*

Then, the policy of parameter adjustment for each input point is determined as (B1) - (B3) if the adjustment rules are used. However, this is not the case. The reason will be discussed in Step 2.

(B1) Since p_1 and p_2 belong to U^1, the consequent parameters of L^1 should be adjusted. <type (A1)>.

(B2) Since p_3 belongs to U^4, the premise parameters of $A^3(x_2)$ and $A^4(x_2)$ and should be adjusted. <type (A2)>.

(B3) Similarly, since p_4 belongs to U^4 and U^5, the premise parameters $A^1(x_1)$, $A^2(x_1)$, $A^3(x_2)$ and $A^4(x_2)$ should be adjusted <types (A2) and (A3)>

Step 2: *Extension to fuzzy adjustment rules.* In many model fitting problems, input-output data generated by a real complex system are often contaminated by noise. So we must consider the influence of noise. For input-output data with noise, $R^1 - R^5$ may be influenced considerably. To avoid the influence of noise, the adjustment rules are extended to fuzzy adjustment rules (FAR) as follows:

FR^1: IF (x_1, x_2) is $\tilde{U}^1$ THEN adjust the consequent parameters of L^1,
FR^2: IF (x_1, x_2) is $\tilde{U}^2$ THEN adjust the consequent parameters of L^2,
FR^3: IF (x_1, x_2) is $\tilde{U}^3$ THEN adjust the consequent parameters of L^3,
FR^4: IF (x_1, x_2) is $\tilde{U}^4$ THEN adjust the premise parameters of $A^3(x_2), A^4(x_2)$,
FR^5: IF (x_1, x_2) is $\tilde{U}^5$ THEN adjust the premise parameters of $A^1(x_1), A^2(x_1)$,
where, $\tilde{U}^1 - \tilde{U}^5$ are fuzzy sets on $X_1 \times X_2$,

Let us show the validity of the extended fuzzy adjustment rules for input output data contaminated by noise. Now, assume that the premise parameters of $A^3(x_2)$ and $A^4(x_2)$ are adjusted at time instance t as shown in Figure 4(b). The details of premise parameter adjustment will be discussed in Section 3.2.2. It is seen from Figure 4 that p_1 belongs to U^1 at time instance t as usual, whereas p_2 belongs to U^4. We may interpret from this fact that the grade of membership of p_1 in U^1 differs from that of p_2 in U^1 at time instance $t - 1$ when input-output data are contaminated by noise. In other words, though p_1 and p_2 completely belong to U^1 at time instance $t - 1$, that is,

$$U^1(x_{12}^0, x_{22}^0) = U^1(x_{11}^0, x_{21}^0) = 1, \tag{10}$$

we should interpret at time instance t -1 as follows:

$$0 \leq U^1(x_{12}^0, x_{22}^0) < U^1(x_{11}^0, x_{21}^0) \leq 1, \tag{11}$$

where $U^1(x^0{}_{11}, x^0{}_{21})$ and $U^1(x^0{}_{12}, x^0{}_{22})$ denote the grades of membership of p_1 and p_2 in U^1, respectively.

It is obvious that U^1 must be a fuzzy set since a crisp set cannot satisfy (11).

The membership values of $\tilde{U}^1$ ($i = 1, 2, \ldots, 5$) will be utilized as weights of the WRLSA in the adjustment level. These weights can be regarded as safety coefficients for the observed data contaminated by noise. It is easily guessed that $R^1 - R^5$ may be unadaptable for input-output data contaminated by noise since the weight of WRLSA is 0 or 1.

Thus, we can expect that the parameters of the fuzzy model are adjusted adaptively for the contaminated data using FAR.

However, we must determine $\tilde{U}^1 - \tilde{U}^5$. The derivation of $\tilde{U}^1 - \tilde{U}^5$ is given below.

(B) *Derivation of* $\tilde{U}^1 - \tilde{U}^5$: We determine the fuzzy sets $\tilde{U}^1 - \tilde{U}^5$ such that the influence of noise disturbances is reduced. Since the identification algorithm proposed here is successive identification, these fuzzy sets are recursively modified every time input-output data are given. This means that the weights for applying the WRLSA are modified.

The deviation is automatically realized using an extended concept of the contrast intensification which was proposed by Zadeh [10].

Zadeh's contrast intensification is defined as

$$\text{INT}(A)(x) = \begin{cases} 2A^2(x), & 0 \leq A(x) \leq 0.5, \\ 1 - 2(1 - A(x))^2, & 0.5 \leq A(x) \leq 1, \end{cases} \tag{12}$$

where A is a fuzzy set, and $A(x)$ is the grade of membership of x in A.

The operator suggested here is defined as:

$$\text{INT}(A)(x) = \begin{cases} 2^{m-1} A^m(x), & 0 \le A(x) \le 0.5, \\ 1 - 2^{m-1}(1 - A(x))^m, & 0.5 \le A(x) \le 1, \end{cases} \qquad (13)$$

where m is a real number. Equation (13) is equal to Equation (12) when $m = 2$. Therefore (13) can be regarded as a generalization of (12). The fuzzy set INT(A, m) is reduced to a crisp set when $m \to \infty$.

$\tilde{U}^1 - \tilde{U}^5$ are generated from $U^1 - U^5$ by modifying the parameter m in (13). The main idea to modify the parameter m is based on the following:

(D1) When the premise parameters, which partition the input space of a fuzzy model, have converged to a fixed value, we can adjust the parameters of the fuzzy model without considering the influence of noise disturbances.

(D2) Conversely, when the premise parameters change or oscillate, we must consider the influence of noise disturbances.

(D2) requires that the safety coefficient for the contaminated data should be used. In this situation, we should decrease the parameter m of the contrast intensification. Conversely, we should increase the parameter m in the situation of (D1). These aspects will be utilized at Step 3 in the calculation processes of $\tilde{U}^1 - \tilde{U}^5$, which consists of five steps. These steps are executed every time input-output data are given. Figure 5 shows the calculation process of $\tilde{U}^1$.

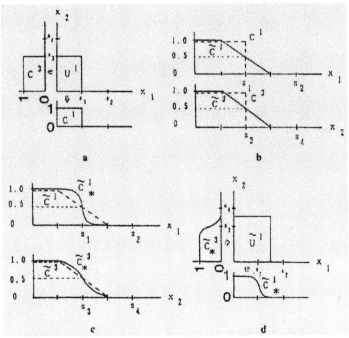

Fig. 5. *Calculation process of $\tilde{U}^1$. (a) Step 1 (projection), (b) Step 2 (fuzzification), (c) Step 3 and 4 (contrast intensification), (d) Step 5 (reconstruction of $\tilde{U}^1$).*

Step 1: From $U^1 - U^5$ in $R^1 - R^5$ and the definition of projection, we can obtain

$$C^1(x_1) = \text{proj}(U^1 : X_1) = \text{proj}(U^2 : X_1), \quad C^2(x_1) = \text{proj}(U^3 : X_1),$$
$$C^3(x_2) = \text{proj}(U^1 : X_2) \quad\quad\quad C^4(x_2) = \text{proj}(U^2 : X_2), \quad (14)$$
$$N^1(x_2) = \text{proj}(U^5 : X_1) \quad\quad\quad N^2(x_2) = \text{proj}(U^4 : X_2),$$

where $\text{proj}(U^i:X_1)$ denotes the projection onto X_1 of the relation U_i on $X_1 \times X_2$. It is obvious that $C^1 - C^4$ and $N^1 - N^2$ of (14) are crisp sets since $U^1 - U^4$ are crisp relations on $X_1 \times X_2$

Step 2: The crisp sets $C^1 - C^4$ and $N^1 - N^2$ are transformed into the following fuzzy sets:

$$\widetilde{C}^1(x_1) = \begin{cases} C^1(x_1), & x_1 < s_1 - f, \\ -x_1/2f + (s_1 + f)/2f, & s_1 - f \leq x_1 \leq s_1 + f, \\ 0, & s_1 + f < x_1, \end{cases}$$

$$\widetilde{C}^2(x_1) = \begin{cases} C^2(x_1), & s_2 + f < x_1, \\ x_1/2f + (f - s_2)/2f, & s_2 - f \leq x_1 \leq s_2 + f, \\ 0, & x_1 < s_2 - f, \end{cases} \quad (15)$$

$$\widetilde{N}^1(x_1) = \begin{cases} 0, & x_1 < s_1 - f, \\ x_1/2f + (f - s_1)/2f, & s_1 - f \leq x_1 \leq s_1 + f, \\ -x_1/2f + (s_2 + f)/2f, & s_2 - f \leq x_1 \leq s_2 + f, \\ 0, & s_2 + f < x_1, \end{cases}$$

$$f = (s_2 - s_1)/2 \neq 0,$$

where f denotes the fuzzification of the crisp sets. $\widetilde{C}^3$, $\widetilde{C}^4$ and $\widetilde{N}^2$ are calculated in the same manner as in (15).

For the contrast intensification, it is convenient to use the transformation of (15), because each transformed fuzzy set ($\widetilde{C}^1(x_1)$, $\widetilde{C}^2(x_1)$ and $\widetilde{N}^1(x_1)$) can be reduced to the original crisp sets ($C^1(x_1)$, $C^2(x_1)$ and $N^1(x_1)$), respectively, when $m \to \infty$.

Step 3: Next, we calculate the parameter m of (13). The parameter m is recursively calculated as follows:

$$m_p(t) = m_p(t-1) + \Delta m_p(t), \quad p = 1, 2, 3, 4,$$

$$\Delta m_p(t) = \begin{cases} 1 - \alpha \dfrac{|s_p(t) - s_p(t-1)|}{|s_p(t-1) - s_p(t-2)|}, & s_p(t-1) \neq s_p(t-2), \\ -1, & s_p(t) \neq s_p(t-1), s_p(t-1) = s_p(t-2), \\ 1, & s_p(t) = s_p(t-1), s_p(t-1) = s_p(t-2), \end{cases} \quad (16)$$

where

$1 < m_p(t) < 10$ and $-1 \leq \Delta m_p(t) \leq 1$
for $p = 1, 2, 3, 4$. $S_p(t)$ denotes the value of the p-th break point s_p at time instance t. $m_p(t)$ is the parameter of the contrast intensification for the membership function which has the p-th break point s_p. α is a constant value and is used to control the range of $\Delta m_p(t)$. $\Delta m_p(t)$, defined by (16), is calculated only when there exists an integer p such that the grade of the translated fuzzy set with the p-th break point is more than 0. Otherwise, assume that $\Delta m_p(t) = 0$.

We can see from (16) that $\Delta m_p(t) > 0$ when s_p converging to a value and that $\Delta m_p(t) < 0$ when s_p changes or oscillates.

Step 4: Fuzzy set $C_*^1 - C_*^4$ and $N_*^1 - N_*^4$ are generated from the previous $C_*^1 - C_*^4$ and $N_*^1 - N_*^4$ by the contrast intensification with m_p determined in Step 3.

The membership functions of $C_*^1 - C_*^4$ and $N_*^1 - N_*^4$ are

$$\tilde{C}_*^1(x_1) = \text{INT}(\tilde{C}^1, m_1)(x_1),$$
$$\tilde{C}_*^2(x_1) = \text{INT}(\tilde{C}^2, m_2)(x_1),$$
$$\tilde{N}_*^1(x_1) = \begin{cases} \text{INT}(\tilde{N}^1, m_1)(x_1), & x_1 < \tfrac{1}{2}(s_1 + s_2), \\ \text{INT}(\tilde{N}^1, m_2)(x_1), & x_1 \geq \tfrac{1}{2}(s_1 + s_2), \end{cases}$$
$$\tilde{C}_*^3(x_2) = \text{INT}(\tilde{C}^3, m_3)(x_2), \qquad (17)$$
$$\tilde{C}_*^4(x_2) = \text{INT}(\tilde{C}^4, m_4)(x_2),$$
$$\tilde{N}_*^2(x_2) = \begin{cases} \text{INT}(\tilde{N}^2, m_3)(x_2), & x_2 < \tfrac{1}{2}(s_3 + s_4), \\ \text{INT}(\tilde{N}^2, m_4)(x_2), & x_2 \geq \tfrac{1}{2}(s_3 + s_4), \end{cases}$$

The fuzzy set $\tilde{C}_*^1$ is reduced to the crisp set C^1 when $m_1 \to \infty$ and is reduced to the fuzzy set $\tilde{C}^1$ when $m_1 = 0$. Similarly, $\tilde{N}_*^1$ is reduced to the crisp set N^1 when $m_1 \to \infty$ and $m_2 \to \infty$, and is reduced to the fuzzy set $\tilde{N}^1$ when $m_1 = 1$ and $m_2 = 1$. Consequently $FR^1 - FR^5$ are reduced to $R^1 - R^5$, respectively, when $m_p \to \infty$ for $p = 1, ..., 4$. In other words, WRLSA is reduced to the recursive least square method.

Step 5: Finally $\tilde{U}^1 - \tilde{U}^5$ in FAR are reconstructed using and $\tilde{C}_*^1 - \tilde{C}_*^4$, $\tilde{N}_*^1 - \tilde{N}_*^4$, C^1 and N^1. The membership functions of $U^1 - U^5$ are

$$\tilde{U}^1(x_1, x_2) = \tilde{C}_*^1(x_1) \cdot \tilde{C}_*^3(x_2),$$
$$\tilde{U}^2(x_1, x_2) = \tilde{C}_*^1(x_1) \cdot \tilde{C}_*^4(x_2),$$
$$\tilde{U}^3(x_1, x_2) = \tilde{C}_*^2(x_1), \qquad (18)$$
$$\tilde{U}^4(x_1, x_2) = (C^1(x_1) \vee \cdot N^1(x_1)) \cdot \tilde{N}_*^2(x_2),$$
$$\tilde{U}^5(x_1, x_2) = \tilde{N}_*^1(x_1),$$

where $\cdot$ denotes the algebraic product, and $\vee$ denotes the max operator.

3.2.2. Adjustment level

The adjustment level consists of four parts, A, B, C and D. First, the consequent parameters of the linear equations are adjusted in part A by WRLSA. Second, it is checked in part B whether the premise parameter adjustment is possible or not. Third, data for the premise parameter adjustment are calculated in part C. The data are called 'premise data'. Finally, the premise parameters of fuzzy sets, that is, the linear oblique lines of the membership functions, are also adjusted using the premise data in part D by WRLSA.

Let us consider the fuzzy model shown in Figure 3 again. Now, assume that a given input-output data is $(x^0{}_1, x^0{}_2, y^0)$. At this time, the perimeter adjustment is performed is follows.

Part A (consequent parameter adjustment). The consequent parameters are adjusted by referring to the fuzzy adjustment rules FR^1 - FR^3.

For example, if we have $\tilde{U}^1(x^0{}_1, x^0{}_2) = 0.8$ for the input data $(x^0{}_1, x^0{}_2)$, then the consequent parameters of L^1 are calculated using the input-output data by WRLSA with the weight 0.8.

Part B (checking). Next, it is checked in this part whether the premise parameter adjustment is possible or not. The premise parameters are adjusted if the output y^1 calculated from each implication satisfies the following condition:

$$\min_i y^i \leq y^0 \leq \max_i y^i \tag{19}$$

It is possible to make the output of the fuzzy model coincide with that of a real system by adjusting the premise parameters if (19) is satisfied. Whereas if (19) is not satisfied, then both the following parts C and D are omitted.

Part C (calculation of premise data). The premise data are circulated from the input-output data $(x^0{}_1, x^0{}_2, y^0)$ in this part since the input-output data cannot be directly utilized in the adjustment of the parameters of the fuzzy sets. The premise parameters are adjusted using the premise data.

If we have $A^1(x^0{}_1) = m^0$ and $A^3(x^0{}_2) = k^0$ for the input data $(x^0{}_1, x^0{}_2)$ as shown in Figure 3, then the inferred output is

$$y = (w^1 y^1 + w^2 y^2 + w^3 y^3)/(w^1 + w^2 + w^3),$$

where

$$w^1 = m^0 k^0, \quad w^2 = m^0(1 - k^0), \quad w^3 = 1 - m^0$$
$$y^1 = x_1^0 + x_2^0, \quad y^2 = 2x_1^0 + x_2^0, \quad y^3 = x_1^0 + 2x_2^0 \tag{20}$$

The inferred output y is not always equal to the output data y^0. We assume that the output y of (20) is equal to y^0, when $A^1(x^0{}_1) = m$ and $A^3(x^0{}_2) = k$, where m and k are unknown. Then m and k must satisfy (21):

$$y^0 = (mky^1 + m(1-k) y^2 + (1-m) y^3)/(mk + m(1-k) + (1-m)). \tag{21}$$

From (21), we can obtain the equation

$$k = a/m + b, \tag{22}$$

where

$$a = (y^0 - y^3)/(y^1 - y^2), \quad b = (y^3 - y^2)/(y^1 - y^2)$$

However, (22) has infinitely many solutions for m and k. So we find the nearest value (m^*, k^*) to (m, k) on the line expressed by (22). From (m^*, k^*), we can obtain two premise data (x_1^0, m^*) and (x_2^0, k^*) for the premise variables x_1 and x_2, respectively. These premise data are utilized in the next part.

Part D (premise parameter adjustment). The premise parameters of the membership functions are adjusted using the premise data by referring to the fuzzy adjustment rules FR^4 - FR^5.

If $\tilde{U}^4(x_1^0, x_2^0) = 0.2$ and $\tilde{U}^5(x_1^0, x_2^0) = 0.6$ for the input data (x_1^0, x_2^0) then the premise parameters of $A^1(x_1)$ and $A^3(x_2)$ are calculated using the premise data (x_1^0, m^*) and (x_2^0, k^*) by WRLSA with the weights 0.2 and 0.6, respectively. In other words, the parameters of the oblique lines of $A^1(x_1)$ and $A^3(x_2)$ are adjusted by WRLSA. Because of relations such as

$$A^2(x_1) = 1 - A^1(x_1) \text{ and } A^4(x_2) = 1 - A^3(x_2),$$

it is sufficient to identify the parameters of $A^1(x_1)$ and $A^3(x_2)$.

4. Applications to prediction of complex system

4.1. Prediction CO2 concentration (gas furnace)

Successive fuzzy modeling is applied to real data generated by a gas furnace system [1]. The gas furnace data are well known and have been used in many modeling exercises. The data consist of 296 pairs of input-output measurements. The input $u(k)$ is gas flow rate into the furnace and output $y(k)$ is CO_2, concentration in outlet gas. The sampling interval is 9 s.

To begin with, an initial model is identified from 50 input-output data pairs by the off-line fuzzy modeling method.

Next, the parameters of the initial model are identified by successive fuzzy modeling. At this time, the remaining data consisting of 246 pairs are utilized.

Equation (23) shows the performance index of the model:

$$J = \frac{1}{246} \sum_{k=51}^{296} (y(k) - \hat{y}(k))^2, \quad (23)$$

where $y(k)$ is the output of the furnace at the k-th pair, and $\hat{y}(k)$ is the output of the model.

Table 1 compares the results of the successive fuzzy modeling with other models identified from the same data: the performance index (J), the number of predictor variables (PV) and the number of fuzzy implications (FI). The other fuzzy models are all represented by fuzzy relations.

The numbers in Table 1 with ° are those of the premise subspaces which are partitioned by referential fuzzy sets [4, 9], instead of the number of FI. This is because those models which consist of the referential fuzzy sets do not have fuzzy implications. FM and FM' denote the results of the successive fuzzy modeling, where FM has PV = 6 and FI = 2, and FM' has PV = 2 and FI = 2.

Table 1. Comparison results of successive fuzzy modeling and other models (gas furnace data)

Tong [8]	0.469	2	19
Pedrycz [4]	0.320	2	81ᶜ
Xu [9]	0.328	2	25ᶜ
Box [1]	0.202	6	-
FM	0.068	6ᵇ	2
FM'	0.359	2	2

a Non fuzzy model
b Existence of a constant term.
c Number of partitioned premise subspaces (see text)

The result of FM is far superior to that of the other fuzzy models. Here, we must notice that FM has PV = 6, whereas the other fuzzy models have $PV = 2$. Therefore, we cannot directly compare FM with the other fuzzy models. However, since FM' has PV = 2, we can directly compare FM' with the other fuzzy models. The performance of FM' is as good as those of the other fuzzy models. We should notice that FI of FM' is a very small number. As for the number of FI, FM and FM' are particularly excellent. A more accurate fuzzy model can be realized if we increase the number of FI. It is obvious that the results are produced by the effect of fuzzy implications described in (5).

Figure 6 shows the results of prediction by FM. The measured output data are solid dots and the predictions of the fuzzy model are a continuous line.

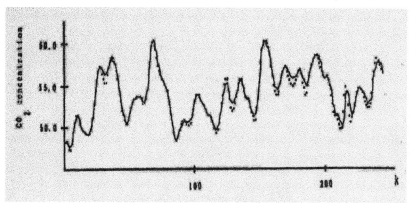

Fig. 6. Predictions of FM.

We can point out from Table 1 that the fuzzy model based on (5) can express the behavior of a system with a high accuracy in spite of a small number of FI (FI = 2). Figures 7 and 8 show the identification results of FM and FM', respectively.

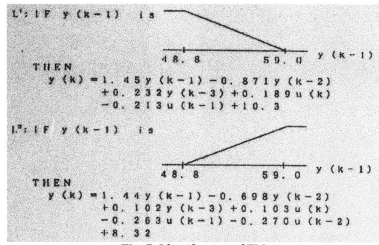

Fig. 7. *Identification of FM.*

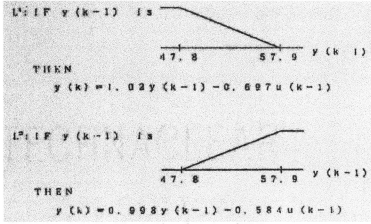

Fig. 8. *Identification of FM'.*

4.2. Prediction of river water flow

The second example is a prediction of water flow rate of a river in Japan. The flow of the river with two dams is illustrated in Figure 9. Arrows denote the direction of the water flow and Q_1 and Q_2 denote the rate of water flow discharged from dam A and dam B, respectively. Q_3 and Q_T denote the rate of water flow at points C and E, respectively. R denotes the rainfall at point D. The sampling interval is 30 min. We will predict $Q_T(t + 6)$ at point E from the following 13 predictor variables:

$$Q_1(t), Q_1(t-5), Q_1(t-6), Q_2(t), Q_2(t-5), Q_2(t-6),$$
$$Q_3(t), Q_3(t-5), Q_3(t-6), Q_T(t), R(t), R(t-5), R(t-6),$$

where $Q_1(t)$ denotes the value of Q_1 at time instance t.

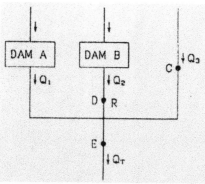

Fig. 9. Flow of river

Four kinds of input-output data for identification are provided:
DATA A: June 21 - July 8 (1985), $n = 847$,
DATA B: July 10-July 18 (1985), $n = 437$,
DATA C: September 20-September 27 (1985), $n = 319$,
DATA D: October 11-October 14 (1985), $n = 170$,
where n is the number of input-output data. DATA A is the data set for identification of the initial model. DATA B, C and D are the data sets for the successive fuzzy modeling.

Equation (24) is the performance index of the model:

$$J = \frac{1}{n}\sum_{t=1}^{n} \frac{|Q_T(t+6) - \hat{Q}_T(t+6)|}{Q_T(t+6)} \times 100\%, \qquad (24)$$

where $Q_T(t + 6)$ and $\hat{Q}_T(t + 6)$ are the outputs of the fuzzy model and the real system at time instance $t + 6$, respectively.

It is very difficult to identify a fuzzy relational model since this system has many predictor variables. In this case, the fuzzy relational model becomes 14 dimensions. We try to identify a fuzzy model for this complex real system by the successive fuzzy modeling.

First, we identify the initial model with DATA A. Next, the parameters of the initial model are identified by successive fuzzy modeling. The successive fuzzy modeling is separately performed for DATA B, C and D. Figure 10 shows the identification result of the successive fuzzy modeling for DATA B.

Table 2 shows the values of performance index J for the predictions of the fuzzy model (PV = 9, FI = 2) and a linear model (PV = 7) for DATA B, C and D. This linear model is used in the prediction of the water flow rate at present. These results show the validity of the successive fuzzy modeling. It is seen from Table 2 that a more accurate fuzzy model can be realized by adjusting its parameters successively.

Figure 11 shows the predictions of the fuzzy model and the linear model for DATA C. In Figure 11, the measured output data are solid dots, and the predictions of the fuzzy model and the linear model are a continuous line and a broken line, respectively

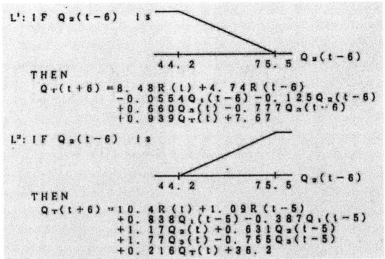

Fig. 10. Identification result of successive fuzzy modeling (DATA B).

Table 2. Comparison results of successive fuzzy modeling and linear model (prediction of river water flow)

Model		Data B	Data C	Data D
Linear		4.166	7.326	10.044
Fuzzy	Initial	3.113	5.848	6.694
Fuzzy	Modified	3.021	4.325	4.165

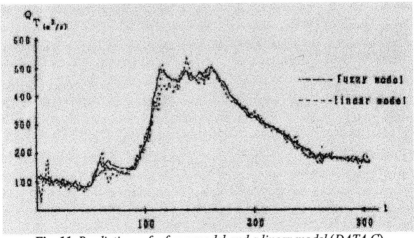

Fig. 11. Predictions of a fuzzy model and a linear model (DATA C)

Here we explain the identification algorithm of a fuzzy model [5, 6, 7]. Figure 12 shows the identification algorithm of a fuzzy model. The identification procedure can be mainly classified into three parts:

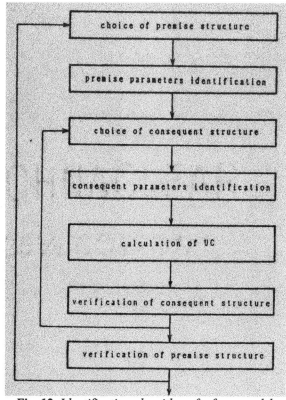

Fig. 12. Identification algorithm of a fuzzy model.

5. Conclusion

We have suggested a method of successive identification of a fuzzy model which was proposed by Takagi and Sugeno. The advantages of this method are:
(1) The fuzzy model can express a highly nonlinear relation inspite of a small number of fuzzy implications.
(2) It is easy to adjust the parameters of a multi-input/multi-output fuzzy model.
(3) More accurate fuzzy models can be successively improved by this method.

Point (1) is not directly concerned with successive fuzzy modeling, but shows that a new type of fuzzy implication is useful in the field of system modeling. Point (2) is very important when this method is applied to real complex systems. In this case, the fuzzy relational model becomes multidimensional, which requires a huge computer memory. Point (3) shows that this method can be applied to time varying systems.

The successive fuzzy modeling suggested in this paper may be applicable to many other real complex systems.

Appendix A: Outline of the identification algorithm of a fuzzy model

Here we explain the identification algorithm of a fuzzy model [5, 6, 7]. Figure 12 shows the identification algorithm of a fuzzy model. The identification procedure can be mainly classified into three parts:

(1) determination of the premise structure and the consequent structure;
(2) estimate of the parameters of the structure determined in (1);
(3) verification of structure of the fuzzy model.

The premise structure is concerned with the fuzzy partition of the input space which is directly related to the number of necessary fuzzy implications.

We apply the least squares method to the identification of the consequent parameters since each of the consequent parts is described by a linear equation. In the identification of the premise parameters, the well-known complex method is used since the problem of finding the optimum premise parameters minimizing a performance index is reduced to non-linear programming.

As a proper criterion of the verification of the structure of the fuzzy model, we utilize the unbiasedness criterion (UC). To calculate UC, we must divide the observed data into two sets NA and NB. UC is calculated as

$$UC = \left[\sum_{i=1}^{n_A} (y_i^{AB} - y_i^{AA})^2 + \sum_{i=1}^{n_B} (y_i^{BA} - y_i^{BB})^2 \right]^{1/2},$$

where n_A, is the number of the data set N_A, y_i^{AA} the estimated output for the data set N_A, from the model identified using the data set N_A and y_i^{AB} the estimated output for the data set N_A from the model identified using the data set N_B. This criterion is based on the fact that the parameters of the fuzzy model with the true structure are the least sensitive to the changes of the observed data which are used for identifying the parameters.

For further details of the structure identification of the fuzzy model, the readers should refer to [6].

Appendix B: Weighted recursive least squares algorithm (WRLSA)

The standard WRLSA is given by

$$\hat{\theta}_k = \hat{\theta}_{k-1} + \frac{P_{k-1} z_k}{1/w_k + z_k^T P_{k-1} z_k} (y_k - z_k^T \hat{\theta}_{k-1}), \tag{B.1}$$

$$P_k = P_{k-1} + \frac{P_{k-1} z_k z_k^T P_{k-1}}{1/w_k + z_k^T P_{k-1} z_k}, \tag{B.2}$$

with z_k, k, $\theta_k \in R^n$, $P_k \in R^{n \times n}$, y_k, $w_k \in R$, where θ_k is the estimated vector at k, z_k is an input vector, w_k is a weight for z_k, and n is the number of parameters.

We apply the grades of membership of input points in U^1 - U^5 to the weight w_k. The initial values of θ_0 and P_0 are set is follows: θ_0 = the consequent parameters of initial model, $P_0 = \alpha I$, where α is a large number and I is the identity matrix. k is recursively calculated by (B.1) and (B.2).

References

[1]. G.E.P. Box and G.M. Jenkins. *Time Series Analysis, Forecasting and Control* (Holden Day. San Francisco. 1970).

[2]. E. Czogala and W. Pedryez. On identification in fuzzy systems and its applications in control problem. *Fuzzy Sets and Systems* **6** (1981) 73-83.

[3]. E.H. Mamdani, Applications of fuzzy algorithms for control of a simple dynamic plant, *Proc. IEE* **121**(12) (1974) 1585-1588.

[4]. W. Pedryez, An identification algorithm in fuzzy relational systems. *Fuzzy Sets and Systems* **13** (1984) 153-167.

[5]. M. Sugeno and G.T. Kang. Fuzzy modeling and control of multilayer incinerator, *Fuzzy Sets and Systems* **18** (1986) 329-346.

[6]. M. Sugeno and G.T. Kang. Structure identification of fuzzy model, *Fuzzy Sets and Systems* **28** (1998) 15-33.

[7]. T. Takagi and M. Sugeno. Fuzzy identification of systems and its applications to modeling and control, *IEEE Trans. Systems Man Cybernet.* **15** (1985) 116-132.

[8]. R.M. Tong, Synthesis of fuzzy models for industrial processes. *Int. J. General Systems* **4** (1978) 143-162.

[9]. C. Xu and Y. Lu, Fuzzy model identification and self learning for dynamic systems. *IEEE Trans. System Man Cybernet.* **17** (1987) 683-689.

[10]. L.A. Zadeh, A fuzzy-set-theoretic interpretation of linguistic hedges, *J. Cybernetics* **2** (1972) 4-34.

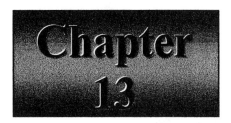

A Fuzzy-Logic-Based Approach to Qualitative Modeling[*]

Michio Sugeno
*Department of Systems Science, Tokyo Institute of Technology,
4259 Nagatsuta. Midoriki, Yokohama 227, Japan.*

Takahiro Yasukawa
*Department of Control Engineering,
Tokyo Institute of Technology,
4259 Nagatsuta, Midoriku, Yokohama 227, Japan.*

*Manuscript received August 2, 1991
Revised April 12, 1992*

Abstract

This paper discusses a general approach to qualitative modeling based on fuzzy logic. The method of qualitative modeling is divided into two parts: fuzzy modeling and linguistic approximation. It proposes to use a fuzzy clustering method (fuzzy c-means method) to identify the structure of a fuzzy model. To clarify the advantages of the proposed method, it also shows some examples of modeling, among them a model of a dynamical process and a model of a human operator's control action.

1. Introduction

In this paper we discuss a method of qualitative modeling based on fuzzy logic. Related terminologies for qualitative modeling in fuzzy theory are known as fuzzy modeling and linguistic modeling. Qualitative modeling is not so popular in general but this concept is indirectly implied by fuzzy modeling or linguistic modeling. Here we deal with a qualitative model as a system model

---

[*] © 1993 IEEE. Retyped with written permission from
IEEE Transactions on Fuzzy Systems, Vol.1, No. 1, (1993) 7-31.

based on linguistic descriptions just as we use them in sociology or psychology.

Though the terminology of "linguistic modeling" may be straightforward and more appropriate, we use "qualitative modeling" partly because we like to discuss certain problems in comparison with the "qualitative reasoning" approach in artificial intelligence and also because "qualitative modeling" has been one of the most important issues since the very beginning of fuzzy theory. Here artificial intelligence will be abbreviated to AI as usual and the terminology "fuzzy modeling" will be used in a narrow sense.

Before entering the main subject, we give an overview of fuzzy modeling in fuzzy theory and qualitative reasoning in AI; then we state the problems concerning qualitative modeling based on fuzzy logic from a systems theory point of view.

After these discussions, we propose a method of qualitative modeling in a general framework known as the black box approach in systems theory. That is, we build a qualitative model of a system without *a priori* knowledge about a system provided that we are given numerical input-output data.

1.1. *Fuzzy Modeling*

Though the term "fuzzy modeling" has not been used so often, fuzzy modeling is the most important issue in fuzzy logic or more widely in fuzzy theory. In fact we can find the seminal ideas of fuzzy modeling in the early papers of Zadeh. Thus, the research concerned with fuzzy modeling has a history of more than 20 years. There are many interpretations of fuzzy modeling. For instance, we can consider a fuzzy set as a fuzzy model of a human concept. In this study, we simply understand the fuzzy modeling to be an approach to form a system model using a description language based on fuzzy logic with fuzzy predicates. In a broader sense we can interpret the fuzzy modeling as a qualitative modeling scheme by which we qualitatively describe system behavior using a natural language. The fuzzy modeling in a narrow sense is a system description with fuzzy quantities. Fuzzy quantities are expressed in terms of fuzzy numbers or fuzzy sets associated with linguistic labels where a fuzzy set usually does not have a tight relation with a linguistic label; either a fuzzy number does not need to have a linguistic interpretation or a linguistic label is expressed as if it were a frill of a membership function.

On the other hand, what we mean by a qualitative model is a generalized fuzzy model consisting of linguistic explanations about system behavior. Linguistic terms in linguistic explanations are found such that they linguistically approximate the fuzzy sets in an underlying fuzzy model. Furthermore, when a relation of state variables in a system is to be expressed linguistically in a qualitative model, we can use a method of linguistic approximation to accomplish this aim.

In this paper, we focus our attention on qualitative modeling since we want to stress that one of the final targets of fuzzy modeling is "qualitative modeling." We could even say that fuzzy sets and/or fuzzy logic were suggested with qualitative modeling in mind.

Let us go back to Zadeh's early ideas. In his paper in 1968 [1] following the

first paper, "Fuzzy Sets," in 1965, he suggested using an idea of fuzzy algorithm such as
   a) set y approximately equal to 10 if x is approximately equal to 5;
   b) if x is large, increase y by several units.

Those fuzzy algorithms are nothing but qualitative descriptions of a human action, or decision making. As examples, he shows cooking recipes, directions for repairing a TV set, instructions on how to treat a disease, and instructions for parking a car. As for the necessity of fuzzy algorithms, he notes that "most realistic problems tend to be complex, and many complex problems are either algorithmically unsolvable or, if solvable in principle, are computationally infeasible."

The most remarkable paper related to qualitative modeling is his paper of 1973 [2] on linguistic analysis, where he states "the principle of incompatibility," according to which "as the complexity of a system increases, our ability to make precise and yet significant statements about its behavior diminishes until a threshold is reached beyond which precision and significance (or relevance) become almost mutually exclusive characteristics. It is in this sense that precise quantitative analyses of the behavior of humanistic systems are not likely to have much relevance to the real world societal, political, economic, and other types of problems which involve humans either as individuals or in groups."

Based on the above considerations Zadeh suggested linguistic analysis in place of quantitative analysis. If we look at his idea of linguistic analysis from the viewpoint of modeling, it is seen to be precisely qualitative modeling. As the main characteristics of this approach, he suggests "1) use of so-called linguistic variables in place of or in addition to numerical variables; 2) characterization of simple relations between variables by conditional fuzzy statements; and 3) characterization of complex relations by fuzzy algorithms."

It is well known that, motivated by these ideas of "fuzzy algorithm" and "linguistic analysis," Mamdani first applied fuzzy logic to control [3]. This topic has come to be known as fuzzy algorithmic control or linguistic control. Fuzzy control can be viewed in a certain sense as the result of the qualitative modeling of a human operator working at plant control. For example, fuzzy control rules may be as follows:
   1) if the *error* is *positive big* and the *change of error* is *positive medium*, then the *change of control* is *positive big*;
   2) if the error is *positive small* and the *change of error* is positive small, then the *change of control* is *zero*,

where the *error* is the difference between the reference input and the plant output, the *change of error* is the derivative of the error, and the *change of control* is the derivative of the control input.

As seen in the above example, we use, to describe control rules, linguistic variables which take linguistic values such as "*positive big*", "*positive medium*", "*zero*", and "*negative small*." These are not merely symbols but rather the linguistic labels with quantitative semantics given by underlying fuzzy sets which are associated with membership functions. So we may call a label "*positive big*" a fuzzy quantity. In fact we sometimes use the term "fuzzy number" or "fuzzy interval."

The main problem of fuzzy control is to design a fuzzy controller where we usually take an expert-system-like approach. That is, we derive fuzzy control rules from a human operator's experience and/or engineer's knowledge, which are mostly based on their qualitative knowledge of an objective system.

The design procedure is thus something like the following: first, we build linguistic control rules; second, we adjust the parameters of fuzzy sets by which the linguistic terms in the control rules are quantitatively interpreted. For example, let us look at the linguistic rules for controlling a cement kiln [4], as shown in Fig. 1. From these, we can easily derive fuzzy control rules. In this sense, we may say that a set of fuzzy control rules is a linguistic model of human control actions, which is not based on a mathematical description of human control actions but is directly based on a human way of thinking about plant operation.

Case	Condition	Action to be taken
1	BZ low OX low BE low	When BZ is drastically low: (a) reduce kiln speed (b) reduce fuel When Bz is slightly low: (c) increase I. D. speed (d) decrease fuel rate
2	BZ low OX low BE O. K.	(a) reduce kiln speed (b) reduce fuel rate (c) reduce I. D. fan speed
3	BZ low OX low BE high	(a) reduce kiln speed (b) reduce fuel rate (c) reduce I. D. fan speed

BE = back end temperature, BZ = burning zone temperature, OX= percentage of oxygen gas in kiln exit gas.
Total of 27 rules.

Fig. 1. Operator's manual of cement kiln control.

Apart from fuzzy control, we have many studies on fuzzy modeling. Those are divided in two groups. The studies of the first group deal with fuzzy modeling of a system itself or a fuzzy modeling for simulation [5]-[8]. Some of those are considered as examples of qualitative modeling. The studies of the second group deal with fuzzy modeling of a plant for control [9]-[13]. Just as with the modern control theory, we can design a fuzzy controller based on a fuzzy model of a plant if a fuzzy model can be identified. Fuzzy modeling in the latter sense is not necessarily viewed as qualitative modeling unless the derivation of a qualitative model from the identified fuzzy model is discussed. It is, of course, quite interesting to derive qualitative control rules based on the qualitative model of a plant.

For example [14], assume that we have a dynamical plant model such as

1) if u_n is *positive* and y_{n-1} is *positive*, then y_n is *positive big*;
2) if u_n is *negative* and y_{n-1} is *positive*, then y_n is *positive small*;
3) if u_n is *positive* and y_{n-1} is *negative*, then y_n is *negative small*;
4) if u_n is *negative* and y_{n-1} is *negative*, then y_n is *negative big*.

Then given a reference input $r = 0$, we can derive the following control rules based on the above model:

1) if y_{n-1} is *positive* and y_n is *positive big*, then u_n is *negative big*;
2) if y_{n-1} is *positive* and y_n is *negative*, then u_n is *positive*;
3) if y_{n-1} is *negative* and y_n is *positive*, then u_n is *negative*;
4) if y_{n-1} is *negative* and y_n is *negative* big, then u_n is *positive big*.

Here the error e_n is equal to $-y_n$ since the reference input (the set point) is zero.

In fuzzy modeling, the most important problem is the identification method of a system. The identification for the fuzzy modeling has two aspects as usual: structure identification and parameter identification. This problem will be discussed in Section 1.4.

1.2. Qualitative Reasoning

Outside the research area of fuzzy logic, the concept of qualitative reasoning has been suggested in AI. Earlier studies of qualitative reasoning were found on mechanical systems in 1975 [15], and also on naive physics in 1979 [16]. We can identify as typical studies of qualitative reasoning the following:

1) qualitative physics or naive physics [17],
2) qualitative process theory [18],
3) qualitative simulation [19].

Here we call these ideas qualitative reasoning. The aims and characteristics of qualitative reasoning may be summarized as follows. According to de Kleer [17], "the behavior of a physical system can be discussed by the exact values of its variables (forces, velocities, positions, pressures, etc.) at each time instant. Such a description, although complete, fails to provide much insight into how the system functions. Our long-term goal is to develop an alternate physics in which these same concepts are derived from a far simpler, but nevertheless formal qualitative basis ... our proposal is to reduce the quantitative precision of the behavioral descriptions but retain the crucial distinctions."

Also if we refer to Kuipers [19], we find "an expert system is often a shallow model of its application domain, in the sense that conclusions are drawn directly from observable features of the presented situation. One major line of research toward the representation of deep models is the study of qualitative causal models. Research on qualitative causal models differs from more general work on deep models in focusing on qualitative descriptions of the deep mechanism, capable of representing incomplete knowledge of the structure and behavior of the mechanism."

As we find in the above statements, there are small distinctions and big similarities between fuzzy modeling and qualitative reasoning. One distinction is that fuzzy modeling starts from the fact that a precise mathematical model of a complex system cannot be obtained, whereas qualitative reasoning starts from the fact that, although a complete model may be available, it cannot provide insight into the system; a description based on deep knowledge is needed. On the other hand, similarities are found in the confidence of the advantage of qualitative expressions, in the goals, and in some parts of description languages for modeling. The common idea of qualitative system theory should be emphasized.

Qualitative reasoning makes use of a quantity space on which "landmarks" are defined. Usually one landmark "0" is set, and then three values {+, 0, -} are used which are qualitative in nature, similar to fuzzy values {*positive, zero, negative*} in fuzzy modeling. The difference between them is that qualitative values are crisp, acting as symbols, though they call {+, 0, -} semantics, while fuzzy values are not symbols but quantitative semantics represented by their membership functions. In qualitative reasoning such phenomena as:

1) temperature $\propto Q_+$ pressure
2) pressure $\propto Q_-$ volume

are interpreted as:

1) if the temperature increases (decreases), then pressure increases (decreases);
2) if pressure increases (decreases), then volume decreases (increases).

As we have seen, this sort of description was found in Zadeh's paper in 1968, which shows our motivation to use fuzzy concepts. As far as reasoning is concerned, in qualitative reasoning the following arithmetic is defined:

$$[x+y] = [x] + [y] \tag{1}$$
$$[x \times y] = [x] \times [y] \tag{2}$$
$$[x_n] = [x_{n-1}] + \partial x \tag{3}$$

where $[x]$ means a qualitative value of a physical quantity x, n is a discrete time, and ∂x is the qualitative derivative of x with the values of {+, 0, -}. Eqs. (1) and (2) describe binary operations on qualitative values; Eq. (3) describes the dynamical relation of a system.

We must point out that, in fuzzy theory, the arithmetic of fuzzy numbers is in a much more general setting. For example, we have

$$\underset{\sim}{a+b} = \underset{\sim}{a} + \underset{\sim}{b} \tag{4}$$

$$\underset{\sim}{a \times b} = \underset{\sim}{a} \times \underset{\sim}{b} \tag{5}$$

where $\underset{\sim}{a}$ is a fuzzy number, "approximately a," associated with a membership function, e.g., $h_a(x) = e^{-(x-a)^2}$. (6)

The arithmetic in Eqs. (4) and (5) is a generalization of the ordinary arithmetic, which of course precisely includes Eqs. (1) and (2).

In fuzzy modeling the fuzzy arithmetic is used for computation, and "*if-then*" rules are used for inference together with fuzzy reasoning based on fuzzy logic. Fuzzy reasoning is interpreted as approximate reasoning based on vague or incomplete knowledge.

As far as the tools for model description are concerned, those in qualitative reasoning seem to be very simple compared with those in fuzzy modeling. We can recall a dentist's way of description: a dentist uses the symbols {++, +, ±, -, --}, and a control engineer uses, in the bang-bang control, the symbols {+1, -1}, where the symbol zero can be added as an ideal case in the middle of + 1 and -1 from a theoretical point of view. The dentist's symbols are fuzzy in nature, much more similar to fuzzy values, while the control symbols are crisp and look similar to qualitative reasoning.

Now let us look at the basic approach of qualitative modeling. As far as the authors can determine, this approach suggests 1) to make a model by observations of phenomena based on, for instance, empirical and/or physical knowledge or 2) to derive a qualitative model based on a differential equation of a system. This approach is effective if a system is simple, for example a mechanical system with a small number of state variables and simple interactions among state variables. But, of course, even if a system is composed of many state variables, this approach is useful provided that the system is a lumped-parameter system such as an electric circuit. As a conclusion it seems that qualitative modeling is concerned with well-structured systems; in most studies on qualitative modeling, they deal with toy systems, if we look at those from a systems theory point of view.

In fuzzy modeling we do not take this approach to modeling. First of all, if a mathematical model is available, we can use it for the purpose of engineering (analysis, control and prediction). This fact of course does not prevent us from making a qualitative model of a system, for instance, in order to explain its general structure and behavior to students.

Second, even if we can find the local mechanisms of a system, it is often the case that we cannot build the whole system model by aggregating the local mechanisms. This is why we take a so-called black box approach to systems identification in systems engineering. We have to note that what we mean by a complex system is a system for which we cannot have a relevant mathematical model; in other words, we cannot build a global system model by integrating local models.

There is another reason for a black box approach. It is often the case that we cannot measure the states of a dynamical system. We can observe only the inputs and the outputs of a system in many cases.

Summing up the above discussions on qualitative reasoning and fuzzy modeling, we may conclude: 1) motivations are different, 2) goals look similar, 3) tools are much more powerful in fuzzy modeling, and 4) methods of approach to modeling are different and those in fuzzy modeling are more varied and applicable. In any case, however, we see similarities between qualitative reasoning and fuzzy modeling rather than differences.

The research domain of fuzzy logic started adopting a qualitative approach to systems analysis in the late 1960s. We are quite sure that fuzzy modeling can be directly applied to problems considered by a qualitative reasoning method. Despite these facts, it is a great pity that we can hardly find any reference to fuzzy logic in the papers related to qualitative reasoning.

1.3. Qualitative Modeling

What we imply here by a qualitative model [20]-[23] is a linguistic model. A linguistic model is a model that is described or expressed using linguistic terms in the framework of fuzzy logic instead of mathematical equations with numerical values or conventional logical formula with logical symbols. For example, as we shall see later, a linguistic model of a two input-single output system is something like the following:

1) if x_1 is *more than medium* and x_2 is *more than medium*, then y is *small*, $\partial y/\partial x_1$ is *sort of negative*, and $\partial y/\partial x_2$ is *sort of negative*.
2) If x_1 is *small* and x_2 is *more or less small or medium small*, then y is *big*, $\partial y/\partial x_1$ is *very negative*, and $\partial y/\partial x_2$ is *very negative*.

We can regard most fuzzy models and fuzzy control rules [24], [25] as qualitative models.

We distinguish, however, in this paper, a qualitative model from a fuzzy model. The terminology "a fuzzy model" is used in a narrow sense. A qualitative model is considered a fuzzy model with something more, i.e., with more linguistic expressions.

For example, a fuzzy model of the type

1) if x is *approximately* 3, then y is *approximately* 5;
2) if x *increases by approximately* 5, then y *decreases by approximately* 4

should not be called a qualitative model. We can call the above model a fuzzy number model to distinguish it from a qualitative model.

Further, in an ordinary fuzzy model that is used in fuzzy control such as

1) if x is *positive small*, then y is *negative small*,
2) if x is *positive medium*, then y is *positive small*,

the terms "*positive small*," "*negative small*," etc. are the labels conventionally attached to fuzzy sets, where the fuzzy sets play an important role, not the labels.

We go beyond this stage by utilizing the concept of "linguistic approximation." That is, given a conventional fuzzy model with fuzzy sets, we improve its qualitative nature by using linguistic approximation techniques in fuzzy logic. In other words, we deal with a model in which we focus our attention on how to linguistically or qualitatively explain a system behavior as we shall see in what follows.

Now let us discuss available sources, i.e., information or data, for qualitative modeling. We find the following classification of the sources:

1) conventional mathematical models
2) observation based on knowledge and/or experience
3) numerical data
4) image data
5) linguistic data

In qualitative reasoning as we have seen, a model is built based on 1) and 2). A conventional mathematical model is usually identified based on 2) and 3). A fuzzy model is also based on 2) and 3), as in the case of a mathematical model. Sources 4) and 5) are also useful to build a qualitative model. A qualitative model based on 4), image data, can be interpreted as image understanding, that is, a linguistic explanation of an image. As for the examples concerned with 5), we can consider translation of one language into another or making a summary of a story.

In this paper we deal with qualitative modeling based partly on 2) and mainly on 3), as usual, a black box approach. The reason for this is that system identification is a key issue in the black box approach and all the basic matters in model building appear in the context of system identification. Indeed we think that any method of modeling should be able to solve the problems discussed from now on.

Let us first consider the problems in modeling. Modeling is classified from the viewpoint of a description language. However, there are some common problems to be solved in modeling independently of both the description language and the data type. Thus, we refer to the systems theory. The most complicated problems arise when we take a black box approach to modeling. In the black box approach, we have to build a dynamical model using only input-output data. This stage of modeling is usually referred to as identification. It is worth noting that the concept of system identification and its formal definition were introduced by Zadeh in the early 1960s. We may suppose that Zadeh suggested the idea of fuzzy sets since he found difficulties in the identification of a complex system based on differential equations with numerical parameters.

According to Zadeh's definition [26], given a class of models, system identification involves finding a model which may be regarded as equivalent to the objective system with respect to input-output data.

1.4. Identification in Fuzzy Modeling

Now let us look at the problems in the identification of a fuzzy model. The identification is divided into two kinds: 1) *structure identification* and *2) parameter identification*.

Structure identification can be divided into two types, called, in this paper, type I and type II, where each type is also divided into two subtypes Ia and Ib, respectively. Before going into details, we show a classification of identification in Table 1. Later we shall consider a multi-input and single-output system.

Table 1. Classification of identification

Structure Identification I	a: Input candidates
	b: Input variables
Structure Identification II	a: Number of rules
	b: Partition of input space
Parameter identification	

1) *Structure Identification:* Generally speaking, the structure identification of a system has to solve two problems: in type I, one is to find input variables and in type II, one is to find input-output relations. Type I consists of Ia and Ib. In Ia, we find possible input candidates for the inputs to a system. There are, of course, an infinite number of possible candidates, which should be restricted to a

certain number. This type of identification, like induction, i.e., selection of a finite number out of an infinite number, cannot be solved in general. Let us just recall Newton's law of kinetic motion. There is no systematic way to find the exact causes of an unknown phenomenon. We have to take a heuristic method based on experience and/or common sense knowledge; we do not discuss the identification of type Ia in this study.

In the structure identification of type Ib, given the possible input candidates, we find a set of input variables to a system which affect the output. In this case we select a finite number out of a finite number and so there are some systematic ways to solve this problem. For example, we are familiar with this problem in multivariate analysis. In a conventional black box approach in systems theory, this type of identification is, however, not explicitly discussed, and models are based on preassigned input-output variables. Generally speaking, we need some criterion for identification to evaluate the performance of a model. For instance, we can use the output error, i.e., the difference between the model output and the real output. However, as far as the structure identification is concerned, it is well known that we cannot use the output error. We need a special criterion for the type Ib.

The structure identification of type II, which is concerned with input-output relations, is further divided into the subtypes IIa and IIb. Let us first consider IIa. In this type we have to determine the number of fuzzy rules in a fuzzy model. By structure identification in ordinary systems theory, what we mean is to find the relations between the inputs and the outputs. Given a description language for modeling, it simply means the determination of the order of a model. Let us consider an input-output model of a static system:

$$y = a_0 + a_1 x + a_2 x^2 + \cdots + a_n x^n \qquad (7)$$

and also consider an input-output model of a dynamical system:

$$\begin{aligned} y(t) + a_1 y(t-1) + a_2 y(t-2) + \cdots + a_n y(t-n) \\ = b_1 u(t-1) + \cdots + b_n u(t-n), \end{aligned} \qquad (8)$$

where t is a discrete time. In both cases, the order, n, must be identified.

We know that AIC [27], Akaike's information criterion, can be used to this aim in particular when a system is linear. The idea of the identification of the order, n, can be explained as follows.

We consider the model in Eq. (7). Given a set of the input-output data indicated by 'O', Fig. 2(a) shows an identified model with a high order which has a good fitting for the data. This model minimizes the output error. However, since the data is contaminated by noise, we might also observe another set of data indicated by 'Δ' in the figure for which the model has no fitting. *What is a model which fits both sets of data, independently of noise?* Fig. 2(b) shows a model with a lower order which has a better fitting on the average than the model in Fig. 2(a): this model does not minimize the output error.

So the identification of the order, n, is very important in modeling: usually n is determined so as to minimize the value of AIC, which, roughly speaking, means minimizing both the output error and the order.

In a fuzzy model, the structure identification of this kind is stated in a different way. A fuzzy model consists of a number of "*if-then*" rules. The number of rules n in a fuzzy model corresponds to the order n in a conventional model. This identification of n is called type IIa.

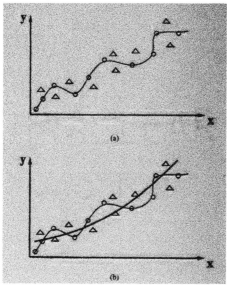

Fig. 2. Input-output data and model.

There are two parts of an "*if-then*" rule: the premise part and the consequent part. So the rules have two structures in principle: the premise structure and the consequent structure. If we use a fuzzy model (see Appendix II), the consequent of which is of a functional type [12] as in (8), we have the same structure problem as in a conventional model. We will not use this sort of a fuzzy model in this paper. Below we shall discuss only the premise structure, which is concerned with IIb.

Suppose we obtain the following fuzzy model with three rules by the identification of IIa:

    1) if x_1 is *small*, then y is *small*;
    2) if x_1 is *big* and x_2 is *small*, then y is *medium*;
    3) if x_1 is *big* and x_2 is *big*, then y is *big*.

The premise space of the above model, which is two-dimensional space of the inputs x_1 and x_2, is partitioned into three fuzzy subspaces as we see; the number of rules corresponds to the number of subspaces. This partition is called the premise structure in a fuzzy model. So the identification IIb implies determining how the input space should be partitioned.

Fig. 3(a) shows the partition of the input space in the above model. As we find in Fig. 3(b) and (c), there are a number of different partitions. The problem is combinatorial: we need a heuristic method to find an optimal partition together with a criterion [12]. However, in our approach, this problem is automatically solved in the process of the structure identification of IIa, where we use a fuzzy clustering method.

As we discussed in the above, we deal with the problems of selecting the input variables (Ib) and finding the number of rules (IIa).

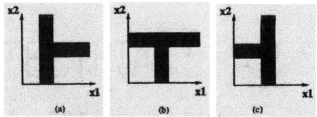

Fig. 3. *Partition of the input space.*

2) *Parameter Identification:* In ordinary system identification, parameters are the coefficients in a functional system model. In a fuzzy model, the parameters are those in the membership functions of the fuzzy sets. There is not a big difference between the two except in the number of the parameters, there being many more in a fuzzy model. We can use a conventional criterion, i.e., the output error, for parameter identification also in a fuzzy model. We have to notice that the structure identification and the parameter identification cannot be separately performed in principle. This fact makes the identification very complicated. In our approach, the parameter identification can, however, be separately performed after the structure identification.

By summing up the above-mentioned tasks to be done for modeling, we could say that the ratio of importance of the structure identification of type I to that of type II and the parameter identification would be, moderately speaking, 100:10:1. In any event, if we know the input candidates to a system, our problem is almost solved. After this, we can certainly find an algorithm of the identification depending on a description language. Unfortunately, we do not have a systematic approach to structure identification of type Ia: it is a problem of induction. Structure identification of type Ib is a combinatorial problem. Type IIa is similar to the identification of the model order in ordinary identification. Type IIb is also a combinatorial problem which appears only in the case of *"if-then"* rules. The parameter identification is merely an optimization problem with an objective function.

In this paper we will mainly deal with structure identification since parameter identification is neither valuable for discussions nor theoretically important.

2. Fuzzy Model

In this paper we use the following type of a fuzzy model for a multi-input and single-output system:

$$R^i: \text{if } x_1 \text{ is } A_1^i \text{ and } x_2 \text{ is } A_2^i \cdots \text{and } x_n \text{ is } A_n^i \text{ then } y \text{ is } B^i, \qquad (9)$$

where R^i is the *i*th rule ($1 \leq i \leq m$), $x_j (1 < j < n)$ are input variables, y is the output, and A_j^i and B^i are fuzzy variables.

We can simply rewrite the form (9) as:

R^i: if x is A^i then y is B^i (9a)

where $x = (x_1, x_2, \cdots, x_n)$ and $A = (A_1, A_2, \cdots, A_n)$.

For simplicity, the membership function of A^i_j is denoted $A^i_j(\bullet)$. In this section we regard the fuzzy variables A and B as those taking as values fuzzy numbers which are not necessarily associated with linguistic labels. Our method of qualitative modeling consists of two steps. In the first step, we deal with a fuzzy model in terms of fuzzy numbers. In the second step, we give linguistic interpretations to this fuzzy model. That is, we deal with a fuzzy model with linguistic terms which we call a qualitative model. Occasionally we let B^i take singletons, i.e., real numbers, instead of fuzzy numbers. As far as fuzzy control is concerned, it is well known that we can use fuzzy rules with singletons in the consequents without losing the performance of the control. However, in order to derive a qualitative model from the fuzzy model in Eq. (9), it is better to use fuzzy numbers in the consequents.

As for reasoning, we modify partly the ordinary method as seen in the following steps 2) and 3):

1) Given the inputs $x_1^0, x_2^0, \cdots,$ and x_n^0, calculate the degree of match, w^i, in the premises for the ith rule, $1 \leq i \leq m$ as

$$w^i = A_1^i(x_1^0) \times A_2^i(x_2^0) \times \cdots \times A_n^i(x_n^0) \qquad (10)$$

2) Then defuzzify B^i in the consequents by taking the center of gravity:

$$b^i = \int B^i(y) y \, dy \Big/ \int B^i(y) \, dy \qquad (11)$$

3) Calculate the inferred value, $\hat{y}$, by taking the weighted average of b^i with respect to w^i

$$\hat{y} = \sum_{i=1}^{m} w^i b^i \Big/ \sum_{i=1}^{m} w^i, \qquad (12)$$

where m is the number of rules.

As we find in the process of reasoning, the rule R^i translates to the form

R^i: if x is A^i then y is b^i. (13)

In what follows we shall call a fuzzy model in the form of Eq. (9), a position type model.

Next we propose occasionally to use a position-gradient type model. It is often the case that we cannot build a fuzzy model over the whole input space because we lack data. In this case we need to take an extrapolation for estimating the output using local fuzzy rules. Note that in conventional fuzzy reasoning we take an interpolation using some number of rules.

A position-gradient model is of the following form:

R^i: if x is A^i then y is B^i and $\partial y / \partial x$ is C^i, (14)

where $\partial y/\partial x = (\partial y/\partial x_1, \cdots, \partial y/\partial x_n)$, $C^i = (C_1^i, \cdots, C_n^i)$, and $\partial y/\partial x_j$ is the partial derivative of y with respect to x_j.

Using the rules shown in Eq. (14), we can infer the output for a given input for which no rule of the position type is available; if $w^i = 0$ in all the rules for the inputs, $\hat{y}$ cannot be inferred in the position type model. The reasoning algorithm of the position-gradient rules is the following:

1) Defuzzify B^i and C^i, and let those values be b^i and c^i respectively, where $c^i = (c_1^i, \cdots, c_n^i)$. We can rewrite Eq. (13) as

$$R^i: \text{if } x \text{ is } A^i \text{ then } y \text{ is } b^i \text{ and } \partial y/\partial x \text{ is } c^i, \qquad (15)$$

2) Calculate the distance d^i between the input and the core region of the ith rule as shown in Fig. 4. As we see, the core region, in a two-dimensional case, of the ith rule: "if x_1 is A_1^i and x_2 is A_2^i then y is B^i, is the crisp region determined by the fuzzy sets A_1^i and A_2^i, where membership grades A_1^i and A_2^i are 1.

3) Calculate the inferred output, y, by

$$\hat{y} = \sum_{i=1}^{m}\left\{w(d^i)*\left(b^i + \sum_{j=1}^{n}(d_j^i \times c_j^i)\right)\right\} \bigg/ \sum_{i=1}^{m} w(d^i) \qquad (16)$$

where n is the number of inputs, m the number of rules, d_j^i the component of d^i on the x_j coordinate axis, and $w(d^i) = \exp(-d^i)$ is the weight of the ith rule depending on the distance d^i.

As is seen in Eq. (16), the term $b^i + \sum_{j=1}^{n}(d_j^i \times c_j^i)$ is the extrapolated value of the output using the value of its partial derivatives $\partial y/\partial x_j$. We shall see below that the membership functions in this study are almost always of a trapezoidal type as shown in Fig. 4.

2.1. Structure Identification

Let us now discuss the method of structure identification. As stated in the Introduction (Section 1) we have to deal with structures of type Ib: to find a set of input variables among the possible candidates, and with structures of type II: to find the number of rules and a fuzzy partition of the input space (see Table I). We shall use an ordinary method for Ib but propose a new method based on fuzzy clustering for type II. In general, it seems that we can neither separate the identification of type Ib from that of type II, nor separate the structure identification from the parameter identification; these are mutually related. However, we can separate these by using our method. This is a great advantage of the new method.

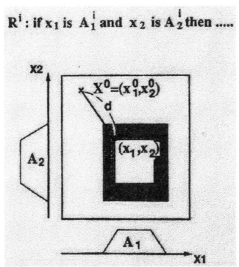

Fig. 4. Distance from the input X to the rule.

In principle, the algorithm for the identification is of the iterative type. We present the algorithm through numerical examples. Let us consider the following nonlinear static system with two inputs, x_1 and x_2, and a single output, y:

$$y = (1 + x_1^{-2} + x_2^{-1.5})^2, \quad 1 \le x_1, x_2 \le 5. \tag{17}$$

We show a three-dimensional input-output graph of this system in Fig. 5. From this system equation, 50 input-output data are obtained as illustrated in Table 2. The data of x_3 and x_4 are put as dummy inputs to check the appropriateness of the identification method.

Fig. 5. Input-output relation of nonlinear system.

Table 2. Input-output data of nonlinear system

	Group A				
No.	x1	x2	x2	x3	y
1	1.40	1.80	3.00	3.80	3.70
2	4.28	4.96	3.02	4.39	1.31
3	1.18	4.29	1.60	3.80	3.35
4	1.96	1.90	1.71	1.59	2.70
5	1.85	1.43	4.15	3.30	3.52
6	3.66	1.60	3.44	3.33	2.46
7	3.64	2.14	1.64	2.64	1.95
8	4.51	1.52	4.53	2.54	2.51
9	3.77	1.45	2.50	1.86	2.70
10	4.84	4.32	2.75	1.70	1.33
11	1.05	2.55	3.03	2.02	4.63
12	4.51	1.37	3.97	1.70	2.80
13	1.84	4.43	4.20	1.38	1.97
14	1.67	2.81	2.23	4.51	2.47
15	2.03	1.88	1.41	1.10	2.66
16	3.62	1.95	4.93	1.58	2.08
17	1.67	2.23	3.93	1.06	2.75
18	3.38	3.70	4.65	1.28	1.51
19	2.83	1.77	2.61	4.50	2.40
20	1.48	4.44	1.33	3.25	2.44
21	3.37	2.13	2.42	3.95	1.99
22	2.84	1.24	4.42	1.21	3.42
23	1.19	1.53	2.54	3.22	4.99
24	4.10	1.71	2.54	1.76	2.27
25	1.65	1.38	4.57	4.03	3.94
	Group B				
No.	x1	x2	x3	x4	y
26	2.00	2.06	2.25	2.37	2.52
27	2.71	4.13	4.38	3.21	1.58
28	1.78	1.11	3.13	1.80	4.71
29	3.61	2.27	2.27	3.61	1.87
30	2.24	3.74	4.25	3.26	1.79
31	1.81	3.18	3.31	2.07	2.20
32	4.85	4.66	4.11	3.74	1.30
33	3.41	3.88	1.27	2.21	1.48
34	1.38	2.55	2.07	4.42	3.14
35	2.46	2.12	1.11	4.44	2.22
36	2.66	4.42	1.71	1.23	1.56
37	4.44	4.71	1.53	2.08	1.32
38	3.11	1.06	2.91	2.80	4.08
39	4.47	3.66	1.23	3.62	1.42
40	1.35	1.76	3.00	3.82	3.91
41	1.24	1.41	1.92	2.25	5.05
42	2.81	1.35	4.96	4.04	1.97
43	1.92	4.25	3.24	3.89	1.92
44	4.61	2.68	4.89	1.03	1.63
45	3.04	4.97	2.77	2.63	1.11
46	4.82	3.80	4.73	2.69	1.39
47	2.58	1.97	4.16	2.95	2.29
48	4.14	4.76	2.63	3.88	1.33
49	4.35	3.90	2.55	1.65	1.40
50	2.22	1.35	2.75	1.01	3.39

1) *Structure Identification of Type I:* In the proposed algorithm, the identification of type Ib is done between the identification of type II and the parameter identification. Let us, however, first consider the identification Ib.

We have four candidates, x_1 - x_4, for the inputs to the system and have to find among them the actual inputs affecting the output, y. This is a combinatorial problem. For instance, we can count 15 cases in this example: four cases if the system has only one input, six cases if it has two inputs, and so on. In general, let X be a set of possible input candidates $x_1, x_2, \cdots, x_n$, then the total number of cases is the number of subsets except an empty subset of X, i.e., $2^n - 1$. Here we take a heuristic method to select some inputs from among the candidates; we increase the number of inputs one by one, watching a criterion.

First we divide the data into two groups, A and B, as in Table 2. As a criterion to this purpose, we use the so-called regularity criterion, RC [28], in GMDH (group method of data handling), which is defined as follows:

$$RC = \left[\sum_{i=1}^{k_A} \left(y_i^A - y_i^{AB} \right)^2 / k_A + \sum_{i=1}^{k_B} \left(y_i^B - y_i^{BA} \right)^2 / k_B \right] \bigg/ 2, \qquad (18)$$

where k_A and k_B are the number of data in groups A and B;

y_i^A and y_i^B are the output data of the groups A and B;

y^{AB} is the model output for the group A input estimated by the model identified using the group B data;

y^{BA} is the model output for the group B input estimated by the model identified using the group A data.

As we can guess from the form of RC, we build two models for two data groups at each stage of the identification. Notice that we have to make the structure identification of type II and the parameter identification in order to calculate RC. We can easily find the meaning of RC in Fig. 2.

Now we show the outline of a heuristic algorithm for Ib. First, we begin with a fuzzy model with one input. We make four models: one model for one particular input. After the identification of the structure II and parameter identification, which will be described later, we calculate RC of each model and select one model to minimize RC from among the one-input models. Next we fix the one input selected above and add another input to our fuzzy model from among the remaining three candidates. Our fuzzy model has two inputs at this stage. We select the second input as we do at the first step, according to the value of RC.

We continue the above process until the value of RC increases. The result is shown in Table 3. As is shown, x_1 is selected at the first step, x_2 at the second step. At the third step, however, both the values of RC for the third inputs, x_3 and x_4, are bigger than the minimal RC at the second step. So the search is terminated at this stage. As a result we have evaluated 9 of the 15 cases and succeeded in finding the true inputs, x_1 and x_2.

Table 3. Structure identification Ib of nonlinear system

	Input Variables	RC	
Step 1	$r1$	0.630	O
	$r2$	0.863	
	$r3$	0.830	
	$r4$	0.937	
Step 2	$r1 - r2$	0.424	OO
	$r1 - r3$	0.571	
	$r1 - r4$	0.583	
Step 3	$r1 - r2 - r3$	0.483	x
	$r1 - r2 - r4$	0.493	x

Now, let us look at Fig. 6 to understand the algorithm.

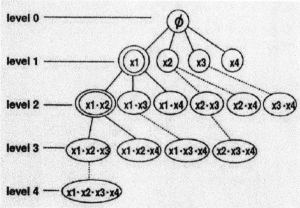

Fig. 6. Search tree in identification of type Ib.

This figure shows a tree structure of this combinatorial problem. As we see, the tree in this case consists of four levels. Each node of the tree corresponds to a subset of the set of input candidates: the node at level 0 corresponds to an empty set. Thus, the total number of meaningful nodes is $2^n - 1$ for a case of n input candidates. The nodes x_1 and $x_1 \bullet x_2$ with double circles are selected in searching the tree as shown in Table 3. Only one node at each level is selected and so those nodes connected by dotted lines are not evaluated. Since the values of RC at level 3 become bigger than that of $x_1 \bullet x_2$ at level 2, the search is terminated at level 2. If not, the search is continued until the final level. As a result of this algorithm, we evaluate at most $n(n + 1)/2$ nodes in this algorithm out $2^n - 1$.

We can further reduce the number of evaluations. For example, suppose the RC of the node $x_1 \bullet x_4$ at level 2 is not improved compared with that of the best node, x_1, at level 1. Then we consider that the input x_4 should be eliminated and thus we stop evaluating the node $x_1 \bullet x_2 \bullet x_4$ at level 3. This case will be shown later in the example of modeling a gas furnace (see Table 7).

2) *Structure Identification of Type* II: Usually in the design of a fuzzy controller we first pay attention to rule premises and find an optimal partition based on a certain criterion. Here we propose a different method; that is, we first pay attention to the consequents of the rules and then find a partition concerning the premises. Also in our method, we do not take an ordinary fuzzy partition of the input space, as is shown in Fig. 7, for if we take this kind of partition, the number of rules increases exponentially with the number of inputs.

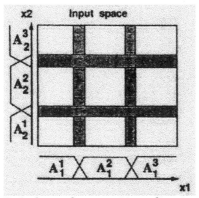

Fig. 7. Ordinary fuzzy partition of input space.

For this reason, we introduce the fuzzy c-means method, abbreviated FCM, for the structure identification of type IIb. The algorithm of FCM will be shown in Appendix I. Using FCM, we make fuzzy clustering of the output data; we use all the data. As a result, every output y is associated with the grade of membership belonging to a fuzzy cluster B. Notice that we now have the following data associated with the grade of the membership of

$$y^i \text{ in } B^j (1 \le j \le c):$$
$$(x^i, y^i), B^1(y^i), B^2(y^i), \cdots, B^c(y^i). \tag{19}$$

We can induce a fuzzy cluster A in the input space as is shown in Fig. 8. By making the projection of the cluster A onto the axes of the coordinates x_1 and x_2 we obtain the fuzzy sets A_1 and A_2 as shown in Fig. 9. As is easily seen, we have at this stage the following relation:

$$A_1(x_1^i) = A_2(x_2^i) = B(y^i), \tag{20}$$

where B is the output cluster. Now this cluster gives a fuzzy rule:
if x_1 is A_1 and x_2 is A_2, then y is B.

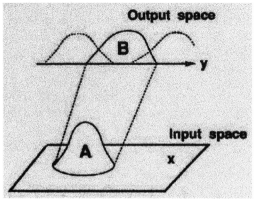

Fig. 8. *Fuzzy Cluster in the input space.*

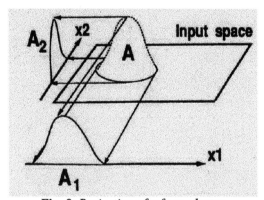

Fig. 9. *Projection of a fuzzy cluster.*

Remark 1: Although the output cluster B is convex, the input cluster A corresponding to B might not be convex: for instance we might obtain A_1, as is shown in Fig. 10(a). In this case we approximate the input cluster in Fig. 10(a) with a convex fuzzy set as in Fig. 10(b). Finally we approximate this convex fuzzy set and B as well, with a fuzzy set of trapezoidal type as shown in Fig. 10(c), which is used in the fuzzy model.

Remark 2: The next problem is that we might have more than two fuzzy clusters, A^1 and A^2 in the input space which corresponds to the same fuzzy cluster B in the output space. In this case we carefully form two convex fuzzy clusters as illustrated in Fig. 11. We obtain the following two rules with the same consequent:

$$R^1: \text{ if } x_1 \text{ is } A_1^1 \text{ and } x_2 \text{ is } A_2^1 \text{ then } y \text{ is } B$$
$$R^2: \text{ if } x_1 \text{ is } A_1^2 \text{ and } x_2 \text{ is } A_2^2 \text{ then } y \text{ is } B$$

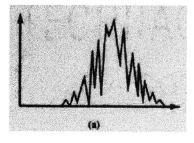

(a) Input cluster   *(b) Approximated convex fuzzy set*

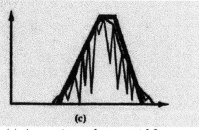

(c) Approximated trapezoid fuzzy set

Fig. 10. *Construction of a membership function.*

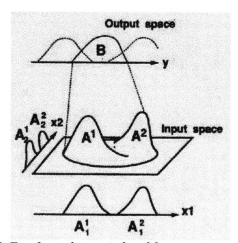

Fig. 11. *Two fuzzy clusters induced from one output cluster.*

As we can easily understand, a fuzzy partition of the input space is obtained as a direct result of fuzzy clustering; we do not have to consider the structure identification of type IIb.

Let us discuss a method to determine the number of rules which is related to the number of fuzzy clusters in fuzzy clustering. Note that, because of *Remark 2*, the number of fuzzy rules is not exactly the same as that of fuzzy clusters in the output space, as we have just discussed above.

The determination of the number of clusters is the most important issue in clustering. There are many studies on this issue [29]-[31]. Here we use the following criterion [31] for this purpose:

$$S(c) = \sum_{k=1}^{n}\sum_{i=1}^{c}(\mu_{ik})^{m}\left(\left\|x_{k}-v_{i}\right\|^{2}-\left\|v_{i}-\bar{x}\right\|^{2}\right), \qquad (21)$$

where

- n: number of data to be clustered;
- c: number of clusters, $c \geq 2$;
- x_k: kth data, usually vector;
- $\bar{x}$: average of data: $x_1, x_2, \cdots, x_n$
- v_i: vector expressing the center of ith cluster;
- $\|\cdot\|$: norm;
- μ_{ik} grade of the kth data belonging to the ith cluster
- m adjustable weight (usually $m=1.5 \sim 3$)

The number of clusters, e, is determined so that $S(c)$ reaches a minimum as c increases: it is supposed to be a local minimum as usual. As is seen in Eq. (21), the first term of the right-hand side is the variance of the data in a cluster and the second term is that of the clusters themselves. Therefore, the optimal clustering is considered to minimize the variance in each cluster and to maximize the variance between the clusters. Fig. 12 shows the change of $S(c)$ in fuzzy modeling of the system shown in Eq. (17). In this case, the optimal number of fuzzy clusters is found to be 6.

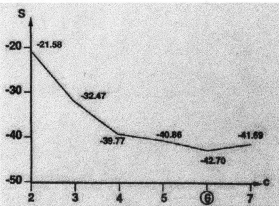

Fig. 12. *Behavior of $S(c)$ in nonlinear system model.*

From this, the number of rules is identified by taking account of the case where a fuzzy cluster in the input space is divided into two fuzzy clusters. As it is easily seen, in the process of fuzzy clustering, we refer neither to the input nor to the parameters in premises. Therefore, we can separate the identification of type II from that of Ib and the parameter identification. Further, as we find in the above discussion, a fuzzy partition of the input space IIb as well as the parameters A^i in the premises and B^i in the consequents are obtained as by-products of IIa. We use these A^i and B^i to calculate RC in the identification of type Ib.

After identifying the structure of type Ib (input variables), we combine the two groups of the data and induce fuzzy clusters in the input space from the output clusters to find the parameters A. We use those parameters obtained in the process of fuzzy clustering as an initial guess in the parameter identification. Fig. 13 shows the fuzzy model of Eq. (17). Below we shall use the mean square error of output as a performance index of a fuzzy model:

$$PI = \sum_{i=1}^{m} (y^i - \hat{y}^i)^2 / m, \qquad (22)$$

where m is the number of data, y^i is the ith actual output, and $\hat{y}^i$ is the ith model output. In this example, we have $PI = 0.318$.

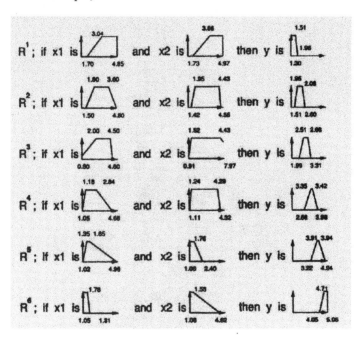

Fig. 13. Model of the nonlinear system.

3) *Identification of a Position-Gradient Model*: Finally let us discuss the identification of a position-gradient type model of the form shown in Eq. (14): the determination of the term "$\partial y / \partial x_j$ is C_j". Since the partial derivative $\partial y / \partial x_i$ is not given as data, we have to estimate it from the given input-output data.

We assume an equation around the input-output data $(x_1^0, \cdots, x_n^0, y^0)$:

$$y = y^0 + \partial y / \partial x_1 (x_1 - x_1^0) + \cdots + \partial y / \partial x_n (x_n - x_n^0). \tag{23}$$

We estimate the coefficients $\partial y / \partial x_j$ in Eq. (23) by using the method of weighted least squares using the other input-output data. As weights, we use

$$w^k = \exp(-|x_1^0 - x_1^k| - 2\sum_{j=2}^{n} |x_j^0 - x_j^k|) \tag{24}$$

where $x_j^k (1 \le j \le n)$ is the kth data. Then we can obtain a fuzzy set C_j^i in the form of Eq. (14) from the output cluster B^i just as we do to obtain the fuzzy sets A_j^i.

Summing up the above method of identification, we arrive at the algorithm given in Fig. 14.

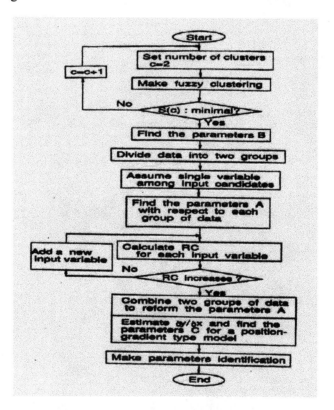

Fig. 14. Algorithm for identification.

We should notice as pointed out in the preceding section that the parameter identification can be done separately from the structure identification in our method since the parameters obtained as by-products in fuzzy clustering are available to calculate *RC* in the process of the structure identification Ib. In ordinary algorithms of identification, the parameter identification must be performed in the process of the structure identification, which makes an algorithm complicated. Our method also enables us to omit the identification IIb of the partition of the input space.

2.2. *Parameter identification*

As discussed in the preceding sections, we determine at the stage of parameter identification the values of parameters in a system model. In the case of a fuzzy model, the parameters are those concerned with membership functions.

In this paper, we approximate a convex fuzzy set with a trapezoidal fuzzy set. A trapezoidal fuzzy set has four parameters, as shown in Fig. 15, where $p1 < p2 < p3 < p4$.

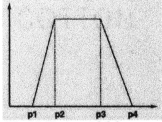
Fig.15. Trapezoidal fuzzy set.

As mentioned, we have already found the parameters by making the projection of the clusters onto the axes of coordinates. We can derive a qualitative model from a fuzzy model with these parameters. However, it is better to improve the parameters in order to use a fuzzy model for simulation. So we adjust the parameters as we do in the ordinary parameter identification.

Since parameter identification is a problem of nonlinear optimization, there are studies on fuzzy modeling using the nonlinear optimization method for parameter identification, e.g., the complex method [12].

In the method we propose, we have a great advantage in that we can find, as we did in our example, the approximate values of the parameters; we can use these values as an initial guess in the parameter identification.

Now we show an algorithm for the identification process.

1) Set the value f of adjustment.

2) Assume that the *k*th parameter of the *j*th fuzzy set is p_j^k.

3) Calculate $p_j^k + f$ and $p_j^k - f$. If $k = 2,3,4$, and $p_j^k - f$ is smaller than p_j^{k-1}, then $\hat{p}_j^k = p_j^{k-1}$; else $\hat{p}_j^k = p_j^k - f$. Also if $k = 1,2,3$ and $p_j^k + f$ is bigger than p_j^{k+1}, then $\hat{p}_j^k = p_j^{k+1}$; else $\hat{p}_j^k = p_j^k + f$. f is a constraint for adjusting parameters, as shown in Fig. 16.

4) Choose the parameter which shows the best performance *P1* in Eq. (21) among $\{\hat{p}_j^k, p_j^k, \hat{\hat{p}}_j^k\}$ and replace p_j^k with it.
5) Go to step 2 while unadjusted parameters exist.
6) Repeat step 2 until we are satisfied with the result.

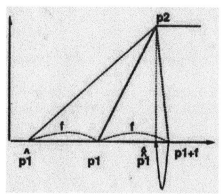

Fig. 16. Constraint for adjusting parameter.

We use 5% of the width of the universe of discourse as the value of f and we repeat 20 times steps 1 to 6. Note that we do not adjust the parameters in the consequents of the rules.

3. Qualitative Model

3.1 *Linguistic Approximation*

In this section we discuss a way to derive a qualitative model from a fuzzy model. To this aim we use a method of linguistic approximation [32], [33] of fuzzy sets. We can state our problem as follows: "Given a proposition with fuzzy predicates, find a word or a phrase out of a given set of words to linguistically approximate it, with hedges and connectives." After this procedure, we can obtain a qualitative model with linguistic rules from the identified fuzzy model.

We use the following two indices to measure the matching degree of two fuzzy sets [34]:

1) *Degree of similarity*:
$$S(N, L) = \|N \cap L\| / \|N \cup L\| \qquad (25)$$

where, $\|\cdot\|$ is the cardinality of the fuzzy set, and N and L are fuzzy sets.

2) *Degree of inclusion*:
$$I(N, L) = \|N \cap L\| / \|L\| \qquad (26)$$

In the above setting, N is a fuzzy set considered as a fuzzy number and L is a fuzzy set associated with a linguistic label.

In order to choose the closest L to a given N, we may refer only to $S(N, L)$. In some cases, however, it is better to refer also to $I(N, L)$. In Fig. 17, for two fuzzy sets P and Q, we have

$$S(N, P) = S(N, Q) \text{ and } I(N,P) < I(N,Q).$$

Fig. 17. Fuzzy set and linguistic approximation P and Q.

As Q is included in N, we can say that Q approximates more appropriately N than P. We suggest leaving the exact use of these to the user's preference.

In this paper we consider linguistic approximation on three levels.
    Level 1: approximation with linguistic terms.
    Level 2: approximation with linguistic terms and hedges.
    Level 3: approximation with linguistic terms, hedges and connectives.

We shall use the following hedges [35]: *very, more or less, slightly, sort of not, more than, less than*, as shown in Fig. 18. We apply certain constraints in the use of hedges. For instance, we may say *"very big,"* or *"more than middle,"* but we do not say *"more than big."*

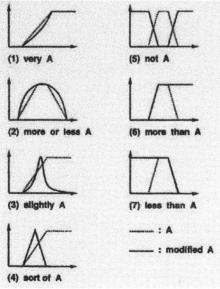

Fig. 18. Linguistic hedges.

As an example, let us approximate two fuzzy numbers, A and B, concerning the strength of wind in Fig. 19(a) with the linguistic terms in Fig. 19(b). The top and bottom parts of Table 4 show the linguistic approximations of A and B according to three levels where the values of two indices are shown.

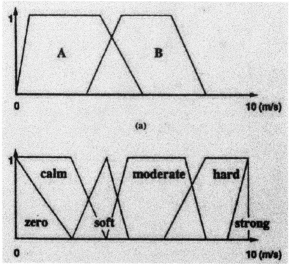

Fig. 19. *Fuzzy numbers and words for linguistic approximation: (a) fuzzy numbers and (b) wind scale.*

Table 4. *Results of linguistic approximation*

Level	Linguistic Expression	S	I
1	calm	0.62	0.96
2	less than soft	0.87	0.97
3	more or less calm or more or less soft	0.94	0.94

(A)

Level	Linguistic Expression	S	I
1	moderate	0.76	0.92
2	more or less moderate	0.81	0.83
3	not strong but more or less moderate	0.86	0.98

(B)

3.2. Control of Number of Linguistic Rules

It is often the case that we are asked to explain phenomena "*more precisely*" or "*more simply*." We have a limitation in responding to this sort of request if we merely try to refine or simplify the linguistic approximation. We can manage this problem by increasing or decreasing the number of linguistic rules. Two features contribute to this flexibility, namely, hedges and connectives.

As we adopt a fuzzy clustering method for the identification of Iia, i.e., the number of rules, we can simply control the number of linguistic rules by adjusting the number of fuzzy clusters. This is also a great advantage of the proposed method.

4. Illustrative Examples

We deal with six illustrative examples in this section. The first example involves qualitative modeling of a nonlinear system, which has been partly discussed in the previous section. The second is also a numerical example in which we discuss the performance of our fuzzy modeling method. In the third example we make a qualitative model of a dynamical process. The fourth and fifth examples are concerned with qualitative modeling of human operation at process control. In the final example we make a qualitative model to explain the trend in the time series data of the price of a stock.

In order to understand the examples in connection with the preceding discussions, let us list the key points:
    a) fuzzy model of position type,
    b) fuzzy model of position-gradient type,
    c) qualitative model,
    d) structure of type Ib: input variables,
    e) control of the number of linguistic rules,
    f) structure of type IIa: number of rules,
    g) parameter identification.

In what follows, we shall attach symbols (a)-(e) to each example. Throughout the examples, (f), the structure identification of type IIa, and (g), the parameter identification, are performed.

4.1. Nonlinear System (a, b, c, d)

This example deals with the explanation of the proposed method of identification. Let us recall a static and nonlinear two input-single output system of the type shown in Eq. (17):

$$y = (1 + x_1^{-2} + x_2^{-1.5})^2, \quad 1 \le x_1, \quad x_2 \le 5.$$

This system shows the nonlinear characteristic illustrated in Fig. 5 and we use the data in Table 2. The process of the identification of type IIa has been shown in Table 3 and we have found the true inputs x_1 and x_2. The optimal number of rules has been found to be 6. In this example, the number of rules is found to be equal to the number of clusters. After the parameter identification, we obtain a fuzzy model as in Fig. 20, where the performance index of the model is $PI = 0.079$. Before the parameter identification, we have obtained the fuzzy model with unadjusted parameters as we see in Fig. 13, where $PI = 0.318$. We can certainly improve the model by adjusting its parameters.

In this example, we also obtain a fuzzy model of position-gradient type as illustrated in Fig. 21. In the case of the latter model, the performance is greatly improved. We have $PI = 0.010$.

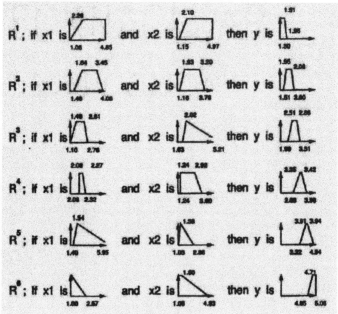

Fig. 20. Model of nonlinear system after parameter identification.

Rule	Premise	y	∂y/∂x1	∂y/∂x2
1		1.42	-0.71	-0.79
2		2.02	-0.58	-0.57
3	the same as Fig. 20	2.58	-0.43	-0.43
4		3.39	-0.50	-0.56
5		3.94	-0.59	-1.04
6		4.90	-1.20	-1.35

★ Consequents are singletons

Fig. 21. Position-gradient of nonlinear system.

Finally, we show a qualitative model of the position-gradient type.
1) If x_1 is more than MEDIUM and x_2 is more than MEDIUM, then y is SMALL, $\partial y/\partial x_1$ is sort of NEGATIVE, and $\partial y/\partial x_2$ is sort of NEGATIVE.
2) If x_1 is not SMALL but less than MEDIUM BIG and x_2 is not SMALL but less than MEDIUM BIG, then y is MEDIUM SMALL, $\partial y/\partial x_1$ is sort of NEGATIVE, and, $\partial y/\partial x_2$ is sort of NEGATIVE.

3) If x_1 is not SMALL but less than MEDIUM BIG and x_2 is more or less MEDIUM BIG, then y is more or less MEDIUM SMALL and MEDIUM, $\partial y/\partial x_1$ is sort of NEGATIVE, and $\partial y/\partial x_2$ is sort of NEGATIVE.
4) If x_1 is less than MEDIUM and x_2 is less than MEDIUM BIG, then y is MEDIUM and more or less MEDIUM BIG, $\partial y/\partial x_1$ is sort of NEGATIVE, and $\partial y/\partial x_2$ is sort of NEGATIVE.
5) If x_1 is more or less MEDIUM SMALL or MEDIUM and x_2 is more or less SMALL, then y is MEDIUM BIG, $\partial y/\partial x_1$ is sort of NEGATIVE, and $\partial y/\partial x_2$ NEGATIVE.
6) If x_1 is SMALL and x_2 is more or less SMALL or MEDIUM SMALL, then y is BIG, $\partial y/\partial x_1$ is very NEGATIVE, and $\partial y/\partial x_2$ is very NEGATIVE.

We can understand the performance of the model by referring to the characteristics of the system shown in Fig. 5.

4.2. Fuzzy System Model (a, d)

In this example we investigate the reproductivity of the proposed method of system identification. That is, we obtain input-output data from a fuzzy system consisting of fuzzy rules and make its fuzzy model using the data.

Fig. 22 shows a fuzzy system with nine rules in the form of Eq. (13) which has singletons in the consequents.

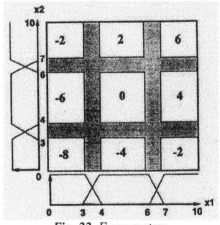

Fig. 22. Fuzzy system.

We use 100 data points of x_1, x_2, x_3, x_4, and y taken from the system where x_3 and x_4 are dummy input variables; some data are shown in Table 5.

After the fuzzy clustering, we find the optimal number of clusters to be 8. From this we derive nine rules; one of eight clusters is divided into two rules because of the nonconvexity of an input cluster.

The structure identification of type Ib proceeds as is shown in Table 6. In step 2 of this table, the two inputs, x_1 and x_2, with the minimal RG are correctly selected.

Table 5. Some input-output data of fuzzy system

$x1$	$x2$	$x3$	$x4$	y
8.85	9.32	0.69	5.84	6.00
3.89	9.19	2.49	9.00	1.56
4.43	3.67	2.95	5.17	-1.32
1.98	6.74	7.08	6.62	-3.04
1.69	7.20	5.41	9.23	-2.00
6.23	7.26	6.50	6.25	2.92

Table 6. Structure identification Ib of fuzzy system

	Input Variables	RC	
Step 1	$x1$	9.810	O
	$x2$	11.335	
	$x3$	19.820	
	$x4$	18.249	
Step 2	$x1 - x2$	3.827	OO
	$x1 - x3$	10.904	
	$x1 - x4$	10.476	

After parameter identification, we obtain a fuzzy model of the type shown in Fig. 23. We can see a good reproductivity by comparing the identified model with the original system.

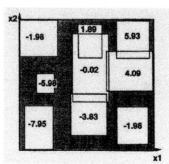

Fig. 23. Fuzzy system model.

4.3. Box and Jenkins's Gas Furnace (b, c, d)

We discuss the qualitative modeling of a dynamical process using a famous example of the system identification given by Box and Jenkins [36]. The process is a gas furnace with single input $u(t)$ and single output $y(t)$: gas flow rate and CO_2 concentration, respectively. We omit to show data which are found in [36].

Since the process is dynamical, we consider as candidates the 10 variables $y(t-1), \cdots, y(t-4), u(t-1), \cdots, u(t-6)$ to affect the present output $y(t)$. Using 296 data points, we make fuzzy clustering and find six clusters. Table 7 shows the structure identification process of type Ib. As we find, we can select $y(t-1)$, $u(t-4)$, and $u(t-3)$ by referring to the values of RG.

In Table 7, some sets of input candidates, for example, $\{y(t-1), y(t-2)\}$, $\{y(t-1), c(t-3)\}$ and $\{y(t-1), y(t-4)\}$ in step 2 are associated with the mark 'x'. The mark 'x' shows that the values of RC for those input sets are bigger than the minimal RG at the previous step, i.e., RC for $y(t-1)$. As discussed in "*Structure Identification of Type I*" of Section 2.1, we eliminate the input candidates $y(t-2), y(t-3)$, and $y(t-4)$ after step 2 since we may not have good prospects by considering those as candidates at proceeding steps. So we evaluate only five sets of input candidates at step 3 as seen in Table 7.

Table 7. Structure identification Ib of gas furnace

	Input Variables	RC	
Step 1	$y(t-1)$	0.851	O
	$y(t-2)$	2.276	
	$y(t-3)$	4.151	
	$y(t-4)$	5.720	
	$u(t-1)$	6.724	
	$u(t-2)$	4.816	
	$u(t-3)$	3.257	
	$u(t-4)$	1.869	
	$u(t-5)$	1.465	
	$u(t-6)$	2.040	
Step 2	$y(t-1) - y(t-2)$	0.973	x
	$y(t-1) - y(t-3)$	1.067	x
	$y(t-1) - y(t-4)$	1.020	x
	$y(t-1) - u(t-1)$	0.745	
	$y(t-1) - u(t-2)$	0.598	
	$y(t-1) - u(t-3)$	0.439	
	$y(t-1) - u(t-4)$	0.418	OO
	$y(t-1) - u(t-5)$	0.565	
	$y(t-1) - u(t-5)$	0.846	
Step 3	$y(t-1) - u(t-4) - u(t-1)$	0.439	x
	$y(t-1) - u(t-4) - u(t-2)$	0.429	x
	$y(t-1) - u(t-4) - u(t-3)$	0.382	OOO
	$y(t-1) - u(t-4) - u(t-5)$	0.454	x
	$y(t-1) - u(t-4) - u(t-6)$	0.499	x

After the parameter identification, we obtain a fuzzy model of the position-gradient type as shown in Fig. 24. Fig. 25 shows the model behavior in comparison with the actual process. In Table 8, we compare our fuzzy model with other fuzzy models where we see a high performance of our model; the number of rules are remarkably reduced, and so is the value of *P1*.

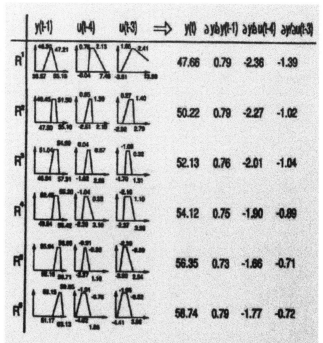

Fig. 24. Fuzzy model of gas furnace (position-gradient type).

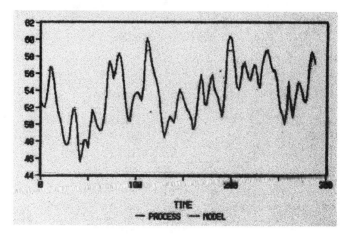

Fig. 25. Output of fuzzy model of gas furnace.

Table 8. Comparison of fuzzy models with other models

Model Name	Inputs	Number of Rules	Model Error
Tong's model [5]	$y(t-1)$, $u(t-4)$	19	0.469
Pedryc's model [6]	$y(t-1)$, $u(t-4)$	81	0.320
Xu's model [7]	$y(t-1)$, $u(t-4)$	25	0.328
Linear model	$y(t-1)$, $y(t-2)$, $u(t-3)$, $u(t-4)$, $u(t-5)$	—	0.193
Takagi Sugeno model [35]	$y(t-1)$, $y(t-2)$, $y(t-3)$, $u(t-1)$, $u(t-2)$, $u(t-3)$	2	0.068
Position-gradient model	$y(t-1)$, $u(t-4)$, $u(t-3)$	6	0.190

Now let us derive a qualitative model from a fuzzy model of position type using linguistic approximation at level 3. We obtain the following model.

1) If $y(t-1)$ is sort of SMALL or MEDIUM SMALL $u(t-4)$ is MEDIUM BIG or sort of BIG and $u(t-3)$ is more or less MEDIUM BIG or sort of BIG, then $y(t)$ is SMALL.
2) If $y(t-1)$ is not SMALL but more or less MEDIUM SMALL, $u(t-4)$ is not BIG but more or less MEDIUM BIG, and $u(t-3)$ is more or less MEDIUM, then $y(t)$ is MEDIUM SMALL.
3) If $y(t-1)$ is not MEDIUM BIG but more or less MEDIUM, $u(t-4)$ is not MEDIUM SMALL but more or less MEDIUM, and $u(t-3)$ is more or less MEDIUM, then $y(t)$ is MEDIUM SMALL and MEDIUM.
4) If $y(t-1)$ is not MEDIUM SMALL but more or less MEDIUM, $u(t-4)$ is MEDIUM SMALL or more or less MEDIUM, and $u(t-3)$ is not small but less than MEDIUM BIG, then $y(t)$ is MEDIUM or MEDIUM BIG.
5) If $y(t-1)$ is not BIG but more or less MEDIUM BIG, $u(t-4)$ is MEDIUM SMALL or MEDIUM, and $u(t-3)$ is not SMALL but more or less MEDIUM SMALL, then $y(t)$ is MEDIUM BIG and sort of BIG.
6) If $y(t-1)$ is MEDIUM BIG or more or less BIG, $u(t-4)$ is more or less MEDIUM SMALL or MEDIUM, and $u(t-3)$ is less than MEDIUM, then $y(t)$ is very BIG.

As is well known, we can make a linear model of this process. The performance index of the identified linear model is $PI = 0.193$. The PI value of 0.190 of our fuzzy model is almost the same as that of the linear model. We should remark that we do not aim at a good numerical model by fuzzy modeling in this study. If we want to do it by fuzzy modeling, we can use Takagi-Sugeno's fuzzy model and we can improve the performance quite a bit more such that $PI = 0.068$ [37].

We will show this fuzzy model as well as a linear model in Appendix II.

4.4. *Human Operation at a Chemical Plant* (a. c, d, e)

We deal with a model of an operator's control of a chemical plant. The plant is for producing a polymer by the polymerization of some monomers. Since the start-up of the plant is very complicated, a man has to make the manual operation at the plant. The structure of the human operation is shown in Fig. 26.

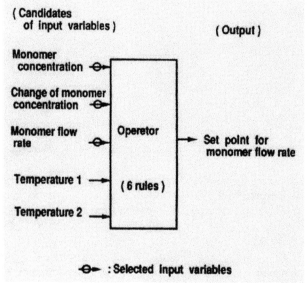

Fig. 26. Structure of plant operation.

There are five input candidates which a human operator might refer to for his control, and one output, i.e., his control. These are the following:

    $u1$ monomer concentration,
    $u2$:   change of monomer concentration,
    $u3$:   monomer flow rate.
    $u4, u5$: local temperatures inside the plant,
    y set point for monomer flow rate.

where an operator determines the set point for the monomer flow rate and the actual value of the monomer flow rate to be put into the plant is controlled by a PID controller. We obtain 70 data points of the above six variables from the actual plant operation as shown in Appendix III. First we find six clusters by fuzzy clustering; we obtain six rules in this case. Fig. 27 shows the change of $S(c)$ in the process of IIa.

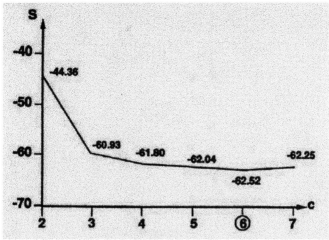

Fig. 27. Behavior of $S(c)$ in plant operation model.

The identification process of type Ib is as shown in Table 9. As a result, we conclude that the operator must refer to the three informations $u1$ (monomer concentration), $u2$ (change of $u1$) and $u3$ (monomer flow rate) to decide his control action.

Table 9. Structure identification of Ib of chemical plant operation

	Input Variables	RC	
Step 1	$u1$	602715	
	$u2$	6077539	
	$u3$	⌜60756⌝	O
	$u4$	6663660	
	$u5$	5570199	
Step 2	$u3 - u1$	46178	
	$u3 - u2$	⌜41418⌝	OO
	$u3 - u4$	64124	×
	$u3 - u5$	60277	
Step 3	$u3 - u2 - u1$	⌜38950⌝	OOO
	$u3 - u2 - u5$	41846	×

236 *Fuzzy Modeling and Control: Selected Works of M. Sugeno*

We show the obtained fuzzy model in Fig. 28. Fig. 29 shows the excellent performance of the fuzzy model.

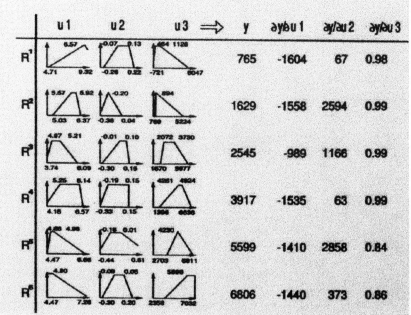

Fig. 28. Fuzzy model of plant operation (position-gradient type).

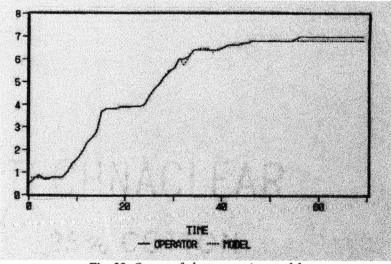

Fig. 29. Output of plant operation model.

From the fuzzy model, we can derive the following qualitative model.

1) If $u1$ is more or less BIG, and $u2$ is not INCREASED, and $u3$ is SMALL, then y is SMALL or MEDIUM SMALL.
2) If $u1$ is more or less MEDIUM, and $u2$ is DECREASED, and $u3$ is SMALL or MEDIUM SMALL, then y is MEDIUM SMALL.
3) If $u1$ is MEDIUM, and $u2$ shows NO CHANGE, and $u3$ is MEDIUM SMALL or MEDIUM, then y is MEDIUM.
4) If $u1$ is more or less MEDIUM, and $u2$ is ANY VALUE, and $u3$ is MEDIUM, then y is MEDIUM or MEDIUM BIG.
5) If $u1$ is more or less SMALL, and $u2$ is very INCREASED, and $u3$ is MEDIUM BIG, then y is BIG.
6) If $u1$ is more or less SMALL, and $u2$ is sort of INCREASED, and $u3$ is BIG, then y is very BIG.

We have shown this qualitative model to some operators and obtained their agreements with the model. It is worth noticing that we can make a fuzzy controller using the obtained fuzzy model to automate the operator's control.

In general it is difficult to evaluate this sort of a qualitative model since a model error in an ordinary sense is not available. One method taken here is to ask an expert about the performance of a model. Also in such a case as Subsection 4.1, we can qualitatively evaluate a model by comparing a model with the original shape in Fig. 5.

If we are asked to explain the operator's action in a simple form, we can present a qualitative model, for instance, with three linguistic rules by adjusting the number of clusters. Those are the following.

M1) If $u1$ is MEDIUM SMALL, and $u2$ it less than NO CHANGE, and $u3$ is sort of SMALL, then y is sort of SMALL or more or less MEDIUM SMALL
M2) If $u1$ is more or less MEDIUM, and $u2$ is ANY, and $u3$ is more or less MEDIUM, then y is more or less MEDIUM.
M3) If $u1$ is sort of SMALL or more or less MEDIUM SMALL, and $u2$ is more or less NO CHANGE. And $u3$ is more than MEDIUM BIG, then y is more than MEDIUM BIG.

By comparing those **M1-M3** with 1-6 above, we can find that **M1** is a summary of 1, 2, and 3, **M2** is that of 3 and 4, and **M3** is that of 4, 5, and 6. That is, the original six rules are fuzzily summarized in three rough rules.

4.5. Human Operation at Water Purification Process (a, c, d)

A water purification process is a process to produce clean water for civil use. Turbid water, called raw water, flows into the process from rivers or lakes and then a chemical product, called a flocculating agent, is put into the process so that the turbid part of the water form flocks by cohesion which settle out in the settling basin. After sterilization by chlorine, we get clean water.

The chemical reactions in this process are very complicated and so a human operator has to decide the amount of the flocculating agent by referring to some state variables. In principle, his control is of a feedforward type; it is generally difficult to automate such a human operator behavior. The referred variables are:

x_1: change of turbidity of raw water
x_2: turbidity of raw water
x_3: turbidity of water in the settling basin
x_4: alkalinity
x_5: PH
y: amount of flocculating agent

We use 624 data points. Making the identification of the operator's control action, we obtain a fuzzy model with seven rules as is shown in Fig. 30. Fig. 31 shows the performance of the model. From this fuzzy model, we derive a qualitative model as is shown in Table 10.

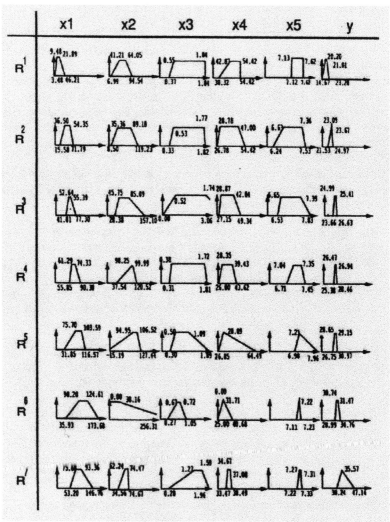

Fig. 30. Fuzzy model of water purification process operation.

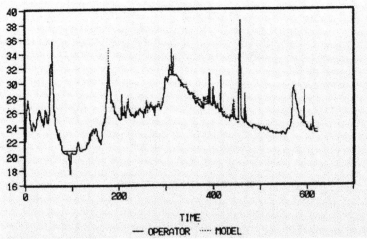
Fig. 31. Output of operation model at water purification process.

Finally we deal with the trend data of stock prices. We use the daily data of two different stocks: 100 data points of the first stock, A, and 150 data points of the second stock, B. The data of stock A are shown in Appendix IV. Both data consist of 10 inputs and 1 output. These are

- x_1: past change of moving average (1) over a middle period;
- x_2: present change of moving average (I) over a middle period;
- x_3: past separation ratio (1) with respect to moving average over a middle period;
- x_4: present separation ratio (1) with respect to moving average over a middle period;
- x_5: present change of moving average (2) over a short period
- x_6: past change of price (1), for instance, change on one day before;
- x_7: present change of price (1):
- x_8: past separation ratio (2) with respect to moving average over a short period;
- x_9: present change of moving average (3) over a long period;
- x_{10}: present separation ratio (3) with respect to moving average over a short period;
- y: prediction of stock price.

Here the separation ratio is a value concerning the difference between a moving average of a stock price and price of a stock. Making fuzzy clustering, we obtain five rules in both cases. Here we omit fuzzy models. Figs. 32 and 33 show a comparison of the price with the actual price; we can see the good performance of the models. The qualitative models of the trend data of the two stock prices are shown in Tables 11 and 12. This sort of study is quite interesting. It implies that we induce "qualitative laws", by observing data, to which economic phenomena are subject.

Table 10. Qualitative model of water purification process operation

Rule	Change of Turbidity of Raw Water	Turbidity of Raw Water	Turbidity of Water in the Settling Basin	Alkalinity	pH	Amount of Flocculating Agent
1	very small	slightly small		more than medium	more than medium	small
2	slightly small	slightly small		less than medium	sort of small	medium-small
3	slightly small	slightly big		sort of small		slightly medium-small
4	medium	medium		sort of small	medium	medium
5	medium	medium	less than medium	more or less small	more or less big	slightly medium-big
6	more or less big	sort of small	slightly small	small	medium	medium-big
7	medium	slightly small	more than medium	slightly small	slightly big	big

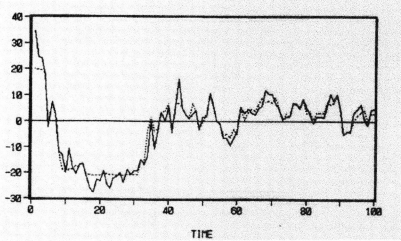

Fig. 32. Output of stock price model A.

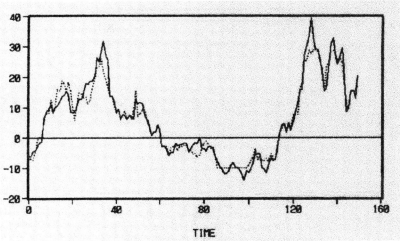

Fig. 33. Output of stock price model B.

5. Conclusions

We have discussed an approach to qualitative modeling based on fuzzy logic. A qualitative model is derived from a fuzzy model using the linguistic approximation method. We have proposed the use of a fuzzy clustering method for the structure identification of a fuzzy model. The proposed method has been examined in several case studies. What we have to do further in qualitative modeling is to improve the linguistic approximation method.

Table 11. Qualitative model of trend of stock price A.

Rule	Change of Moving Average 1		Separation Ratio 1		Change of Moving Average 2	Change of Price 1		Separation Ratio 2	Change of Moving Average 3	Separation Ratio 3	Prediction
	Past	Now	Past	Now	Now	Past	Now	Past	Now	Now	
1	more or less rise	level or rise	sort of plus	less than zero	level	level	rise	more or less zero	level	more less zero	rise
2		level or decline		minus	more or less level		level	more or less zero	level	more or less zero	sort of rise
3		level or decline	more or less zero	sort of minus	more or less level	level or rise	sort of decline	more or less zero	more or less level	more or less zero	level
4	more or less rise	level	plus	more or less zero	more or less level	sort of rise	level	more or less zero	more or less level	more or less zero	sort of decline
5	level or rise	rise	more than zero	plus	sort of rise	sort of rise	level	sort of plus	level or rise	more than zero	decline

Table 12. Qualitative model of trend of stock price B

Rule	Change of Moving Average 1		Separation Ratio 1		Change of Moving Average 2	Change of Price 1		Separation Ratio 2	Change of Moving Average 3	Separation Ratio 3	Prediction
	Past	Now	Past	Now	Now	Past	Now	Past	Now	Now	
1	sort of rise	level or decline	less than zero	sort of minus	level	sort of decline	more or less level	sort of minus	sort of decline	more or less zero	rise
2	level or rise	level or decline	less than zero	more or less zero	level or rise	sort of decline	level or rise		sort of rise	more than zero	sort of rise
3	level or rise	level	more or less zero	more or less zero	level or rise	sort of rise	level or rise		sort of rise		level
4	more or less rise	more or less rise	more or less zero	slightly plus	more or less level	sort of decline	level or rise	more or less zero	level or decline	less than zero	sort of decline
5	more or less rise	more or less rise	plus	more than zero	level	sort of decline	level	sort of minus	decline	more or less zero	decline

References

[1]. L. A. Zadeh, "Fuzzy algorithm," *Information and Control*, vol.12, pp. 94-102, 1968.

[2]. L. A. Zadeh, "Outline of a new approach to the analysis of complex systems and decision processes," IEEE *Trans. Syst., Man. Cybern.*, vol. 3, pp. 28-44, 1973.

[3]. E. H. Mamdani and S. Assilian, "Applications of fuzzy algorithms for control of simple dynamic plant." *Proc. Inst. Elec. Eng.*, vol.121, pp. 1585-1588, 1974.

[4]. L. P. Holmblad and J. J. Ostergaard, "Control of a cement kiln by fuzzy logic," in *Fuzzy Information and Decision Processes*, M. M. Gupta and E. Sanchez, Eds. North-Holland, 1982.

[5]. R. M. Tong. "The evaluation of fuzzy models derived from experimental data," *Fuzzy Sets and Systems*, vol.4, pp. 1-12, 1980.

[6]. W. Pedrycz, "An identification algorithm in fuzzy relational systems," *Fuzzy Sets and Systems*, vol.13, pp.153-167,1984.

[7]. C. W. Xu and Z. Yong, "Fuzzy model identification and self-learning for dynamic Systems," IEEE *Trans. Syst., Man. Cybern.*, vol.17, no.4. pp. 683-689, 1987.

[8]. D. Filev, "Fuzzy modeling of complex systems," *Int. J. Approximate Reasoning*, vol.5, pp. 281-290, 1991.

[9]. D. Willaeys, N. Malvache, and P. Hammond, "Utilization of fuzzy sets for systems modeling and control," in *Proc. 16th* IEEE *Conf Decision and Control* (New Orleans, LA), Dec.1975.

[10]. E. Czogala and W. Pedrycz, "On identification in fuzzy systems and its applications in control problem," *Fuzzy Sets and Systems*, vol.6, pp. 73-83, 1981.

[11]. T. Takagi and M. Sugeno, "Fuzzy identification of systems and its application to modeling and control," IEEE *Trans. Syst., Man, Cybern.*, vol. 15, pp. 116-132, 1985.

[12]. M. Sugeno and G. T. Kang, "Fuzzy modeling and control of multilayer incinerator," *Fuzzy Sets and Systems*, vol.18, pp. 329-346, 1986.

[13]. K. Tanaka and M. Sugeno, "Stability of analysis of fuzzy systems using Lyapunov's direct method and construction procedure for Lyapunov functions," in *Proc. 6th Fuzzy Systems Symposium* (Tokyo), 1990, pp. 353-356.

[14]. M. Sugeno and T. Takagi, "A new approach to design of fuzzy controller," *Advances in Fuzzy Sets, Possibility Theory, and Applications*, P.P. Wang and S. K. Chang, Eds. New York: Plenum Press, 1983, pp. 325-334.

[15]. J. De kleer, "Qualitative and quantitative knowledge in classical mechanics," Artificial Intelligence Laboratory, TR-352, MIT, Cambridge, MA, 1975.

[16]. Hayes, P. J., "The naive physics manifesto," *Expert in the Micro-electronic Age*, D. Michie, Ed. Edinburgh: Edinburgh University Press, 1979.

[17]. J. De Kleer and J. S. Brown, "A qualitative physics based on confluences,' *Artijicial Intelligence*, vol.24, pp. 7-83, 1984.

[18]. K. D. Forbus, "Qualitative process theory," *Artificial Intelligence.* vol. 24 pp. 85-168, 1984.

[19]. B. Kuipers, "Qualitative simulation," *Artificial Intelligence.* vol.29, pp. 289-338, 1986.

[20]. F. Wentop, "Deductive verbal models of organizations," *Int. J. Man-Machine Studies*, vol.8, pp. 293-311, 1976.

[21]. W. J. M. Kickert, "An example of linguistic modeling: the case of Mulder's theory of power," in *Advances in Fuzzy Set Theory and Applications*, M. M. Gupta, R. K. Ragade, and R. R, Yager, Eds. North-Holland, 1979, pp. 519-540.

[22]. W. J. M. Kickert, "Towards an analysis of linguistic modeling," *Fuzzy Sets and Systems*, vol.2, pp. 293-307, 1979.

[23]. B. D'Ambrosio, *Qualitative Process Theory Using Linguistic Variables*. New York: Springer-Verlag, 1989.

[24]. E. H. Mamdani. "Application of fuzzy logic to approximate reasoning using linguistic synthesis," IEEE *Trans. Comput.* vol.26, pp. 1182-1191, 1977.

[25]. T. J. Procyk and E. H. Mamdani, "A linguistic self-organizing process controller," *Automatica*, vol.15, pp. 15-30, 1979.

[26]. L. A. Zadeh, "From circuit theory to system theory," in *Facets of Systems Science*, G. J. Klir, Ed. New York: Plenum Press, 1991.

[27]. H. Akaike, "A new look at the statical model identification," IEEE *Trans. Automat. Contr.*, vol.19 pp. 716-723, 1974.

[28]. J. Ihara, "Group method of data handling towards a modeling of complex systems-IV," *Systems and Control* (in Japanese), vol.24, pp. 158-168, 1980.
[29]. J. C. Dunn, "Well-separated cluster and optimal fuzzy partition," *J Cybern.*, vol.4, pp. 95-104, 1974.
[30]. J. C. Bezdek, *Pattern Recognition with Fuzzy Objective Function Algorithm*. New York: Plenum Press. 1981.
[31]. Y. Fukuyama and M. Sugeno, "A new method of choosing the number of clusters for fuzzy c-means method." in *Proc. 5th Fuzzy System Symposium* (in Japanese), 1989, pp. 247-250.
[32]. F. Eshragh and E. H. Mamdani, "A general approach to linguistic approximation," *Int. J. Man-Machine Studies*, vol. II, pp. 501-519, 1979.
[33]. A. Ralescu and H. Narazaki, "Integrating artificial intelligence techniques in linguistic modeling from numerical data," *Proc. Int. Fuzzy Engineering Symp.* (Yokohama), 1991.
[34]. D. Dubois and H. Prade, *Fuzzy Sets and Systems-Theory and Applications*, New York: Academic Press, 1980.
[35]. L. A. Zadeh, "A fuzzy-set-theoretic interpretation of linguistic hedges," *J. Cybern.*, vol.2, pp. 4-34, 1972.
[36]. G. E. P. Box and G. M. Jenkins, *Time Series Analysis, Forecasting and Control*, San Francisco: Holden Day, 1970.
[37]. K. Tanaka, "*An approach to modeling, analysis and design of fuzzy control systems,*" Ph.D. thesis, Tokyo Institute of Technology, 1990.

Appendix I

Algorithm of Fuzzy C-Means Methods [30]

Given a set of vector data x_k, $1 \leq k \leq n$ we classify those into a certain number of fuzzy clusters. A fuzzy cluster is characterized by μ_{ik}, which shows the grade of membership of the kth data, x_k, belonging to the ith cluster. We assume that

$$\sum_{i=1}^{c} \mu_{ik} = 1 \qquad (27)$$

where c is the number of clusters, and we define a matrix U consisting of μ_{ik}.

Our problem now is to find c and to determine U. The algorithm is as follows:

1) Set $c = 2$, an initial value $U^{(1)}$ of U and $l = 1$.
2) Calculate the center v_i of the fuzzy cluster by

$$v_i = \sum_{k=1}^{n}(\mu_{ik})^m x_k \bigg/ \sum_{k=1}^{n}(\mu_{ik}), \quad 1 \leq i \; c. \qquad (28)$$

We define the distance from the kth data to the center of the ith cluster by

$$d_{ik} = \|x_k - v_i\| \qquad (29)$$

3) Calculate the element of a new $U^{(l)}$ as follows for $l = l + 1$:

$$I_k \triangleq \{i | 1 \; i \; c; d_{ik} = \|x_k - v_i\| = 0\}$$

$$\widetilde{I}_k \triangleq \{1, 2, \cdots, c\} - I_k$$

$$I_k = \emptyset \Rightarrow \mu_{ik} = 1 \bigg/ \left[\sum_{j=1}^{c}(d_{ik}/d_{kj})^{2/(m-1)}\right]$$

$$I_k = \emptyset \Rightarrow \mu_{ik} = 0 \; \forall i \in \tilde{I}_k \text{ and } \sum_{i \in I_k} \mu_{ik} = 1. \tag{30}$$

4) If $\left\|U^{(l-1)} - U^{(l)}\right\| \leq \varepsilon$, then stop; otherwise go to step 2.

In this algorithm, m is an adjustable parameter, which is set as 2 in this paper.

Appendix II

Takagi-Sugeno's Fuzzy Model And A Linear Model Of Gas Furnace
Takagi-Sugeno's fuzzy model is of the following form:

$$R^i: \quad \text{if } x_1 \text{ is } A_1^i \text{ and } \cdots \text{ and } x_n \text{ is } A_n^i \text{ then}$$

$$y = y^i = c_0^i + c_1^i x_1 + \cdots + c_n^i x_n \tag{31}$$

where y^i is the output from the ith rule, A_j^i is a fuzzy set, and c_j^i is a consequent parameter.

Given an input $(x_1^0, x_2^0, \cdots, x_n^0)$, the final output of this model is inferred by taking the weighted average of the y^i's as follows:

$$y = \sum_{i=1}^{n} w^i y^{i0} \bigg/ \sum_{i=1}^{n} w^i, \tag{32}$$

where y^{i0} is calculated for the input by the consequent equation of the ith implication and the weight w^i implies the degree of match in the premise as of the ith rule calculated as in Eq. (10).

This fuzzy model can express a highly nonlinear functional relation in spite of a small number of rules. The simple algorithm of the identification of this model is as follows:

1) Choose the premise structure and the consequent structure.
2) Estimate the parameters of the structure determined in 1).
3) Evaluate the model.
4) Repeat 1) until we are satisfied with the result.

For further details of the identification of Takagi-Sugeno's model, the reader is referred to [12].

Using the above type of modeling method, we obtain a fuzzy model of Box and Jenkins's gas furnace with two rules as follows [37]:

R^1 : if $y(t-1)$ is [trapezoidal fuzzy set with breakpoints 48.8 and 59.0]

then $y(t) = 1.45y(t-1) - 0.87y(t-2) + 0.23y(t-3)$
$+ 0.19u(t-1) - 0.21u(t-2) + 10.3$

R^2 : if $y(t-1)$ is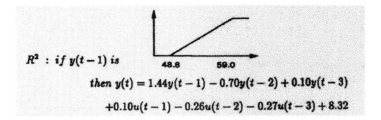

then $y(t) = 1.44y(t-1) - 0.70y(t-2) + 0.10y(t-3)$
$+ 0.10u(t-1) - 0.26u(t-2) - 0.27u(t-3) + 8.32$

On the other hand, we can also obtain a linear model of this system using the conventional system modeling method. The linear model is:

$$y(t) = 0.57y(t-1) + 0.02y(t-2) - 0.53u(t-3)$$
$$- 0.33u(t-4) - 0.51u(t-5) + 21.87.$$

Appendix III

Data of Human Operation at a Chemical Plant

u1	u2	u3	u4	u5	y
6.80	-0.05	401.00	-0.20	-0.10	500.00
6.59	-0.21	464.00	-0.10	0.10	700.00
6.59	0.00	703.00	-0.10	0.10	900.00
6.50	-0.09	797.00	0.10	0.10	700.00
6.48	-0.02	717.00	-0.10	0.10	700.00
6.54	0.06	706.00	-0.20	0.10	800.00
6.45	-0.09	784.00	0.00	0.10	800.00
6.45	0.00	794.00	-0.20	0.10	800.00
6.20	-0.25	792.00	0.00	0.00	1000.00
6.02	-0.18	1211.00	0.00	0.10	1400.00
5.80	-0.22	1557.00	-0.20	0.00	1600.00
5.51	-0.29	1782.00	-0.10	0.00	1900.00
5.43	-0.08	2206.00	-0.10	0.10	2300.00
5.44	0.01	2404.00	-0.10	-0.10	2500.00
5.51	0.07	2685.00	0.10	0.00	2800.00
5.62	0.11	3562.00	-0.40	0.10	3700.00
5.77	0.15	3629.00	-0.10	0.00	3800.00
5.94	0.17	3701.00	-0.20	0.10	3800.00
5.97	0.03	3775.00	-0.10	0.00	3800.00
6.02	0.05	3829.00	-0.10	-0.10	3900.00
5.99	-0.03	3896.00	0.20	-0.10	3900.00
5.82	-0.17	3920.00	0.20	-0.10	3900.00
5.79	-0.03	3895.00	0.20	-0.10	3900.00
5.65	-0.14	3887.00	-0.10	0.00	3900.00
5.48	-0.17	3930.00	0.20	0.00	4000.00
5.24	-0.24	4048.00	0.10	0.00	4400.00
5.04	-0.20	4448.00	0.00	0.00	4700.00
4.81	-0.23	4462.00	0.00	0.10	4900.00
4.62	-0.19	5078.00	-0.30	0.30	5200.00
4.61	-0.01	5284.00	-0.10	0.20	5400.00
4.54	-0.07	5225.00	-0.30	0.10	5600.00
4.71	0.17	5391.00	-0.10	0.00	6000.00
4.72	0.01	5668.00	0.00	-0.10	6000.00
4.58	-0.14	5844.00	-0.20	0.10	6100.00
4.55	-0.03	6068.00	-0.20	0.00	6400.00
4.59	0.04	6250.00	-0.20	-0.10	6400.00
4.65	0.06	6358.00	-0.10	-0.10	6400.00
4.70	0.05	6368.00	-0.10	0.00	6400.00
4.81	0.11	6379.00	-0.30	0.00	6400.00
4.84	0.03	6412.00	-0.10	-0.10	6400.00
4.83	-0.01	6416.00	0.10	-0.10	6500.00
4.76	-0.07	6514.00	0.00	0.00	6600.00
4.77	0.01	6587.00	-0.10	0.10	6600.00
4.77	0.00	6569.00	0.00	-0.10	6600.00
4.77	0.00	6559.00	0.00	0.00	6700.00
4.73	-0.04	6672.00	0.00	0.00	6700.00
4.73	0.00	6844.00	-0.10	0.00	6800.00
4.74	0.01	6775.00	-0.20	0.00	6800.00
4.77	0.03	6779.00	0.00	-0.10	6800.00
4.71	-0.06	6783.00	0.00	0.00	6800.00
4.66	-0.05	6816.00	0.00	0.00	6800.00
4.70	0.04	6812.00	0.00	0.00	6800.00
4.63	-0.07	6849.00	0.00	0.00	6800.00
4.61	-0.02	6803.00	0.00	0.00	6800.00
4.57	-0.04	6832.00	0.00	0.10	6800.00
4.56	-0.01	6832.00	-0.10	0.10	6900.00
4.54	-0.02	6862.00	-0.10	-0.10	7000.00
4.51	-0.03	6958.00	0.10	-0.10	7000.00
4.47	-0.04	6998.00	0.00	0.10	7000.00
4.47	0.00	6986.00	-0.10	0.10	7000.00
4.48	0.01	6975.00	0.00	0.00	7000.00
4.48	0.00	6973.00	0.00	0.00	7000.00
4.50	0.02	7006.00	0.00	0.10	7000.00
4.50	0.00	7027.00	0.00	0.00	7000.00
4.48	-0.02	7032.00	0.00	0.00	7000.00
4.54	0.06	6995.00	0.00	0.00	7000.00
4.57	0.03	6986.00	0.10	-0.10	7000.00
4.56	-0.01	7009.00	-0.10	0.10	7000.00
4.56	0.00	7022.00	0.00	0.00	7000.00
4.57	0.01	6998.00	-0.10	0.00	7000.00

Appendix IV

Daily Data of Stock A

x1	x2	x3	x4	x5	x6	x7	x8	x9	x10	y
0.239	0.227	8.307	2.820	-0.498	0.870	-3.281	-2.563	-2.523	-3.265	34.612
0.279	0.246	16.771	5.222	-0.250	2.358	1.696	-4.568	0.848	-0.502	24.390
0.319	0.273	21.253	5.819	-0.418	0.841	-0.834	-3.265	4.352	0.756	24.187
0.346	0.270	19.867	4.654	-0.420	-0.834	-0.841	-0.502	2.588	0.338	17.662
0.351	0.303	19.447	7.848	0.084	3.364	-2.545	0.756	3.364	3.626	-2.441
0.269	0.268	14.320	5.809	0.253	-1.627	0.870	0.338	0.834	1.682	7.444
0.250	0.267	11.157	5.527	0.084	0.0	2.588	3.626	1.682	1.597	2.481
0.256	0.336	12.787	11.263	0.756	5.790	0.841	1.682	4.068	6.672	-11.728
0.296	0.379	12.453	14.309	1.168	3.127	-0.834	1.597	9.098	8.739	-12.889
0.326	0.400	12.088	16.444	1.649	2.274	3.364	6.672	11.580	9.408	-20.015
0.354	0.350	13.587	14.317	1.379	-1.483	-1.627	8.739	3.909	6.320	-10.534
0.395	0.315	15.971	14.815	1.200	0.752	0.0	9.408	1.516	5.850	-17.924
0.388	0.295	19.283	12.767	0.949	-1.494	5.790	6.320	-2.224	3.289	-20.470
0.433	0.294	22.512	16.698	1.410	3.791	3.127	5.850	3.010	5.714	-16.801
0.466	0.350	28.465	19.689	1.390	2.922	2.274	3.289	5.228	7.311	-16.324
0.466	0.340	26.015	21.823	1.752	2.129	-1.483	5.714	9.098	7.710	-20.848
0.391	0.404	19.058	26.391	2.171	4.170	0.752	7.311	9.496	8.817	-26.017
0.346	0.478	22.512	30.825	2.051	4.003	-1.494	7.710	10.646	11.917	-27.582
0.348	0.401	11.815	29.467	1.651	-0.641	3.791	9.817	7.644	9.393	-22.595
0.357	0.427	10.504	28.084	1.342	-0.646	2.922	11.917	2.668	7.247	-23.392
0.357	0.361	11.021	20.990	0.906	-5.198	2.129	9.393	-6.414	0.760	-19.191
0.314	0.378	10.674	22.186	0.967	1.371	4.170	7.247	-4.519	1.163	-24.341
0.292	0.413	13.067	25.798	1.436	3.381	4.003	0.760	-0.650	3.102	-26.161
0.297	0.457	11.830	26.044	1.146	0.654	-0.641	1.163	5.483	2.600	-22.092
0.266	0.455	9.733	28.734	1.133	2.599	-0.646	3.102	6.761	4.087	-21.533
0.255	0.408	5.862	25.776	0.725	-1.900	-5.198	2.600	1.308	1.374	-20.013
0.264	0.415	7.374	28.491	0.589	2.582	1.371	4.087	3.249	3.383	-23.914
0.249	0.404	6.214	25.557	0.0	-1.888	3.381	1.374	-1.267	1.431	-18.602
0.243	0.394	5.086	24.262	0.0	-0.641	0.654	3.383	0.0	0.781	-20.658
0.223	0.410	2.165	24.554	0.130	0.646	2.599	1.431	-1.888	1.299	-19.243
0.227	0.364	2.820	24.102	0.650	0.0	-1.900	0.781	0.0	0.646	-19.243
0.246	0.318	5.222	17.360	0.0	-5.131	2.582	1.299	-4.519	-4.519	-14.875

(Table continues on next page)

x1	x2	x3	x4	x5	x6	x7	x8	x9	x10	y
0.272	0.238	5.819	17.873	-0.258	0.676	-1.888	0.646	-4.490	-3.625	-16.790
0.270	0.176	4.654	10.554	-0.906	-6.044	-0.641	-4.519	-10.263	-8.622	-13.581
0.303	-0.026	7.848	-5.226	-2.462	-14.296	0.646	-3.625	-18.932	-19.692	0.0
0.268	0.070	5.809	2.607	-1.674	8.340	0.0	-8.622	-12.760	-11.512	-10.778
0.267	0.053	5.527	-2.184	-2.384	-4.619	-5.131	-19.692	-11.437	-13.538	-4.036
0.336	-0.035	11.263	-10.837	-3.001	-8.878	0.676	-11.512	-5.838	-18.777	3.543
0.379	-0.044	14.309	-9.218	-2.878	1.771	-6.044	-13.538	-11.547	-14.889	0.0
0.400	-0.097	16.444	-14.666	-3.556	-6.092	-14.296	-18.777	-12.914	-17.127	5.561
0.350	0.0	14.317	-5.967	-2.842	10.195	8.340	-14.889	5.314	-6.008	-3.364
0.315	-0.070	14.815	-13.023	-3.004	-7.569	-4.619	-17.127	-4.352	-10.432	5.459
0.295	-0.114	12.767	-16.885	-3.586	-4.550	-8.878	-6.008	-2.780	-11.327	16.206
0.294	-0.026	16.698	-9.731	-2.198	8.580	1.771	-10.432	-4.205	-1.556	5.268
0.330	0.026	19.689	-6.585	-0.173	3.512	-6.092	-11.327	7.279	2.078	2.544
0.340	0.053	21.823	-9.802	-1.385	-3.393	10.195	-1.556	8.580	0.0	0.878
0.404	0.035	26.391	-12.209	-1.141	-2.634	-7.569	2.078	-2.634	-1.510	2.705
0.478	0.018	30.825	-10.641	0.0	1.803	-4.550	0.0	-4.241	0.266	4.429
0.401	0.105	29.467	-5.201	0.444	6.200	8.580	-1.510	5.268	6.012	-3.336
0.427	0.123	28.084	-6.897	0.884	-1.668	3.512	0.266	6.312	3.330	1.696
0.361	0.088	20.990	-6.978	-0.088	0.0	-3.393	6.012	4.429	3.421	1.696
0.378	0.061	22.186	-11.766	0.175	-5.089	-2.634	3.330	-6.672	-2.014	10.724
0.413	0.017	25.798	-10.993	0.700	0.894	1.803	3.421	-4.241	-1.826	6.200
0.457	0.061	26.044	-5.533	0.522	6.200	6.200	-2.014	1.696	3.720	0.0
0.455	0.105	28.734	-2.484	0.519	3.338	-1.668	-1.826	10.724	6.626	-0.807
0.408	0.149	25.776	-2.628	0.861	0.0	0.0	3.720	9.743	5.717	-6.457
0.415	0.113	28.491	-5.094	0.853	-2.421	-5.089	6.626	0.834	2.284	-6.617
0.404	0.174	25.557	-0.557	1.184	4.963	0.894	5.717	2.421	6.104	-9.456
0.394	0.148	24.262	-3.834	0.251	-3.152	6.200	2.284	-0.807	2.502	-6.509
0.410	0.174	24.554	-1.658	0.667	2.441	3.336	6.104	4.136	4.308	-3.971
0.364	0.130	24.102	-1.786	0.663	0.0	0.0	2.502	-0.788	3.621	5.560
0.318	0.052	17.360	-1.837	1.152	0.0	-2.421	4.308	2.441	2.441	3.177
0.238	-0.009	17.873	-3.388	0.895	-1.589	4.963	3.621	-1.589	-0.081	4.843
0.176	-0.026	10.554	-6.702	0.081	-2.421	-3.152	2.441	-3.971	-2.579	3.308
-0.026	-0.035	-5.226	-6.450	-0.322	-0.827	2.441	-0.081	-4.766	-3.072	2.502
0.070	-0.026	2.607	-9.547	-0.647	-3.336	0.0	-2.579	-6.457	-5.696	5.177

(Table continues on next page)

x1	x2	x3	x4	x5	x6	x7	x8	x9	x10	y
0.053	0.026	-2.184	-7.230	-0.163	2.588	0.0	-3.072	-1.654	-3.097	5.887
-0.035	-0.009	-10.837	-8.783	-0.815	-1.682	-1.589	-5.696	-2.502	-3.944	11.976
-0.044	-0.026	-9.218	-10.320	-0.657	-1.711	-2.421	-3.097	-0.863	-4.963	10.444
-0.097	-0.035	-14.666	-11.070	-0.993	-0.870	-0.897	-3.944	-4.205	-4.845	10.536
0.0	-0.043	-5.967	-10.250	-0.919	-0.878	-3.336	-4.963	-1.711	-3.120	6.962
-0.070	-0.061	-13.023	-9.414	-0.843	0.870	2.588	-4.845	0.870	-1.446	5.177
-0.114	-0.043	-16.885	-4.663	-0.170	5.177	-1.682	-3.120	7.024	3.833	0.820
-0.026	-0.096	-9.731	-6.157	-0.085	-1.641	-1.711	-1.446	4.353	2.217	1.658
0.026	-0.148	-6.585	-5.234	0.085	0.834	-0.870	3.833	4.314	2.981	1.654
0.053	-0.183	-9.802	-9.772	-0.085	-4.963	0.878	2.217	-5.742	-2.046	6.962
0.035	-0.131	-12.209	-10.440	-0.426	-0.870	0.870	2.981	-5.004	-2.483	7.024
0.018	-0.044	-10.641	-7.255	0.086	3.512	5.177	-2.046	-2.481	0.655	5.089
0.105	-0.052	-5.201	-8.760	0.085	-1.696	-1.641	-2.483	0.870	-0.940	8.628
0.123	-0.009	-6.897	-5.624	0.513	3.451	0.834	0.655	5.268	1.956	5.004
0.088	-0.017	-6.978	-5.607	0.425	0.0	-4.963	-0.940	1.696	1.524	2.502
0.061	0.017	-11.766	-2.475	0.677	3.336	-0.870	1.956	6.902	4.205	-0.807
0.017	-0.044	-10.993	-5.582	-0.168	-3.228	3.512	1.524	0.0	1.011	1.668
0.061	-0.035	-5.548	-5.548	0.0	0.0	-1.696	4.205	0.0	1.011	1.668
0.105	0.009	-2.484	-3.194	0.168	2.502	3.451	1.011	-0.807	3.364	1.827
0.149	-0.017	-2.626	-8.692	0.084	-5.696	0.0	1.011	-3.336	-2.605	6.040
0.113	-0.061	-5.094	-11.001	-0.084	-2.588	3.336	3.364	-5.838	-5.046	10.629
0.174	-0.035	-0.657	-9.393	-0.252	1.771	-3.228	-2.606	-6.509	-3.120	7.833
0.148	-0.026	-3.834	-9.369	-0.084	0.0	0.0	-5.046	-0.863	-3.038	10.444
0.174	0.053	-1.658	-4.666	0.084	5.222	2.502	-3.120	7.086	1.939	4.963
0.130	0.149	-1.786	4.618	1.096	9.926	-5.696	-3.038	15.666	10.843	-5.267
0.052	0.096	-1.837	2.188	0.600	-2.257	-2.588	1.939	13.055	7.801	-3.849
-0.009	0.087	-3.388	2.069	0.830	0.0	1.7711	0.843	7.444	6.914	-3.849
-0.026	0.052	-5.702	-1.911	0.412	-3.849	0.0	7.801	-6.020	2.377	3.203
-0.035	0.0	-6.450	-3.482	0.0	-1.601	5.222	6.914	-5.389	0.738	4.852
-0.026	0.009	-9.547	-4.275	0.492	-0.814	9.926	2.377	-6.159	-0.571	6.563
0.026	0.044	-7.230	-1.177	1.060	3.281	-2.257	0.738	0.801	1.614	0.794
-0.009	0.026	-8.783	2.720	1.291	3.971	0.0	-0.571	6.509	4.303	-1.528
-0.026	-0.044	-10.320	-0.375	0.956	-3.056	-3.849	1.614	4.102	0.158	4.728
-0.035	-0.079	-11.070	-1.062	0.395	-0.788	-1.601	4.303	0.0	-1.022	4.766

Stability Analysis and Design of Fuzzy Control Systems[*]

Kazuo Tanaka and Michio Sugeno
*Department of Systems Science,
Tokyo Institute of Technology,
4259 Nagatsuta, Midori-ku, Yokohanw 227, Japan*

*Received November 1989
Revised May 1990*

Abstract

The stability analysis and the design technique of fuzzy control systems using fuzzy block diagrams are discussed. First, we show the concept of fuzzy blocks and consider the connection problems of fuzzy block diagrams. We derive some theorems and corollaries with respect to two basic types of connections of fuzzy blocks. In order to preserve some properties in a connection of fuzzy blocks, continuous piecewise-polynomial membership functions are defined. Second, a sufficient condition which guarantees the stability of fuzzy systems is obtained in terms of Lyapunov's direct method. We give an important fact based on this condition. Third, we propose a new design technique of a fuzzy controller. The fuzzy block diagrams and the stability analysis are applied to the design problems of a model-based fuzzy controller.

Keywords: control theory, model-based controller, fuzzy control systems, fuzzy block diagrams, continuous piecewise polynomial membership function, stability analysis, Lyapunov's direct method.

---

[*] © *1992 North-Holland. Retyped with written permission from Fuzzy Sets and Systems, 45 (1992) 135-156.*

1. Introduction

Recently, fuzzy control has been applied to many practical industrial applications. From miscellaneous applications, we can see that fuzzy control has the following advantages:
(1) realization of multi-objective control;
(2) realization of expert control;
(3) realization of robust control.

On the other hand, as one of the disadvantages of fuzzy control, it is said that at present we lack analytical tools for fuzzy control systems. In other words, we need a fuzzy systems theory like linear systems theory. We consider three important concepts concerning the establishment of a fuzzy systems theory:
(C1) the connection problems of fuzzy block diagrams;
(C2) the stability analysis of fuzzy control systems;
(C3) the new design technique of a fuzzy controller.

In the linear case, block diagrams, which are represented by transfer functions, are utilized in the analysis of linear control systems. Similarly, we can analyze fuzzy control systems if we have the concepts of fuzzy block diagrams which represent the diagrams of fuzzy control systems.

The stability analysis of fuzzy control systems is one of the important concepts in the analysis of control systems.

Theoretically we can design a model-based fuzzy controller if we have a useful stability criterion for fuzzy control systems. (C3) is concerned with the design technique.

We need at least a fuzzy model of an objective system in the analysis of fuzzy control systems. We have a useful method [6-9] to identify a fuzzy model using input-output data of an objective system. In this paper, it is assumed that a fuzzy model of an objective system has already been identified.

This paper is organized as follows. Section 2 shows the outline of a fuzzy model. Section 3 describes the connection problems of fuzzy block diagrams. Section 4 shows the stability analysis of fuzzy control systems. Section 5 gives the new design technique of a fuzzy controller.

2. Takagi and Sugeno's fuzzy model

In this paper, we deal with Takagi and Sugeno's fuzzy model [9]. This fuzzy model is of the following form:

$$L^i : IF\ x(k)\ is\ A_1^i\ and\ ...\ and\ x(k-n+1)\ is\ A_n^i\ and$$
$$u(k)\ is\ B_1^i\ and\ ...\ and\ u(k-m+1)\ is\ B_m^i \quad (1)$$
$$THEN\ x^i(k+1) = a_0^i + a_1^i x(k) + \cdots + a_n^i x(k-n+1)$$
$$+ b_1^i u(k) + \cdots + b_m^i u(k-m+1),$$

where L^i: ($i = 1, 2, \ldots, l$) denotes the i-th implication, l is the number of fuzzy implications, $x^i(k+1)$ is the output from the i-th implication, a_p^i ($p = 0, 1, \ldots, n$)

and b_q^i ($q = 0, 1, \ldots, m$) are consequent parameters, $x(k), \ldots, x(k-n+1)$ are state variables, $u(k), \ldots u(k-m+1)$ are input variables and A_p^i and B_q^i are fuzzy sets whose membership functions denoted by the same symbols are continuous piecewise-polynomial functions. The membership functions are defined in Definition 2.1.

Given an input $(x(k), x(k-1), \ldots, x(k-n+1), u(k), u(k-1), \ldots, u(k-m+1))$, the final output of a fuzzy model is inferred by taking the weighted average of the $x^i(k+1)$'s:

$$x(k+1) = \sum_{i=1}^{l} w^i x^i(k+1) \bigg/ \sum_{i=1}^{l} w^i, \qquad (2)$$

where $\sum_{i=1}^{l} w^i > 0$, and $x^i(k+1)$ is calculated for the input by the consequent equation of the i-th implication, and the weight w^i implies the overall truth value of the premise of the i-th implication for the input calculated as

$$w^i = \prod_{p=1}^{n} A_p^i(x(k-p+1)) \times \prod_{q=1}^{m} B_q^i(u(k-q+1)). \qquad (3)$$

A set of fuzzy implications shown in (1) can express a highly nonlinear functional relation in spite of a small number of fuzzy implications [8].

Definition 2.1. A fuzzy set A satisfying the following two properties is said to have a continuous piecewise polynomial membership function $A(x)$.

(1) $A(x)$ is a continuous function.

(2) $A(x) = \begin{cases} \phi_1(x), & x \in [p_0, p_1], \\ \vdots & \vdots \\ \phi_s(x), & x \in [p_{s-1}, p_s], \end{cases} \qquad (4)$

Where $\phi_i(x) \in [0,1]$ for $x \in [p_{i-1}, p_i]$, $i = 1, 2, \ldots, s$, and $-\infty = p_0 < p_1 < \cdots < p_{s-1} < p_s = \infty$. The $\phi_i(x)$'s are polynomials of degree n_i, that is,

$$\phi_i(x) = \sum_{k=0}^{n_i} c_k^i x^k. \qquad (5)$$

c_k^i's are parameters of the polynomial $\phi_i(x)$.

The condition of convexity of fuzzy sets is not assumed in Definition 2.1 though it is an important property. The reason will be shown later. Next, we give an example of continuous piecewise polynomial membership functions.

Example 2.1. Let us consider two fuzzy sets of triangular type and trapezoidal type as shown in Figure 1. It is clear that the fuzzy sets satisfy the two properties. For the triangular type, if we choose a fuzzy set A such that

$$A(x) = \begin{cases} 0, & x \in (-\infty, 0], \\ 0.2x & x \in [0, 5], \\ -0.2x + 2, & x \in [5, 10], \\ 0, & x \in [10, \infty), \end{cases}$$

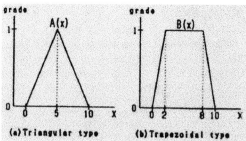

Fig. 1. *Examples of continuous piecewise-polynomial membership functions.*

then $A(x)$ is a continuous piecewise-polynomial membership function. For the trapezoidal type, if we choose a fuzzy set B such that

$$B(x) = \begin{cases} 0, & x \in (-\infty, 0], \\ 0.5x & x \in [0, 2], \\ 1, & x \in [2, 8], \\ -0.5x + 5, & x \in [8, 10], \\ 0, & x \in [10, \infty), \end{cases}$$

then $B(x)$ is a continuous piecewise-polynomial membership function. Moreover, the well-known π-function and s-function are continuous piecewise-polynomial membership functions, whereas the membership function of an exponential type is not a continuous piecewise-polynomial function. If it is, however, approximated by the Taylor expansion, it is a continuous piecewise-polynomial function.

Since (1) has many variables, it is convenient to express this in a vector form such as

L^i: IF $x(k)$ is P^i and $u(k)$ is Q^i

$$\text{THEN } x^i(k+1) = a_0^i + \sum_{p=1}^{n} a_p^i x(k-p+1) + \sum_{q=1}^{n} b_q^i u(k-q+1), \quad (6)$$

where

$x(k) = [x(k), x(k-1), \cdots x(k-n+1)]^{\mathrm{T}}$,
$u(k) = [u(k), u(k-1), \cdots u(k-m+1)]^{\mathrm{T}}$
$P^i = [A_1^i, A_2^i \cdots, A_n^i]^{\mathrm{T}}, Q^i = [B_1^i, B_2^i \cdots, B_m^i]^{\mathrm{T}}$

and

$x(k) = P^i \Leftrightarrow x(k)$ is A_1^i and $\cdots$ and $x(k-n+1)$ is A_n^i.

3. Connection problems of fuzzy block diagrams

In linear systems theory, block diagrams are the most suitable for analyzing control systems. The blocks are usually expressed in terms of transfer functions. On the other hand, fuzzy block diagrams which represent the connections between fuzzy systems appear very complex since a fuzzy system is described by a set of fuzzy implications. In order to establish a fuzzy systems theory, however, we need the concept of fuzzy block diagrams. It plays an important role in the connection problems of fuzzy systems. In this section, we analyze two basic types of connections of fuzzy blocks.

Definition 3.1. A fuzzy block is a block which expresses a fuzzy input-output relation described by (1).

Figure 2 shows the fuzzy block of (1). In the linear case we know that the result of serial connection of two linear systems is linear. Since linearity is preserved, we can apply linear systems theory to the connected linear system. Similarly, we can apply an analytical tool for fuzzy systems to a connected fuzzy system if some properties of fuzzy systems in a connection are preserved. Here the following two properties are under consideration:
(1) All the fuzzy sets in the premise parts are described by continuous piecewise-polynomial membership functions.
(2) All the consequent parts are described by linear equations.

We show an important proposition with respect to the preservation of premise parts.

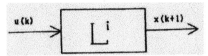

Fig. 2. *A single-input single-output fuzzy block.*

Proposition 3.1. *The product of two continuous piecewise-polynomial membership functions is also a continuous piecewise-polynomial membership function.*

Proof. Consider two fuzzy sets A^1 and A^2 on X. Assume that the fuzzy sets have continuous piecewise-polynomial membership functions such that

$$A^1(x) = \begin{cases} \phi_1^1(x), & x \in [p_0^1, p_1^1], \\ \vdots & \vdots \\ \phi_{s1}^1(x), & x \in [p_{s1-1}^1, p_{s1}^1], \end{cases} \quad A^2(x) = \begin{cases} \phi_1^2(x), & x \in [p_0^2, p_1^2], \\ \vdots & \vdots \\ \phi_{s2}^2(x), & x \in [p_{s2-1}^2, p_{s2}^2], \end{cases}$$

Then, $A^1(x) \times A^2(x)$ is calculated as follows:

$$A^1(x) \times A^2(x) = \phi_i^1(x) \times \phi_j^2(x)$$

for $x \in [p_{i-1}^1, p_i^1] \cap [p_{j-1}^2, p_j^2] (\neq \emptyset)$ where $i = 1, 2, \cdots, s_1$ and $j = 1, 2, \cdots, s_2$. Obviously $A^1(x) \times A^2(x)$ is a piecewise-polynomial membership function. Moreover, the product of two continuous functions is also a continuous function. Therefore, $A^1(x) \times A^2(x)$ is a continuous piecewise-polynomial membership function. □

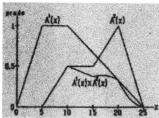

Fig. 3. An example of the product of two continuous piecewise-polynomial membership functions.

Figure 3 shows an example of the product of two continuous piecewise polynomial membership functions. In Figure 3,

$$A^1(x) = \begin{cases} 0, & x \in (-\infty, 0], \\ 0.2x, & x \in [0, 5], \\ 1, & x \in [5, 10], \\ -0.067x + 1.67, & x \in [10, 25], \\ 0, & x \in [25, \infty). \end{cases} \quad A^2(x) = \begin{cases} 0, & x \in (-\infty, 5], \\ 0.1x - 0.5, & x \in [5, 10], \\ 0.5, & x \in [10, 15], \\ 0.1x - 1, & x \in [15, 20], \\ -0.2x + 5, & x \in [20, 25], \\ 0, & x \in [25, \infty). \end{cases}$$

Then,

$$A^1(x) \times A^2(x) = \begin{cases} 0, & x \in (-\infty, 5], \\ 0.1x - 0.5, & x \in [5, 10], \\ -0.034x + 0.84, & x \in [10, 15], \\ -0.0067x^2 + 0.234x - 1.67, & x \in [15, 20], \\ 0.0134x^2 - 0.669x + 8.35, & x \in [20, 25], \\ 0, & x \in [25, \infty). \end{cases}$$

It is clear that $A^1(x) \times A^2(x)$ is a continuous piecewise-polynomial membership function.

Proposition 3.1 shows that the preservation of the premise parts in a connection is guaranteed if we use continuous piecewise-polynomial membership functions as the fuzzy sets in the premise parts. However, we must point out from Figure 3 that $A^1(x) \times A^2(x)$ is not a convex fuzzy set though $A^1(x)$ and $A^2(x)$ are convex fuzzy sets, that is, the convexity of fuzzy sets in a connection is not preserved in general. Therefore, from the viewpoints of the simplification of fuzzy modeling and the preserving properties of connection problems, we recommend continuous piecewise-polynomial membership functions as one of the realizable forms of the fuzzy sets in the premise parts.

Figure 4 shows two basic types of connections of single-input single-output fuzzy blocks, where L, R and S denote the fuzzy model of the objective system, the fuzzy controller, and the total fuzzy system, respectively. We give some theorems and corollaries for two types of connections of single-input single-output fuzzy blocks. The proofs of all the theorems and corollaries in this section will be given in Appendixes A- D.

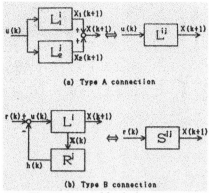

Fig. 4. Fuzzy block diagrams.

3.1. Type A connection
We consider the type A connection of L^i_1 and L^j_2 as shown in Figure 4(a).

Theorem 3.1. *Assume that,* L^i_1 *and* L^j_2 *are the following fuzzy blocks:*

L^i_1: IF $x(k)$ is P^i and $u(k)$ is Q^i

$$\text{THEN } x^i_2(k+1) = a^i_0 + \sum_{p=1}^n a^i_p x(k-p+1) + \sum_{q=1}^n b^i_q u(k-q+1),$$

L^j_2: IF $x(k)$ is G^j and $u(k)$ is H^j

$$\text{THEN } x^i_2(k+1) = c^i_0 + \sum_{p=1}^n c^i_p x(k-p+1) + \sum_{q=1}^n d^i_q u(k-q+1),$$

where $i = 1, 2, \cdots, l_1, j = 1, 2, \cdots, l_2$ and

$$P^i = [A^i_1, A^i_2, \ldots, A^i_n]^T, \quad Q^i = [B^i_1, B^i_2, \ldots, B^i_m]^T,$$
$$G^j = [C^j_1, C^j_2, \ldots, C^j_n]^T, \quad H^j = [D^j_1, D^j_2, \ldots, D^j_m]^T.$$

Then the result of the type A connection of L^i_1 *and* L^j_2 *is equivalent to the following fuzzy block* L^{ij}.

L^{ij}: IF $x(k)$ is (P^i and G^j) and $u(k)$ is (Q^i and H^j)

$$\text{THEN } x^{ij}(k+1) = (a^i_0 + c^j_0) + \sum_{p=1}^n (a^i_p + c^j_p) x(k-p+1)$$

$$+ \sum_{q=1}^n (b^i_q + d^j_q) u(k-q+1),$$

where, $x(k)$ is $(P^i$ and $G^j) \Leftrightarrow x(k)$ is $(A^i_1$ and $C^j_1)$ and $\cdots$ and $x(k-n+1)$ is $(A^i_n$ and $C^j_n)$, and the membership function of the fuzzy set $(A^i_1$ and $C^j_1)$ is defined as $(A^i_1(x(k)) \times C^j_1(x(k))$.

In Theorem 3.1, we notice that all the membership functions of L^{ij} are continuous piecewise-polynomial functions from Proposition 2.1 and that all the consequent equations

$$x^{ij}(k+1) = (a^i_0 + c^j_0) + \sum_{p=1}^{n}(a^i_p + c^j_p)x(k-p+1) + \sum_{q=1}^{n}(b^i_q + d^j_q)u(k-q+1),$$

are linear equations. That is, the properties of the fuzzy blocks are preserved at the type A connection.

Remark. In Theorem 3.1, the number of fuzzy implications in the fuzzy block L^{ij} is $l_1 \times l_2$. However, it is $l_1 \times l_2 - l_e$ when there exists fuzzy implications such that its weight $w^i v^j$ is equal to 0, where l_e is the number of the fuzzy implications such that $w^i v^j = 0$. It is clear that $w^i v^j = 0$ whenever $A^i_1 \cap C^j_1 = \emptyset$ or $\cdots$, or $A^i_n \cap C^j_n = \emptyset =$ or $B^i_1 \cap D^j_1 = \emptyset$ or $\cdots$, or $B^i_m \cap D^j_m = \emptyset$.

Corollary 3.1. *If $l_1 = l_2 = 1$ and $A^i_1 = C^j_1, \ldots, A^i_n = C^j_n, B^i_1 = D^j_1, \ldots, B^i_n = D^j_n$, for $i \in (1, 2, \ldots l)$ in Theorem 3.1, then the result of the type A connection of L^i_1 and L^j_2 is equivalent to the following fuzzy block L^i*

L^i: IF $x(k)$ is P^i and $u(k)$ is Q^i

$$\text{THEN } x^i(k+1) = (a^i_0 + c^i_0) + \sum_{p=1}^{n}(a^i_p + c^i_p)x(k-p+1)$$

$$+ \sum_{q=1}^{n}(b^i_q + d^i_q)u(k-q+1)$$

3.2 Type B connection

Next, we consider the type B connection of a fuzzy model L^i and a fuzzy controller R^i as shown in Figure 4(b). Figure 4(b) is a feedback type. In Figure 4(b), $r(k)$ is a reference input and $x(k)$ is a state vector, where $x(k) = [x(k), x(k-1), \ldots, x(k-n+1)]^T$.

Theorem 3.2. *Assume that L^i and R^i are the following fuzzy blocks:*

L^i: IF $x(k)$ is P^i and $u(k)$ is Q^i

$$\text{THEN } x^i(k+1) = a^i_0 + \sum_{p=1}^{n} a^i_p x(k-p+1) + b^i u(k),$$

R^i: IF $x(k)$ is G^j and $u(k)$ is H^j

$$\text{THEN } h_j(k) = c^i_0 + \sum_{p=1}^{n} c^i_p x(k-p+1),$$

Stability Analysis and Design of Fuzzy Control System 259

where
$$P^i = [A_1^i, A_2^i, \ldots, A_n^i]^T, \quad Q^i = [B_1^i, B_2^i, \ldots, B_m^i]^T,$$
$$G^j = [C_1^j, C_2^j, \ldots, C_n^j]^T, \quad H^j = [D_1^j, D_2^j, \ldots, D_m^j]^T.$$

As shown in Figure 4(b), $u(k) = r(k) - h(k)$, $r(k)$ is a reference input. The result of the type B connection of R^j and L^i is equivalent to the following fuzzy block:

S^{ij}: IF $x(k)$ is (P^i and G^j) and $v^*(k)$ is (Q^i and H^j)

$$\text{THEN } x^{ij}(k+1) = a_0^i - b^i c_0^j + b^i r(k) + \sum_{p=1}^{n} \{a_p^i - b^i c_p^j\} x(k-p+1),$$

where $i = 1, 2, \ldots, l_1, j = 1, 2, \ldots, l_2$,

$$v^*(k) = [r(k) - e^*(x(k)), r(k-1) - e^*(x(k-1)), \ldots, r(k-m+1) - e^*(x(k-m+1))]^T$$

and e^* is the input-output relation of the block R^j such that $h(k) = e^*(x(k))$.

The properties of fuzzy blocks at the type B connection are preserved.
In Theorem 3.2, we must find the function e^* such that $h(k) = e^*(x(k))$. The calculation process of the function e^* is shown in Appendix E.

Corollary 3.2. If $l_1 = l_2 = l$ and $A^i_1 = C^i_1, \ldots, A^i_n = C^i_n, B^i_1 = D^j_1, \ldots, B^i_m = D^j_m$, for $i \in (1, 2, \ldots l)$ in Theorem 3.2, then the result of the type A connection of L^i and R^i is equivalent to the following fuzzy block S^{ij}:

S^{ij}: IF $x(k)$ is (P^i and P^j) and $v^*(k)$ is (Q^i and Q^j)

$$\text{THEN } x^{ij}(k+1) = a_0^i - b^i c_0^j + b^i r(k) + \sum_{p=1}^{n} \{a_p^i - b^i c_p^j\} x(k-p+1),$$

where, $i, j = 1, 2, \ldots, l$.

Example 3.1. (1) Let us consider the type B connection of a fuzzy model L^i and a fuzzy controller R^j.

L^1: IF $x(k)$ is A^1 THEN $x^1(k+1) = 2.178 x(k) - 0.588 x(k-1) + 0.603u(k)$,

L^2: IF $x(k)$ is A^2 THEN $x^2(k+1) = 2.256 x(k) - 0.361 x(k-1) + 1.120u(k)$,

R^1: IF $x(k)$ is C^1 THEN $h^1(k) = k^1_1 x(k) - k^1_2 x(k-1)$,

R^2: IF $x(k)$ is C^2 THEN $h^2(k) = k^2_1 x(k) - k^2_2 x(k-1)$.

As shown in Figure 4(b), $u(k) = r(k) - h(k)$, where $r(k)$ is a reference input. From Theorem 3.2, we can derive S^{ij} as follows:

S^{11} : IF $x(k)$ (is A^1 and C^1)
THEN $x^{11}(k+1) = (2.178 - 0.603k_1^1)x(k)$
$\qquad + (-0.588 - 0.603k_2^1)x(k-1) + 0.603r(k),$

S^{12} : IF $x(k)$ (is A^1 and C^2)
THEN $x^{12}(k+1) = (2.178 - 0.603k_1^2)x(k)$
$\qquad + (-0.588 - 0.603k_2^2)x(k-1) + 0.603r(k),$

S^{21} : IF $x(k)$ (is A^2 and C^1)
THEN $x^{21}(k+1) = (2.256 - 1.120k_1^1)x(k)$
$\qquad + (-0.361 - 1.120k_2^1)x(k-1) + 1.120r(k),$

S^{22} : IF $x(k)$ (is A^2 and C^2)
THEN $x^{22}(k+1) = (2.256 - 1.120k_1^2)x(k)$
$\qquad + (-0.361 - 1.120k_2^2)x(k-1) + 1.120r(k).$

(2) If $A^1 = C^1$ and $A^2 = C^2$ in (1), we get the following S^{ij} from Corollary 3.2:

S^{11} : IF $x(k)$ (is A^1 and A^1)
THEN $x^{11}(k+1) = (2.178 - 0.603k_1^1)x(k)$
$\qquad + (-0.588 - 0.603k_2^1)x(k-1) + 0.603r(k),$

S^{12} : IF $x(k)$ (is A^1 and A^2)
THEN $x^{12}(k+1) = (2.178 - 0.603k_1^2)x(k)$
$\qquad + (-0.588 - 0.603k_2^2)x(k-1) + 0.603r(k),$

S^{21} : IF $x(k)$ (is A^2 and A^1)
THEN $x^{21}(k+1) = (2.256 - 1.120k_1^1)x(k)$
$\qquad + (-0.361 - 1.120k_2^1)x(k-1) + 1.120r(k),$

S^{22} : IF $x(k)$ (is A^2 and A^2)
THEN $x^{22}(k+1) = (2.256 - 1.120k_1^2)x(k)$
$\qquad + (-0.361 - 1.120k_2^2)x(k-1) + 1.120r(k).$

In this case, we notice that S^{12} and S^{21} have the same weights for any $x(k)$. So, we try to simplify the fuzzy system. In other words, we try to decrease the number of fuzzy implications as follows.

S^{11} : IF $x(k)$ (is A^1 and A^1)

   THEN $x^{11}(k+1) = (2.178 - 0.603k_1^1)x(k)$

   $+(-0.588 - 0.603k_2^1)x(k-1) + 0.603r(k)$,

$2S^{12*}$: IF $x(k)$ (is A^1 and A^2)

   THEN $x^{21*}(k+1) = \{2.217 - (1.120k_1^1 + 0.603k_1^2)/2\}x(k)$

   $+\{-0.4745 - (1.120k_2^1 + 0.603k_2^2)/2\}x(k-1)$

   $+0.8615r(k)$,

S^{22} : IF $x(k)$ (is A^2 and A^2)

   THEN $x^{22}(k+1) = (2.256 - 1.120k_1^2)x(k)$

   $+(-0.361 - 1.120k_2^2)x(k-1) + 1.120r(k)$,

where the unified implication of S^{12} and S^{21} is $2S^{12*}$.

The symbol 2 in '$2S^{12*}$' means that the weight for this implication must be doubled in the calculation of the final output. That is, the final output must be calculated as follows:

$$\frac{w^1 x^{11}(k+1) + 2w^2 x^{12*}(k+1) + w^3 x^{22}(k+1)}{w^1 + 2w^2 + w^3}$$

where w^1, w^2 and w^3 denote the weights of S^{11}, $2S^{12*}$, S^{22} for a given input, respectively.

4. Stability Analysis

One of the most important concepts concerning the properties of control systems is stability. We have some studies [1, 2, 4, 5, 10, 11] on stability and also on the analysis of system behavior in fuzzy control systems.

We derive theorems for the stability of a fuzzy system in accordance with the definition of stability in the sense of Lyapunov. A sufficient condition which guarantees the stability of a fuzzy system is obtained in terms of Lyapunov's direct method.

Let us consider the following fuzzy free system:

L^i : IF $x(k)$ is A_1^i and $\cdots$ and $x(k-n+1)$ is A_n^i

   THEN $x^i(k+1) = a_1^i x(k) + \cdots + a_n^i x(k-n+1)$,

where $i = 1, 2, \ldots l$.

The linear subsystems in the consequent part of the i-th implication can be written in the matrix form

$A_i x(k)$,

where $x(k) \in R^n$, $A_i \in R^n \times R^n$,

$x(k) = [x(k), x(k-1), \ldots, x(k-n+1)]^T$, and

$$A_i = \begin{bmatrix} a_1^i & a_2^i & \cdots & a_{n-1}^i & a_n^i \\ 1 & 0 & \cdots & 0 & 0 \\ 0 & 1 & \cdots & 0 & 0 \\ 0 & 0 & & 0 & 0 \\ \vdots & \vdots & \ddots & \vdots & \vdots \\ 0 & 0 & & 0 & 0 \\ 0 & 0 & \cdots & 1 & 0 \end{bmatrix}.$$

The output of the fuzzy system is inferred as follows:

$$x(k+1) = \sum_{i=1}^{l} w^i A^i x(k) \Big/ \sum_{i=1}^{l} w^i , \qquad (7)$$

where l is the number of fuzzy implications.

Theorem 4.1 and Lemma 4.1 are necessary in order to derive Theorem 4.2 which is an important theorem with respect to the stability of a fuzzy system. Theorem 4.1 is the well-known Lyapunov's stable theorem.

Theorem 4.1 [3]. *Consider a discrete system described by*
$$x(k+1) = f(x(k)),$$
where $x(k) \in R^n$, $f(x(k))$ is an $n \times 1$ function vector with the property that $f(0) = 0$ for all k.

Suppose that there exists a scalar function $V(x(k))$ continuous in $x(k)$ such that

(a) $V(0) = 0$,
(b) $V(x(k)) > 0$ for $x(k) \neq 0$,
(c) $V(x(k))$ approaches infinity as $\| x(k) \| \to \infty$
(d) $\Delta V(x(k)) < 0$ for $x(k) \neq 0$.

Then the equilibrium state $x(k) = 0$ for all k is asymptotically stable in the large and $V(x(k))$ is a Lyapunov function.

Lemma 4.1. *If P is a positive definite matrix such that*

$$A^T P A - P < 0 \text{ and } B^T P B - P < 0,$$

where, $A, B, P \in R^{n \times n}$, then

$$A^T P B + B^T P A - 2P < 0.$$

Proof.
$$A^T P B + B^T P A - 2P = -(A - B)^T P(A - B) + A^T P A + B^T P B - 2P$$
$$= -(A - B)^T P(A - B) + A^T P A - P + B^T P B - P.$$

Since P is a positive definite matrix,
$$-(A - B)^T P(A - B) \leq 0.$$
Therefore, the conclusion of the lemma follows.

Theorem 4.2. *The equilibrium of a fuzzy system, shown in (7), is globally asymptotically stable if there exists a common positive definite matrix* P *for all the subsystems such that*

$$A_i^T P A_i - P < 0 \text{ for } i \in \{1, 2, \ldots, l\}. \tag{8}$$

Proof. Consider the scalar function $V(x(k))$ such that
$$V(x(k)) = x^T(k) P x(k),$$
where P is a positive definite matrix. This function satisfies the following properties:
  (a) $V(0) = 0$,
  (b) $V(x(k)) > 0$ for $x(k) \neq 0$,
  (c) $V(x(k))$ approaches infinity as $\|x(k)\| \to \infty$

Next,
$$\Delta V(x(k)) = V(x(k+1)) - V(x(k)) = x^T(k+1) P x(k+1) - x^T(k) P x(k)$$

$$= \left(\sum_{i=1}^{l} w^i A_i x(k) \Big/ \sum_{i=1}^{l} w^i\right)^T P \left(\sum_{i=1}^{l} w^i A_i x(k) \Big/ \sum_{i=1}^{l} w^i\right) - x^T(k) P x(k)$$

$$= x^T(k) \left\{\left(\sum_{i=1}^{l} w^i A_i^T \Big/ \sum_{i=1}^{l} w^i\right)^T P \left(\sum_{i=1}^{l} w^i A_i \Big/ \sum_{i=1}^{l} w^i\right) - P\right\} x(k)$$

$$= \sum_{i,j=1}^{l} w^i w^j x^T(k) \{A_i^T P A_j - P\} x(k) \Big/ \sum_{i,j=1}^{l} w^i w^j,$$

$$= \left[\sum_{i=1}^{l} (w^i)^2 x^T(k) \{A_i^T P A_i - P\} x(k)\right.$$

$$\left. + \sum_{i<j}^{l} w^i w^j x^T(k) \{A_i^T P A_j + A_j^T P A_i - 2P\} x(k)\right] \Big/ \sum_{i,j=1}^{l} w^i w^j,$$

where $w^i \geq 0$ for $i \in \{1, 2, \ldots, l\}$ and $\sum_{i=1}^{l} w^i > 0$..

From Lemma 4.1 and (8), we obtain
  (d) $\Delta(x(k)) < 0$.

By Theorem 4.1, $V(x(k))$ is a Lyapunov function and the fuzzy system of (7) is globally asymptotically stable. □

This theorem is reduced to the Lyapunov stability theorem for linear discrete systems when $l = 1$.

This theorem can be applied to the stability analysis of a nonlinear system which is approximated by a piecewise linear function if the condition in (8) is satisfied under $w^i \geq 0$ and $\sum_{i=1}^{l} w^i > 0$. We can point out that a piecewise linear function can be described as a special case of (2.1) if we use crisp sets instead of fuzzy sets in the premise parts of a fuzzy system. It is easy to divide a nonlinear system into some linearized subsystems on an input-state space. This means that the system is approximated by a piecewise linear function. Since many nonlinear systems can be approximated by piecewise linear functions, this theorem can be widely applied not only to a fuzzy system, represented by (1), but also to nonlinear systems.

Theorem 4.2 gives, of course, a sufficient condition for ensuring the stability of (7). We may intuitively guess that an approximated nonlinear system is stable if all locally approximating linear systems are stable. However, it is not the case in general. Here we notice the following fact.

All the A_i's are stable matrices if there exists a common positive definite matrix P. There does not always exist a common positive definite matrix P even if all the A_i's are stable matrices. Of course, a fuzzy system may be globally asymptotically stable even if there does not exist a common positive definite matrix P. However, we must notice that a fuzzy system is not always globally asymptotically stable even if all the A_i's are stable matrices as shown in Example 4.1.

Example 4.1. Let us consider the following fuzzy system:

L^1: IF $x(k-1)$ is as in Figure5(a) THEN $x^1(k+1) = x(k) - 0.5\, x(k-1)$,

L^2: IF $x(k-1)$ is as in Figure 5(b) THEN $x^2(k+1) = -x(k) - 0.5\, x(k-1)$.

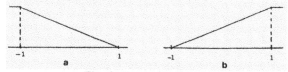

Fig. 5. Example $x(k - 1)$.

From the linear subsystems, we obtain

$$A_1 = \begin{bmatrix} 1 & -0.5 \\ 1 & 0 \end{bmatrix}, \quad A_2 = \begin{bmatrix} -1 & -0.5 \\ 1 & 0 \end{bmatrix}.$$

The initial condition is $x(0) = 0.90$ and $x(1) = -0.70$. Figures 6(a) and (b) illustrate the behavior of the following linear systems for the initial condition, respectively:

$$x(k+1) = A_1 x(k), \quad x(k+1) = A_2 x(k).$$

The linear systems are stable since A_1 and A_2 are stable matrices. However, the fuzzy system which consists of the linear system is unstable as shown in Figure 6(c), where w^1 and w^2 denote the weights of L^1 and L^2, respectively.

Obviously, in the example, there does not exist a common P since the fuzzy system is unstable. Next, we give a necessary condition for ensuring the existence of a common P.

Theorem 4.3. *Assume that A_i is a stable and nonsingular matrix for $i = 1, 2, \ldots, l$. A_iA_j is a stable matrix for $i, j = 1, 2, \ldots, l$ if there exists a common positive definite matrix P such that*

$$A_i^T P A_i - P < 0 \qquad (9)$$

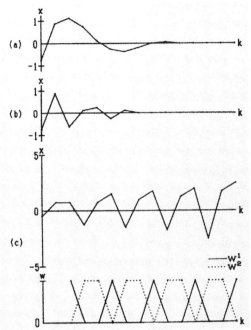

Fig. 6. (a) Behavior of $x(k+1) = A_1 x(k)$. (b) Behavior of $x(k+1) = A_2 x(k)$ (c) Behavior of the fuzzy system.

Proof. From (9), we obtain

$$P - (A_i^{-1})^T P A_i^{-1} < 0,$$

since $(A_i^{-1})^T = (A_i^T)^{-1}$. Therefore, $P < (A_i^{-1})^T P (A_i^{-1})$ for $i = 1, 2, \ldots, l$. Since $A_i^T P A_i < P$ from (9), the following inequality holds for $i, j = 1, 2, \ldots, l$:

$$A_i^T P A_i < (A_j^{-1})^T P (A_j^{-1}).$$

From the inequality, we obtain $A_j^T A_i^T P A_i A_j - P < 0$. Therefore, $A_i A_j$ must be a stable matrix for $i, j = 1, 2, \ldots, l$. □

Theorem 4.3 shows that if one of the $A_i A_j$'s is not a stable matrix, then there does not exist a common P. In Example 4.1, we can see that there does not exist a common P since the eigenvalues of $A_1 A_2$ are -0.135 and -1.865, where

$$A_1 A_2 = \begin{bmatrix} -1.5 & -0.5 \\ -1.0 & -0.5 \end{bmatrix}.$$

In order to check the stability of a fuzzy system, we must find a common positive definite P. It is difficult to find a common positive definite matrix P as effectively as possible. So, the following simple procedure is used. The procedure consists of two steps.

Step 1: We find a positive definite matrix P_i such that

$$A_i^T P_i A_i - P_i < 0$$

for $i = 1, 2, \ldots, l$. It is possible to find a positive definite matrix P_i if A_i is a stable matrix.

Step 2: Next, if there exists P_j in $(P_i \mid i = 1, 2, \ldots, l)$ such that

$$A_i^T P_j A_i - P_j < 0$$

for $i = 1, 2, \ldots, l$, then we select P_j as a common P. If Step 2 has not succeeded, go back to Step 1.

5. Design of a fuzzy controller

We have considered the conditions for the stability of a fuzzy control system by using Lyapunov's direct method in the previous section. In this section, we propose a design method of a model-based fuzzy controller. The controller can be designed so as to guarantee the stability of a fuzzy control system by using the conditions since it is a model-based controller. At this time, the connection theorems discussed in Section 3 are utilized.

5 1. Design procedure of a fuzzy controller

The design procedure of a fuzzy controller consists of three main parts.

Step 1: First, we derive a total fuzzy system by connecting a fuzzy model of an objective system and a fuzzy controller by using the connection theorems.

Let us consider a simple example. Assume that the following fuzzy model is given.

L^1: IF $x(k)$ is A^1 and $u(k)$ is B^1 THEN $x^1(k+1) = a^1 x(k) + b^1 u(k)$,

L^2: IF $x(k)$ is A^2 and $u(k)$ is B^2 THEN $x^2(k+1) = a^2 x(k) + b^2 u(k)$,

For the fuzzy model, we can easily design the following fuzzy controller.

R^1: IF $x(k)$ is A^1 and $u(k)$ is B^1 THEN $u^1(k) = f^1 x(k)$,

R^2: IF $x(k)$ is A^2 and $u(k)$ is B^2 THEN $u^2(k) = f^2 x(k)$.

Here, the subcontrollers of R^1 and R^2 are designed for the subsystems of L^1 and L^3 with the state feedbacks, respectively.

By the type B connection, the following total fuzzy system is obtained:

S^{11} : IF $x(k)$ (is A^1 and A^1) and $u(k)$ (is B^1 and B^1)

THEN $x^{11}(k+1) = (a^1 - b^1 f^1)x(k)$,

$2S^{12*}$: IF $x(k)$ (is A^1 and A^2) and $u(k)$ (is B^1 and B^2)
THEN $x^{12}(k+1) = (a^1 + a^2 - b^1 f^2 - b^2 f^1)x(k)/2$,

S^{22} : IF $x(k)$ (is A^2 and A^2) and $u(k)$ (is B^2 and B^2)
THEN $x^{22}(k+1) = (a^2 - b^2 f^2)x(k)$.

Thus, we can obtain the total system by this method.

Step 2: Next, we focus our attention on determining the parameters f^i ($i = 1$, 2) of the fuzzy controller. These parameters are determined so as to guarantee the stability of the linear subsystems in the total fuzzy system.

Step 3: Last, we check the stability of the total fuzzy system by the procedure to find a common P discussed in Section 4. If the system is not stable, go back to Step 2.

At Step 2, we can easily find the parameters f^i in the linear subsystems by using the linear systems theory. However, we have shown in Example 4.1 that a fuzzy system is not always stable even if all subsystems are stable. So, it is necessary to perform Step 3 in order to check the stability of the total fuzzy system.

We concretely show the design procedure of a fuzzy controller through two examples.

Example 5.1. Let us consider the fuzzy system L^i of Example 3.1 again.

L^1: IF $x(k)$ is as in Figure 7(a)
THEN $x^1(k+1) = 2.178\,x(k) - 0.588\,x(k-1) + 0.603u(k)$.

L^2: IF $x(k)$ is as in Figure 7(b)
THEN $x^2(k+1) = 2.256\,x(k) - 0.361\,x(k-1) + 1.120u(k)$.

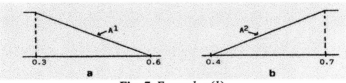

Fig. 7. *Example $x(k)$.*

We try to stabilize the fuzzy system using the linear controller with a proportional gain K. The linear proportional controller can be described as a special case of a fuzzy proportional controller as follows:

R^1:   IF $x(k)$ is any THEN $u(k) = Kx(k)$,

where 'any' is a fuzzy set whose membership functions any $(x(k))$ is 1.0 for all $x(k)$.

Step 1: We can derive the type B connection of L^i and R^i from Theorem 3.2 as follows:

S^1 : IF $x(k)$ is A^1 THEN $x^1(k+1) = (2.178 - 0.603K)x(k) - 0.588x(k-1)$,

S^2 : IF $x(k)$ is A^2 THEN $x^2(k+1) = (2.256 - 1.120K)x(k) - 0.361x(k-1)$.

Here it is assumed that reference input $r(k) = 0$.

Step 2: Next, we utilize the root locus method to determine the parameter K. It is not always necessary to utilize the root locus method. For example, we may use the technique of a Bode diagram or pole assignment.

Figures 8 and 9 show root locus plots for the linear subsystems of S^1 and S^2, respectively, where $0<K<\infty$. It is well known that the stability boundary in the z-plane is the unit circle $|z|=1$. From Figures 8 and 9, we can stabilize the linear subsystems of S^1 and S^2 when we choose a gain K such that $0.980<K<6.25$ and $0.80<K<3.23$, respectively. Therefore, in order to stabilize the fuzzy control system, we must choose a gain K such that $0.98<K<3.23$ at least.

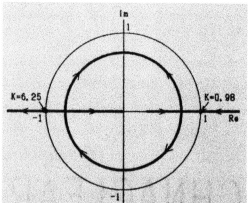

Fig. 8. *Root loci for the consequent equation of S^1.*

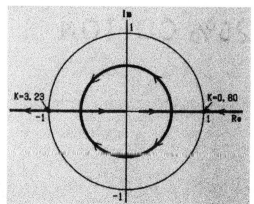

Fig. 9. *Root loci for the consequent equation of S^2.*

Step 3: Last, we must check the stability of the fuzzy control systems using the procedure to find a common P. For the linear subsystems of S^1 and S^2, we obtain

$$A_1 = \begin{bmatrix} 2.178-0.603K & -0.588 \\ 1 & 0 \end{bmatrix}, \quad A_2 = \begin{bmatrix} 2.256-1.120K & -0.361 \\ 1 & 0 \end{bmatrix}.$$

Here, assume that $k = 1.12$. It satisfies the inequality $0.98<K<3.28$. Figure 10 shows a result of the response of the fuzzy control system at $K = 1.12$. In the case of $K = 1.12$, if we choose the positive definite matrix P such that

$$P = \begin{bmatrix} 2.0 & -1.3 \\ -1.3 & 1.0 \end{bmatrix},$$

then the condition $A_i^T P A_i - P < 0$ is satisfied for $i \in \{1, 2\}$, that is, the fuzzy control system is globally asymptotically stable.

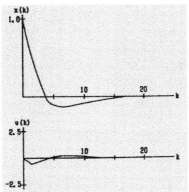

Fig. 10. A result of the response of fuzzy control system.

Example 5.2. Assume that the fuzzy system and the fuzzy controller are those of Example 3.1(2).

L^1:   IF $x(k)$ is as in Figure 7(a)
THEN $x^1(k+1) = 2.178\ x(k) - 0.588\ x(k-1) + 0.603u(k)$,

L^2:   IF $x(k)$ is as in Figure 7(b)
THEN $x^2(k+1) = 2.256\ x(k) - 0.361\ x(k-1) + 1.120u(k)$,

R^1:   IF $x(k)$ is A^1 THEN $u^1(k) = k^1{}_1 x(k) - k^1{}_2 x(k-1)$,

R^2:   IF $x(k)$ is A^2 THEN $u^2(k) = k^2{}_1 x(k) - k^2{}_2 x(k-1)$.

Step 1: The total fuzzy system has been obtained in Example 3.2(2) as follows:

S^{11}:   IF $x(k)$ is $(A^1$ and $A^1)$
THEN $x^{11}(k+1) = (2.178 - 0.603\ k^1{}_1)x(k)$
$+ (-0.588 - 0.603\ k^2{}_1)x(k-1)$,

$2S^{12*}$: IF $x(k)$ is $(A^1$ and $A^2)$
THEN $x^{12*}(k+1) = \{2.217 - (1.120\, k^1_1 + 0.603\, k^2_1)/2\}x(k)$
$+ \{-0.4745 - (1.120\, k^1_2 + 0.603\, k^2_2)/2\}x(k-1)$

S^{22}: IF $x(k)$ is $(A^2$ and $A^2)$
THEN $x^{22}(k+1) = (2.256 - 1.120\, k^2_1)x(k)$
$+ (-0.361 - 1.120\, k^2_2)x(k-1)$

Here it is assumed that $r(k) = 0$, $\forall k$

Step 2: S^{11} has two parameters: k^1_1 and k^1_2, and S^{22} has two parameters: k^2_1 and k^2_2, and $2S^{11*}$ has four parameters: k^1_1, k^1_2, k^2_1 and k^2_2. We determine four parameters of the fuzzy controller so as to satisfy the following design criteria:

(C1) The value ζ of each consequent equation is approximately equal to 0.707, explicitly $0.6 < \zeta < 0.8$, where ζ denotes the damping coefficient,

(C2) the response of the total fuzzy system converges to a set point as quickly as possible.

We determine k^1_1 and k^1_2 so that $\zeta = 0.707$ for the linear subsystem of S^{11}. Similarly, we determine k^2_1 and k^2_2 for S^{22}. These parameters cannot be determined uniquely. So we select two cases.

Case A: $k^1_1 = 1.564$, $k^1_2 = -0.223$, $k^2_1 = 0.912$, $k^2_2 = 0.079$

Case B: $k^1_1 = 2.109$, $k^1_2 = -0.475$, $k^2_1 = 1.205$, $k^2_2 = -0.053$.

However, we must check for $2S^{12*}$ whether these parameters are appropriate or not since these parameters are determined without considering the linear subsystem of $2S^{12*}$. The damping coefficient of $2S^{12*}$ is 0.705 in Case A, and that of $2S^{12*}$ is 0.764 in Case B. Both of them satisfy the design criterion (C1), that is, $0.6 < \zeta < 0.8$. Of course, all the subsystems are stable.

Step 3: Last, we check the stability of the total fuzzy system for Cases A and B. Figure 11 shows the responses of Case A and Case B. Figure 12 shows poles locations in the z-plane for the linear subsystems of S^{11}, S^{22} and $2S^{12*}$. From Figure 11, it is seen that the settling time of Case B is shorter than that of Case A because the poles of Case B are closer to the origin of the z-plane as shown in Figure 12. For the design criterion (C2), the parameters of Case B are better than those of Case A.

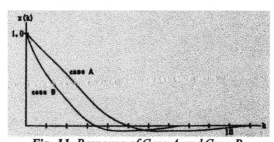

Fig. 11. Response of Case A and Case B.

So we check the stability of the fuzzy control system for Case B. In this case, we obtain

$$A_{11} = \begin{bmatrix} 0.906 & -0.302 \\ 1 & 0 \end{bmatrix}, \quad A_{12*} = \begin{bmatrix} 0.672 & -0.193 \\ 1 & 0 \end{bmatrix}, \quad A_{22} = \begin{bmatrix} 0.906 & -0.302 \\ 1 & 0 \end{bmatrix},$$

The fuzzy control system is stable since we can find a common positive definite matrix P such that

$$A^T_{11} P A_{11} - P < 0, \quad A^T_{22} P A_{22} - P < 0 \quad A^T_{12*} P A_{12*} - P < 0.$$

where A_{11}, denotes a matrix such that $x_{11}(k+1) = A_{11} x(k)$ for the linear subsystem of S^{11}.

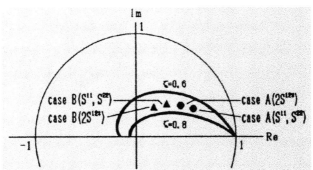

Fig. 12. Pole locations for the consequent equations of S^{11}, S^{11}, and $2S^{12*}$.

By the procedure discussed in Section 4, we find

$$P = \begin{bmatrix} 4.19 & -0.88 \\ -0.88 & 1.38 \end{bmatrix}$$

For Case A also, we can find a common positive definite matrix P such that

$$P = \begin{bmatrix} 9.13 & -3.50 \\ -3.50 & 2.85 \end{bmatrix}.$$

6. Conclusion

We have considered the stability analysis and the design technique of fuzzy control systems using fuzzy block diagrams. These are the most important to establish fuzzy systems theory.

The disadvantage of the connection of fuzzy blocks is to increase the number of fuzzy implications of a connected fuzzy system. This means the complexity of a fuzzy system. It is useful to consider the method of model reduction, that is, the method to reduce the number of fuzzy implications.

We have derived a stability theorem in terms of Lyapunov's direct method. This theorem can be widely used as a tool of stability criterion not only for fuzzy control systems but also for the nonlinear systems which can be approximated by a piecewise linear function. However, we should develop an algorithm to effectively find a common positive definite matrix P.

We believe that a fuzzy system theory must be established in order to improve the method of fuzzy control.

Appendix A: Proof of Theorem 3.1

Let the weights of L^i_1 and L^j_2 for a given input $(x_0(k), \ldots, x_0(k-n+1), u_0(k), \ldots, u_0(k-m+1))$ be w_i and v_j, respectively. Then, the results of fuzzy inference by L^i_1 and L^j_2 are

$$x_1(k+1) = \sum_{i=1}^{l_1} w^i x_1^i(k+1) \bigg/ \sum_{i=1}^{l_1} w^i$$

$$= \sum_{i=1}^{l_1} w^i \left\{ a_0^i + \sum_{p=1}^{n} a_p^i x^0(k-p+1) + \sum_{q=1}^{m} b_q^i u^0(k-q+1) \right\} \bigg/ \sum_{i=1}^{l_1} w^i,$$

$$x_2(k+1) = \sum_{j=1}^{l_2} v^j x_2^j(k+1) \bigg/ \sum_{j=1}^{l_2} v^j$$

$$= \sum_{j=1}^{l_2} v^j \left\{ c_0^j + \sum_{p=1}^{n} c_p^j x^0(k-p+1) + \sum_{q=1}^{m} d_q^j u^0(k-q+1) \right\} \bigg/ \sum_{j=1}^{l_2} v^j,$$

respectively. The result of the type A connection of L^i_1 and L^j_2 is derived as follows:

$$x_1(k+1) = x_1(k+1) + x_2(k+1)$$

$$= \sum_{i=1}^{l_1} w^i \left\{ a_0^i + \sum_{p=1}^{n} a_p^i x^0(k-p+1) + \sum_{q=1}^{m} b_q^i u^0(k-q+1) \right\} \bigg/ \sum_{i=1}^{l_1} w^i$$

$$+ \sum_{j=1}^{l_2} v^j \left\{ c_0^j + \sum_{p=1}^{n} c_p^j x^0(k-p+1) + \sum_{q=1}^{m} d_q^j u^0(k-q+1) \right\} \bigg/ \sum_{j=1}^{l_2} v^j$$

$$= \sum_{i=1}^{l_1} \sum_{j=1}^{l_2} w^i v^j \left\{ (a_0^i + c_0^j) + \sum_{p=1}^{n} (a_p^i + c_p^j) x^0(k-p+1) \right.$$

$$\left. + \sum_{p=1}^{n} (b_q^i + d_q^j) u^0(k-q+1) \right\} \bigg/ \sum_{i=1}^{l_1} \sum_{j=1}^{l_2} w^i v^j.$$

Stability Analysis and Design of Fuzzy Control System 273

This result is equivalent to the fuzzy block L^{ij} since the weight of L^{ij} for the input is $w^i v^j$.

Appendix B: Proof of Corollary 3.1

Let the weights of L^i_1 and L^i_2 for a given input $(x^0(k), \ldots, x^0(k - n + 1), u^0(k), \ldots, u^0(k - m + 1))$ be w_i. The result of the type A connection of L^i_1 and L^i_2 is derived as follows:

$$x(k+1) = x_1(k+1) + x_2(k+1)$$

$$= \sum_{i=1}^{l} w^i \left\{ a_0^i + \sum_{p=1}^{n} a_p^i x^0(k-p+1) + \sum_{q=1}^{m} b_q^i u^0(k-q+1) \right\} \Big/ \sum_{i=1}^{l} w^i ,$$

$$+ \sum_{j=1}^{l} w^i \left\{ c_0^i + \sum_{p=1}^{n} c_p^i x^0(k-p+1) + \sum_{q=1}^{m} d_q^i u^0(k-q+1) \right\} \Big/ \sum_{j=1}^{l} w^j$$

$$= \sum_{i=1}^{l} w^i \left\{ (a_0^i + c_0^i) + \sum_{p=1}^{n} (a_p^i + c_p^i) x^0(k-p+1) + \sum_{q=1}^{m} (b_q^i + d_q^i) u^0(k-q+1) \right\} \Big/ \sum_{i=1}^{l} w^i .$$

This result is equivalent to the fuzzy block L^i since the weight of L^i for the input is w^i.

Appendix C: Proof of Theorem 3.2

Let the weights of L^i and R^j for a given input $(x(k), \ldots, x(k - n + 1), u(k))$ be w^i and v^j, respectively.

Then, the results of fuzzy inference for L^i and R^j are

$$x(k+1) = \sum_{i=1}^{l_1} w^i \left\{ a_0^i + \sum_{p=1}^{n} a_p^i x(k-p+1) + b^i u(k) \right\} \Big/ \sum_{i} w^i , \quad (C.1)$$

$$h(k) = \sum_{j=1}^{l_2} v^j \left\{ c_0^j + \sum_{p=1}^{n} c_p^j x(k-p+1) \right\} \Big/ \sum_{j=1}^{l_2} v^j , \quad (C.2)$$

respectively, where

$$v^j = \prod_{p=1}^{n} C_p^j(x(k-p+1)) \times \prod_{q=1}^{n} D_q^j(r(k-q+1)) - e^*(x(k-q+1))).$$

From (C.2), we obtain
$$u(k) = r(k) - h(k)$$

$$= r(k) - \sum_{j=1}^{l_2} v^j \left\{ c_0^j + \sum_{p=1}^{n} c_p^j x(k-p+1) \right\} \Big/ \sum_{j=1}^{l_2} v^j . \quad (C.3)$$

By substituting (C.3) into (C.1), we can eliminate $u(k)$, that is,

$$x(k+1) = \sum_{i=1}^{l_1} w^i \left[a_0^i + \sum_{p=1}^{n} a_p^i x(k-p+1) \right.$$

$$\left. + b^i \left(r(k) - \sum_{j=1}^{l_2} v^j \left\{ c_0^j + \sum_{p=1}^{n} c_p^j x(k-p+1) \right\} \bigg/ \sum_{j=1}^{l_2} v^j \right) \right] \bigg/ \sum_{i=1}^{l_1} w^i$$

$$= \sum_{i=1}^{l_1} \sum_{j=1}^{l_2} w^i v^j \left[a_0^i - b^i c_0^j + b^i r(k) + \sum_{p=1}^{n} \{a_p^i - b^i c_p^j\} x(k-p+1) \right] \bigg/ \sum_{i=1}^{l_1} \sum_{j=1}^{l_2} w^i v^j .$$

This result is equivalent to the fuzzy block S^{ij} since the weight of S^{ij} for the input is equal to $w^i v^j$.

Appendix D: Proof of Corollary 3.2

If G^j and H^j are replaced by P^j and Q^j in the proof of Theorem 3.2, respectively, we can prove Corollary 3.2 in the same manner as Theorem 3.2.

Appendix E: Calculation process of function e^*

We consider the following fuzzy controller.

R^j: IF $x(k)$ is G^j and $U(k)$ is H^j

THEN $h^j(k) = c_0^j + \sum_{p=1}^{n} c_p^i x(k-p+1)$,

where $j = 1, 2, ..., l$,

$G^j = [C^j_1, ... C^j_n]^T$, and $H^j = [D^j_1, ... D^j_m]^T$

The inputs of the fuzzy controller are $x(k)$, where $x(k) = [x(k), x(k-1), ..., x(k-n+1)]^T$, and the output of that is $u(k)$.

We can rewrite the fuzzy block R^j as follows since $u(k) = r(k) - h(k)$ as shown in Figure 4(b).

R^j: IF $x(k)$ is G^j and $r(k) - h(k)$ is D^j_1 and $\bar{v}(k)$ is $\overline{H}^j$

THEN $h^j(k) = c_0^j + \sum_{p=1}^{n} c_p^j x(k-p+1)$,

where

$\bar{v}(k) = [r(k-1) - h(k-1), ..., r(k-m+1) \ h(k-m+1)]^T$,

and

$\overline{H}^j(k) = [D^j_2, ... D^j_m]^T$

The final output $h(k)$ is inferred as follows:

$$h(k) = \sum_{j=1}^{l} v^j h^j(k) \bigg/ \sum_{j=1}^{l} v^j = \sum_{j=1}^{l} v^j \left\{ c_0^j + \sum_{p=1}^{n} c_p^j x(k-p+1) \right\} \bigg/ \sum_{j=1}^{l} v^j, \quad (E.1)$$

where

$$v^j = \prod_{p=1}^{n} C_p^j(x(k-p+1)) \times D_1^j(r(k) - h(k)) \times \prod_{q=2}^{m} D_q^j(r(k-q-1) - h(k-q+1)),$$

and $C^j(\ )$ and $D^j(\ )$ denote the membership functions of the fuzzy sets C^j and D^j, respectively.

Here we must notice that v^j's depend on $h(k)$. So (E.1) can be rewritten as follows:

$$\sum_{j=1}^{l} t^j D_1^j(r(k) - h(k))h(k) - \sum_{j=1}^{l} t^j D_1^j(r(k) \\ - h(k))\{c_0^j + \sum_{p=1}^{n} c_p^j x(k-p+1)\} = 0, \quad (E.2)$$

where

$$t^j = \prod_{p=1}^{n} C_p^j(x(k-p+1)) \times \prod_{q=2}^{m} D_q^j(r(k-q+1) - h(k-q+1)).$$

Equation (E.2) is locally solved for $h(k)$ since $D^j{}_1(r(k) - h(k))$ is a piecewise-polynomial membership function. For example, (E.2) is a piecewise-polynomial of degree 2 if $D^j{}_1(r(k) - h(k))$ is a piecewise linear function such as a triangular type or a trapezoid type. Now, assume that

$$D_1^j(r(k) - h(k))h(k) = \begin{cases} a_1^j(r(k) - h(k)) + b_1^j, & r(k) - h(k) \in [p_0, p_1], \\ \vdots & \vdots \\ a_s^j(r(k) - h(k)) + b_s^j, & r(k) - h(k) \in [p_{s-1}, p_s], \end{cases}$$

where $a^j{}_r$'s and $b^j{}_r$'s are parameters of the membership functions, and $r = 1, 2, \ldots, s$. Then we obtain

$$\sum_{j=1}^{l} t^j \left\{ a_r^j h(k)^2 - \left[b_r^j + a_r^j \left(r(k) + c_0^j + \sum_{p=1}^{n} c_p^j x(k-p+1) \right) \right] h(k) \right. \\ \left. + a_r^j \left(c_0^j + \sum_{p=1}^{n} c_p^j x(k-p+1) \right) r(k) + b_r^j \left(c_0^j + \sum_{p=1}^{n} c_p^j x(k-p+1) \right) \right\} = 0, \quad (E.3)$$

for $r(k) - h(k) \in [p_{r-1}, p_r]$ $r = 1, 2, \ldots, s$. For each interval $[p_{r-1}, p_r]$, we solve (E.3) for $h(k)$. Therefore, we have s solutions. Let the solution for the r-th interval be $h_r(k)$. We select one value of $h_r(k)$'s such that $r(k) - h_r(k)$ $[p_{r-1}, p_r]$ for $r = 1, 2, \ldots, s$. It is the final output of the fuzzy controller R^j for the input $x(k)$.

Thus, we can find $h(k) = e^*(x(k))$ by solving (E.3) locally for $h(k)$.

Whereas, if $D^j{}_1$ is fuzzy set 'any' for $j = 1, 2, \ldots, l$, that is, $D^j{}_1(r(k) - h(k)) = 1.0$ for any $h(k)$, then we can easily find $h(k) = e^*(x(k))$ such that

$$e^*(x(k)) = \sum_{j=1}^{l} v^j \left\{ c_0^j + \sum_{p=1}^{n} c_p^j x(k-p+1) \right\} \bigg/ \sum_{j=1}^{l} v^j,$$

because v^j's do not depend on $h(k)$ in this case.

References

[1] A. A. Kania, J. F. Kiszka, M. B. Gorzalczany, J.R. Maj and M. S. Stachowicz, On stability of formal fuzziness systems, *Inform, Sci.* **22** (1980) 51-68.
[2] J.B. Kiszka, M. M. Gupta and P.N. Nikiforuk, Energetistic stability of fuzzy dynamic systems, *IEEE Trans. Systems Man Cybernet.* **15**(6) (1985) 783-792.
[3] B. C. Kuo, *Digital Control System* (Holt-Saunders, 1980).
[4] E. H. Mamdani, Advances in the linguistic synthesis of fuzzy controllers, *Internal. J. Man-Machine Stud.* **8**(6) (1976) 669-678.
[5] W. Pedrycz, An approach to the analysis of fuzzy systems, *Internat. J. Control* **34**(3) (1981) 403-421.
[6] M. Sugeno and G.T. Kang, Fuzzy modeling and control of multilayer incinerator, *Fuzzy Sets and Systems* **18** (1986) 329-346.
[7] M. Sugeno and G.T. Kang, Structure identification of fuzzy model, *Fuzzy Sets and Systems* **28** (1988) 15-33.
[8] M. Sugeno and K. Tanaka, Successive identification of fuzzy model and its applications to prediction of complex system, *Fuzzy Sets and Systems* **42** (1991) 315-344.
[9] T. Takagi and M. Sugeno, Fuzzy identification of systems and its applications to modeling and control, *IEEE Trans. Systems Man Cybernet.* **15**(l) (1985) 116-132.
[10] R.M. Tong, Analysis and control of fuzzy systems using finite discrete relations, *Internat. J. Control* **27**(3) (1978) 431-440.
[11] R.M. Tong, Some properties of fuzzy feedback systems, *IEEE Trans. System Man Cybernet.* **10**(6) (1980) 327-330.

An Interpretation of Fuzzy Measures and the Choquet Integral as an Integral with Respect to a Fuzzy Measure*

Toshiaki Murofushi and Michio Sugeno
Department of Systems Science,
Tokyo Institute of Technology,
4259 Nagatsuta, Midori-ku Yokohama 227, Japan

Received December 1986
Revised March 1987

Abstract

In this paper the non-additivity of Sugeno's fuzzy measure is interpreted in terms of addition and the rationality of the Choquet integral is discussed. It is pointed out that a fuzzy measure on a set X expresses the interaction between the subsets of X and can be represented by an additive measure. It is shown through concrete examples that the Choquet integral is reasonable as an integral with respect to a fuzzy measure. It is also found that the Choquet integral is closely related with the representation of a fuzzy measure.

Keywords: fuzzy measure, Choquet integral, representation of fuzzy measure.

1. Introduction

Various integrations with respect to Sugeno's fuzzy measure [10] have been proposed. A fuzzy measure on a measurable space $(X, \mathscr{A})$ is a set function $\mu: \mathscr{A} \to [0, 1]$ with the properties:

(1) $\mu(\emptyset) = 0$, $\mu(X) = 1$,
(2) $A \subset B \Rightarrow \mu(A) \leq \mu(B)$,

---

*© 1989 North-Holland. Retyped with written permission from Fuzzy Sets and Systems, **29** (1989) 201-227.*

(3) $A_n \uparrow A \Rightarrow \mu(A_n) \uparrow \mu(A)$,
(4) $A_n \downarrow A \Rightarrow \mu(A_n) \downarrow \mu(A)$.

Occasionally we call μ not satisfying (4) a fuzzy measure. Monotonicity and non-additivity are the main features of this measure.

The fuzzy integral [10] $\int h \circ \mu$ of a measurable function $h: X \to [0, 1]$ is defined by

$$\int h \circ \mu = \sup_{\alpha \in [0,1]} \min[\alpha, \mu(h^*(\alpha))],$$

where $h^*(\alpha) = \{x \mid h(x) \geq \alpha\} \; \forall \; \alpha \in [0, 1]$. This integral has very nice properties in view of its definition only by order relation. But it is not an extension of the Lebesgue integral while the fuzzy measure is an extension of the probability measure.

Weber [12] pointed out that the functional defined by Choquet [1] can be regarded as an integral with respect to a fuzzy measure. We call it the Choquet integral. The Choquet integral of a measurable function $h: X \to [0, \infty)$ is defined as

$$\int_0^\infty \mu(h^*(\alpha)) \, d\alpha,$$

where $h^*(\alpha) = \{x \mid h(x) \geq \alpha\} \; \forall \; \alpha \in [0, \infty)$. Dempster's upper and lower expected values [2] of a nonnegative function are Choquet integrals. Höhle's integral [4] of a simple function

$$h = \sum_{i=1}^n b_i 1_{B_i} \quad (0 \leq b_1 \leq b_2 \leq \cdots \leq b_n, B_i \cap B_j = \emptyset \text{ for } i \neq j)$$

is defined as

$$\sum_{m=1}^n b_m \left[\mu\left(\bigcup_{i=0}^m B_i \right) - \mu\left(\bigcup_{i=0}^{m-1} B_i \right) \right],$$

where $B_0 = X - \bigcup_{i=1}^n B_i$. We denote the characteristic function of a set B by 1_B, i.e.,

$$1_B(x) = \begin{cases} 1, & x \in B, \\ 0, & x \notin B. \end{cases}$$

Both integrals are extensions of the Lebesgue integral, but the rationality for their definitions was not mentioned in the papers [1] and [4].

A pseudo-additive measure [11] is a special fuzzy measure μ with the property

$$\mu(A \cup B) = \mu(A) \hat{+} \mu(B) \text{ whenever } A \cap B = \emptyset,$$

where $\hat{+}$ is a pseudo-addition, an operation like addition. In the case of $\hat{+}$ being a t-conorm, μ is called $\hat{+}$ - decomposable [12]. Integrals with respect to pseudo-additive measures were proposed by Kruse [7], Weber [12], and Sugeno and Murofushi [11]. These integrals of a nonnegative simple function

$$h(x) = \begin{cases} b_i, & x \in B_i, i = 1, 2, \ldots, n, \\ 0, & \text{otherwise,} \end{cases}$$

are expressed in the same form, except a special case in [12],

$$\sum_{i=1}^{n} b_i \hat{\cdot} \mu(B_i),$$

where $\sum_{i=1}^{n} x_i = x_1 \hat{+} x_2 \hat{+} \cdots \hat{+} x_n$ and $\hat{\cdot}$ is an operation like multiplication.

Though these are also extensions of the Lebesgue integral, they are based on specific operations such as t-conorms and pseudo-additions rather than on general properties of fuzzy measures.

Which integration is proper for the original features of the fuzzy measure: non-additivity and monotonicity?

To answer this question, we should consider the more fundamental question: what does a fuzzy measure express by its non-additivity and monotonicity? We have not yet obtained a universal answer for the fundamental question. Some interpretations have been proposed. Sugeno [10] considered the situation that one guesses whether an ill-located element x in a universe X belongs to a subset A of X, and he regarded a fuzzy measure of A as the grade of "$x \in A$". Höhle [3] interpreted the fuzzy measure in terms of the occurrence of events. In some applications of this measure, it is interpreted as a grade of importance (Sugeno [10], Seif and Aguilar-Martin [9], Ishii and Sugeno [5]).

This paper gives an answer for the above-mentioned questions. We interpret the non-additivity of the fuzzy measure in terms of addition. We discuss the representation of a fuzzy measure and the Choquet integral. The representation was suggested by Höhle [3]. Its essence is that a fuzzy measure μ on a measurable space $(X, \mathcal{A})$ is expressed in terms of a measure on a measurable space $(\Theta, \mathcal{B})$, and a mapping $\eta: \mathcal{A} \to \mathcal{B}$ such that

$$\mu(A) = \lambda(\eta(A)) \quad \forall A \in \mathcal{A}.$$

In Höhle's definition, he sets $\Theta = \{0, 1\}^{\mathcal{A}}$ and $\eta(A) = \{\theta \in \Theta \mid \theta(A) = 1\}$ for the sake of his interpretation. In Section 2 our general interpretation also leads us to this representation. The rationality of the Choquet integral follows from the interpretation, which is discussed in Section 3.

In this paper we deal only with fuzzy measures defined on a non-empty finite set X for the sake of simplicity. In addition, we do not assume that $\mu(X) = 1$. Therefore, by a fuzzy measure, we mean a nonnegative set function $\mu: P(X) \to [0, \infty)$, where $P(X)$ is the power set of X, with the properties:

(1) $\mu(\emptyset) = 0$,
(2) $A \subset B \Rightarrow \mu(A) \leq \mu(B)$,

For a clear distinction between an ordinary measure and a fuzzy measure, we call the former a classical measure.

2. Interpretation of fuzzy measures

A fuzzy measure is a sort of measure. Therefore a fuzzy measure μ on $P(X)$ must measure a certain attribute of $P(X)$, and faithfully represent the properties of the attribute. In this section, on the assumption that addition is a meaningful operation in measuring an attribute, we consider what situation a fuzzy measure expresses. We begin with examples.

Let X be the set of the workers in a workshop, and suppose that they produce the same products. For each $A \in P(X)$, we consider the situation that the members of A work in the workshop. A group A may have various ways to work: various combinations of joint work and divided work. But suppose that a group A works in the most efficient way. Let $\mu(A)$ be the number of the products made by A in one hour. Then μ is a measure of the productivity of a group: the attribute of $P(X)$ in question is the productivity. By the definition of μ, the following statements are natural:

(1) $\mu(\emptyset) = 0$, $\mu(X) = 1$,
(2) $A \subset B \Rightarrow \mu(A) \leq \mu(B)$,

That is, μ is a fuzzy measure.

μ is not necessarily additive. Let A and B be disjoint subsets of X, and let us consider the productivity of the coupled group $A \cup B$. If A and B work separately, then $(A \cup B) = \mu(A) + \mu(B)$. But, since they generally interact with each other, equality may not necessarily hold. The inequality $\mu(A \cup B) > \mu(A) + \mu(B)$ shows effective cooperation of members of $A \cup B$. The converse inequality $\mu(A \cup B) < \mu(A) + \mu(B)$ shows incompatibility between A's operations and B's, that is, the impossibility of separate working. For example, incompatibility is caused by insufficient equipment and/or insufficient workstations in the workshop: sufficient equipment and/or sufficient workstations make separate working possible. As a matter of fact, $A \cup B$ may have both effective cooperation and incompatible operation. Therefore, if the degree of the effective cooperation is greater than that of the incompatible operation, the inequality
$(A \cup B) > \mu(A) + \mu(B)$ holds. If it is not the case, the converse inequality holds.

Now let $X = \{x_1, x_2\}$ for the sake of simplicity. The same argument is applicable to the case where $A = \{x_1\}$ and $B = \{x_2\}$. Let $\lambda_0, \lambda_1, \lambda_2$, and λ_3 be the measures of the productivity corresponding to incompatible operation, x_1's compatible operation, x_2's compatible operation, and effective cooperation, respectively. One's compatible operation means the operation not prevented by the other's operation. We can simply imagine the case where there are two workstations to do a certain job, while incompatible operation means that there is only one workstation for two workers and they have to use it in series.

Then μ can be represented as follows:

$$\mu(\emptyset) = 0,$$
$$\mu(\{x_1\}) = \lambda_0 + \lambda_1, \qquad (1)$$
$$\mu(\{x_2\}) = \lambda_0 + \lambda_2,$$
$$\mu(\{x_1, x_2\}) = \lambda_0 + \lambda_1 + \lambda_3$$

Figure 1 illustrates Eq. (1), where the measures of the productivity are represented by area.

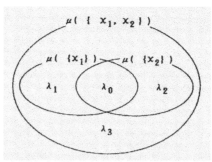

Fig. 1. *Illustration of Eq. (1).*

We denote the index set of λ by Θ, i.e., $\Theta = \{0, 1, 2, 3\}$, and define the subsets $\eta(A)$ $(A \in P(X))$ of Θ as

$$\eta(\emptyset) = \emptyset,$$
$$\eta(\{x_1\}) = \{0, 1\},$$
$$\eta(\{x_2\}) = \{0, 2\},$$
$$\eta(\{x_1, x_2\}) = \{0, 1, 2, 3\}. \qquad (2)$$

Then μ is expressed as

$$\mu(A) = \sum_{i \in \eta(A)} \lambda_i \quad \forall A \in P(X). \qquad (3)$$

We call an element of Θ a feature in respect of the attribute (or a feature for short). In this case 'a feature in respect of the attribute' means 'a feature in respect of productivity', since the attribute in question is productivity. For example, $3 \in \Theta$ is the feature of the effective cooperation of x_1 and x_2. In general if $i \in \eta(A)$, we say that i is a feature of A, or A has the feature i. Then Eq. (3) expresses that the measure of A equals the sum of all the measures i's of the features of A. In addition, we call the set $\eta(A)$ of the features of A 'A in the attribute'.

Note that η as a mapping from $P(X)$ to $P(\Theta)$ does not preserve the operations $\cup$ and $\cap$ but the inclusion $\subset$. This is because of the interactions $0, 3 \in \Theta$ between x_1 and x_2.

The same argument is applied when we discuss weight. Let $X = \{x_1, x_2, \ldots, x_n\}$ be a set consisting of n objects, and let $\mu(A)$ be the weight of $A \in P(X)$. Obviously μ is a classical measure, which is a special fuzzy measure. Since the elements of X are not interactive, the weight of A is characterized only by each element contained in A. Therefore we define the set of the features by $\Theta = (1, 2, \ldots n\}$ where i means a feature 'containing x_i'. Since $\eta(A)$ must satisfy that

$$x_i \in A \Leftrightarrow i \in \eta(A) \quad \forall A \in P(X),$$

$\eta(A)$ is defined by $\eta(A) = \{i \mid x_i \in A\}$. If $\lambda_i = \mu(\{x_i\}) \forall i \in \Theta$, where λ_i represents the increment of weight caused by the feature i, then we have

$$\mu(A) = \sum_{i \in \eta(A)} \lambda_i \ \forall A \in P(X).$$

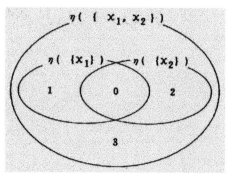

Fig. 2. The subsets $\eta(A)$ of Θ.

The mapping $\eta: P(X) \to P(\Theta)$ is a Boolean algebra isomorphism. That is, $\eta(A)$ is not distinguished from A. In case of classical measures, the distinction between '$P(X)$' and '$P(X)$ in the attribute', i.e., $P(\Theta)$, is unnecessary.

Now we describe a general case. We consider measuring an attribute of $P(X)$ by a set function μ. We assume that the subsets of $P(X)$ interact with each other, and we distinguish '$P(X)$ in the attribute' from '$P(X)$'. That is, we do not assume the mapping $\eta: P(X) \to P(\Theta)$ to be a Boolean algebra isomorphism.

We explain in detail. First suppose it is possible that $\eta(A) \cup \eta(B) \subsetneq \eta(A \cup B)$. This means that $A \cup B$ can have a feature which neither A nor B has. Such a feature can be called a cooperative action of A and B. The set of the cooperative actions is $\eta(A \cup B) - [\eta(A) \cup \eta(B)]$. Second. suppose that the condition $A \cap B = \emptyset$ unnecessarily means that $\eta(A) \cap \eta(B) = 0$. The set $\eta(A) \cap \eta(B)$ represents the overlap of A and B in the attribute, which is interpreted as an incompatibility between A and B or as common features between A and B.

Accordingly we assume that η satisfies only the conditions:

(M1) $\eta(\emptyset) = \emptyset$,
(M2) $A \subset B \Rightarrow \eta(A) \subset \eta(B)$,
(M3) $\eta(X) = \Theta$

That is, η is an order-homomorphism which maps the greatest and least elements of $P(X)$ to the greatest and least elements of $P(\Theta)$, respectively. We call this order-homomorphism η a 0-1 preserving order-homomorphism, where 0 and 1 mean the least and greatest elements, respectively. (M1) means that the empty set has no feature in respect of the attribute. (M2) means that, if $A \subset B$, then B has all the features of A. (M3) means that only the features of the subsets of X are considered. It is not a very essential condition.

For each feature $\theta \in \Theta$, let λ_θ be its measure, and suppose

$$\mu(A) = \sum_{\theta = \eta(A)} \lambda_\theta \quad \forall A \in P(X),$$

i.e., the measure $\mu(A)$ of A is the sum of all the measures of the features of A. Let λ be a classical measure on $P(\Theta)$ defined by

$$\lambda(E) = \sum_{\theta \in E} \lambda_\theta \quad \forall E \in P(\Theta)$$

Then we have that

$$\mu(A) = \lambda(\eta(A)) \quad \forall A \in P(X), \tag{4}$$

that is, μ makes the following diagram commute:

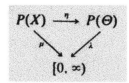

Obviously this set function μ is a fuzzy measure.

We summarize this mathematical result in the following proposition.

Proposition 2.1. *If X and Θ are finite sets, if η: $P(X) - P(\Theta)$ is a 0-1 preserving order-homomorphism, and if λ is a classical measure on $P(\Theta)$, then $\mu = \lambda \circ \eta$ is a fuzzy measure on $P(X)$.*

Given a fuzzy measure μ on $P(X)$, if there exists a set Θ, a classical measure λ on $P(\Theta)$, and a 0 - 1 preserving order-homomorphism η: $P(X) \to P(\Theta)$ such that $\mu = \lambda \circ \eta$, then we call the quadruplet $R = (\Theta, P(\Theta), \lambda, \eta)$ a representation of a fuzzy measure. This definition of the representation is a generalization of Höhle's definition described in Section 1, since Θ and η are not specified in our definition.

The next proposition shows that every fuzzy measure has the interpretation based on Eq. (4). This proposition was pointed out by Höhle [3] in his special setting of Θ and η. Our definition of the representation makes the proof very elementary.

Proposition 2.2. *Every fuzzy measure has its representation.*

Proof. Let $(X, P(X), \mu)$ be a fuzzy measure space. We define the set Θ by
$$\Theta = \{\mu(A) | A \in P(X), \mu(A) \neq 0\},$$
and assume that $\Theta = \{\theta_1, \theta_2, \ldots \theta_n\}$ ($\theta_1 < \theta_2 < \ldots < \theta_n$.) Let η be a mapping from $P(X)$ to $P(\Theta)$ defined by
$$\eta(A) = \{\theta_i | \theta_i \leq \mu(A)\} \quad \forall A \in P(X),$$
and let λ be a classical measure on $P(\Theta)$ defined by
$$\lambda(\{\theta_i\}) = \theta_i - \theta_{i-1}, \quad i = 1,\ldots,n \quad (\theta_0 = 0).$$
Then

$$\lambda(\eta(A)) = \lambda(\{\theta_i | \theta_i \le \mu(A)\})$$
$$= \sum_{\theta_i \le \mu(A)} (\lambda(\{\theta_i\}))$$
$$= \sum_{\theta_i \le \mu(A)} (\theta_i - \theta_{i-1}) \quad (\theta_0 = 0)$$
$$= \mu(A).$$

We have that $\mu = \lambda \circ \eta$. □

Remark 2.1. A fuzzy measure generally has infinitely many representations. We give an example. Let $X = \{x_1, x_2\}$; μ be an arbitrary fuzzy measure on $P(X)$; $\Theta = \{0, 1, 2, 3\}$; and η be a mapping from $P(X)$ to $P(\Theta)$ defined by Eq. (2).

If non-negative numbers λ_i satisfy Eq. (1), then $(\Theta, P(\Theta), \lambda, \eta)$ is a representation of μ, where λ, is defined by

$$\lambda(E) = \sum_{i \in E} \lambda_i \quad \forall E \in P(\Theta).$$

Obviously, under the condition $\lambda_i \ge 0$, $i = 0, 1, 2, 3$, the system of equations (1) generally has infinitely many solutions. Therefore, the fuzzy measure μ generally has infinitely many representations.

Remark 2.2. The representation constructed in the proof of Proposition 2.2 is the most special of all the representations of μ, because the mapping η has a very special property:

$$\eta(A) \subset \eta(B) \Leftrightarrow \mu(A) \le \mu(B) \quad \forall A, B \in P(X).$$

In addition, the set Θ and the mapping η depend on a fuzzy measure μ. But, for every non-empty finite set X, there are a set Θ_x and a mapping η_x such that, for any fuzzy measure on $P(X)$ and for any representation R of μ, a representation equivalent to R is constructed by Θ_x and η_x. Θ_x and η_x are independent of a fuzzy measure μ on $P(X)$. Appendix 1 discusses this universal representation.

Finally we state the difference between fuzzy measures we have discussed and classical measures. The separateness (no interactions) of the subsets of X leads to a classical measure, while the interaction leads to a fuzzy measure. The identity of '$P(X)$' and '$P(X)$ in the attribute' leads to a classical measure, while the distinction leads to a fuzzy measure.

Remark 2.3. We can express this difference in other words. Let us fix a set A. If μ is a classical measure, then for every B for which $A \cap B = \emptyset$, $\mu(A \cup B) - \mu(B)$ is equal to the constant $\mu(A)$. But if μ is a fuzzy measure, this is not true: the value $\mu(A \cup B) - \mu(B)$ depends on B. The value $\mu(A \cup B) - \mu(B)$ is interpreted as the effect of A joining B. Therefore, we can state the difference of a fuzzy measure from a classical measure. For a classical measure, the effect of A joining B does not depend on B. For a fuzzy measure, the effect depends on B.

3. Integration

We again consider the example of productivity at a workshop. Let $X = \{x_1, x_2, \ldots, x_n\}$ be a set of the workers. Suppose that each worker x_i works $g(x_i)$ hours a day from the opening hour. Without loss of generality, we can assume that $g(x_1) \leq g(x_2) \ldots \leq g(x_n)$. We have for $i \geq 2$,
$$g(x_i) - g(x_{i-1}) \geq 0$$
and
$$g(x_i) = g(x_1) + [g(x_2) - g(x_1)] + [g(x_3) - g(x_2)]$$
$$+ \cdots + [g(x_i) - g(x_{i-1})].$$

Now let us aggregate the working hours of all the workers in the following way. First the group X with n workers works $g(x_1)$ hours, next the group $X - \{x_1\} = \{x_2, x_3, \ldots, x_n\}$ works $g(x_2) - g(x_1)$ hours, then the group $X - \{x_1, x_2\} = \{x_3, x_4, \ldots, x_n\}$ works $g(x_3) - g(x_2)$ hours, ..., lastly one worker x_n works $g(x_n) - g(x_{n-1})$ hours. Therefore, since a group $A \subset X$ produces the amount $\mu(A)$ in one hour, the total number of the products produced by them is expressed by

$$g(x_1)\mu(X) + [g(x_2) - g(x_1)]\mu(X - \{x_1\})$$
$$+ [g(x_3) - g(x_2)]\mu(X - \{x_1, x_2\})$$
$$+ \cdots + [g(x_m) - g(x_{m-1})]\mu(\{x_n\}),$$

$$= \sum_{m=1}^{n} [g(x_m) - g(x_{m-1})]\mu(\{x_m, x_{m+1}, \ldots, x_n\}), \tag{5}$$

where $g(x_0) = 0$.

We give another example. Suppose there is a rare book consisting of two volumes. We denote the first volume and the second volume by y_1 and y_2, respectively. Suppose that there is a secondhand bookseller who buys them at the prices: $(v\{y_1\})$ dollars per first volume, $(v\{y_2\})$ dollars per second volume, $v(\{y_1, y_2\})$ dollars per set of two volumes. Since he sets a high value on a complete set, there holds
$$v(\{y_1, y_2\}) > v(\{y_1\}) + v(\{y_2\}).$$
A certain person sells $h(y_1)$ first volumes and $h(y_2)$ second volumes to the secondhand bookseller. We may assume that $h(y_1) \leq h(y_2)$. Then, since he sells $h(y_1)$ complete sets and $h(y_2) - h(y_1)$ second volumes, he gets
$$h(y_1)v(\{y_1, y_2\}) + [h(y_2) - h(y_1)]v(\{y_2\}) \tag{6}$$
dollars.

It is natural that the quantities expressed by (5) and (6) are regarded as the integral of g with respect to μ and that of h with respect to v, respectively. The functions g and h are expressed in

$$g = \sum_{m=1}^{n} [g(x_m) - g(x_{m-1})] 1_{A_m}, \tag{7}$$

where $A_m = \{x_m, x_{m+1}, \ldots, x_n\}$ and $g(x_0) = 0$,
$$h = h(y_1) 1_{\{y_1, y_2\}} + [h(y_2) - h(y_1)] 1_{\{y_2\}}. \tag{8}$$

On the basis of the correspondences of (5) to (7) and (6) to (8), we define an integral with respect to a fuzzy measure.

Now any nonnegative function on X can be expressed in the form

$$f = \bigvee_{i=1}^{n} a_i 1_{A_i}, \qquad (9)$$

where $\vee$ stands for supremum, $0 \leq a_1 \leq a_2 \ldots \leq a_n$ and $A_i = \{x \mid f(x) \geq a_i\}$. Note that $A_1 \supset A_2 \supset \ldots A_n$. This form can be rewritten as

$$f = \sum_{i=1}^{n}(a_i - a_{i-1}) 1_{A_i}, \qquad (10)$$

where $a_0 = 0$.

Definition 3.1. Let $(X, P(X), \mu)$ be a fuzzy measure space. The integral of

$$f = \sum_{i=1}^{n}(a_i - a_{i-1}) 1_{A_i},$$

where $a_0 = 0$, with respect to μ is defined as

$$\sum_{i=1}^{n}(a_i - a_{i-1}) \mu(A_i), \qquad (11)$$

This integral is a Choquet integral because

$$\sum_{i=1}^{n}(a_i - a_{i-1}) \mu(A_i) = \sum_{i=1}^{n} \mu(\{x \mid f(x) \geq a_i\})(a_i - a_{i-1})$$

and it is equal to

$$\int_0^\infty \mu(\{x \mid f(x) \geq \alpha\}) \, d\alpha.$$

In the sequel we shall denote the Choquet integral $\int_0^\infty \mu(\{x \mid f(x) \geq \alpha\}) \, d\alpha$ by $(C) \int f \, d\mu$.

This integral reflects the interaction of subsets. It makes this assertion clear to compare the integral (11) of f with the quantity

$$\sum_{i=1}^{n} a_i \mu(A_i - A_{i+1}),$$

where $A_{n+1} = \emptyset$. This quantity is a classic integral which is based on another expression of f in the form

$$f = \sum_{i=1}^{n} a_i 1_{B_i}, \qquad (12)$$

where $B_i = A_i - A_{i+1}, A_{n+1} = \emptyset$ and $B_i \cap B_j = \emptyset$ for $i \neq j$. Considering that the B_j's are pairwise disjoint, it is not utilized at the above quantity that μ reflects the interaction of subsets. On the contrary, the A_i's used in (10) are not disjoint: they are a monotone sequence $A_1 \supset A_2 \supset \ldots A_n$.

If there are no interactions between the subsets, that is, μ is a classical measure, then the integral based on Eq. (10) coincides with that based on Eq. (12). Taking into account that $\mu(A_i) = \mu(A_{i+1}) + \mu(A_i - A_{i+1})$, we have

$$\sum_{i=1}^{n}(a_i - a_{i-1})\mu(A_i) = \sum_{i=1}^{n} a_i[\mu(A_i) - \mu(A_{i+1})]$$

$$= \sum_{i=1}^{n} a_i \mu(A_i - A_{i+1})$$

That is, for a classical measure, the Choquet integral is equal to the Lebesgue integral:

$$(C)\int f\, d\mu = \int f\, d\mu$$

We can interpret the integral according to Remark 2.3. Since A_i is the disjoint union of A_{i+1} and B_i, $\mu(A_i) - \mu(A_{i+1})$ means the effect of B_i joining to A_{i+1}. Then the integral of f is the sum of all the effects of B_i joining to A_{i+1} of the a_i units:

$$(C)\int f\, d\mu = \sum_{i=1}^{n} a_i[\mu(A_i) - \mu(A_{i+1})], \tag{13}$$

where $A_{n+1} = \emptyset$. We can consider this expression to be based on that of f in the form

$$f = \sum_{i=1}^{n} a_i(1_{A_i} - 1_{A_{i+1}}).$$

Eq. (13) is of course equivalent to Eq. (11). However, if μ is not additive, then

$$\sum_{i=1}^{n} a_i[\mu(A_i) - \mu(A_{i+1})] \neq \sum_{i=1}^{n} a_i\mu(A_i - A_{i+1}).$$

From the foregoing, we state that the Choquet integral is the most proper for the interpretation based on Eq. (4). In Appendix 2 we shall compare the Choquet integral with the others.

The following are some properties of the Choquet integral. Some of them were shown in [1]. Let f and g be nonnegative functions on X.

(1) $(C)\int 1_A\, d\mu = \mu(A)$
(2) $(C)\int af\, d\mu = a(C)\int f\, d\mu \ \forall\ a \geq 0$,
(3) generally $(C)\int (f+g)\, d\mu \neq (C)\int f\, d\mu + (C)\int g\, d\mu$,
(4) $f \leq g \Rightarrow (C)\int f\, d\mu \leq (C)\int g\, d\mu$,
(5) $\mu \leq \nu \Rightarrow (C)\int f\, d\mu \leq (C)\int f\, d\nu$,

Finally we consider the relation between the representation of a fuzzy measure and the Choquet integral.

Proposition 3.1. *Let $(X, P(X), \mu)$ be a fuzzy measure space and let $(\Theta, P(\Theta), \lambda, \eta)$ be its representation. For every nonnegative function f on X, there exists a nonnegative function g on Θ such that*

$$(C) \int f \, d\mu = \int g \, d\lambda,$$

i.e., *the Choquet integral of f with respect to μ is equal to the Lebesgue integral of g with respect to λ. The function g is given as*

$$g(\theta) = \sup\{\alpha \mid \theta \in \eta \circ f^*(\alpha)\} \quad \forall \theta \in \Theta,$$

where $f^*(\alpha) = \{x \mid f(x) \geq \alpha\} \; \forall \alpha \, [0, \infty)$.

Proof. When Y is an arbitrary set, we define $F(Y)$ to be the set of all nonnegative functions on Y, and $\Phi(Y)$ to be the set of all the functions $\phi: \to [0, \infty) \to P(Y)$ with the following properties (P1)-(P3):

(P1) $\phi(0) = Y$.
(P2) $\bigcap_{\beta < \alpha} \phi(\beta) = \phi(\alpha)$,
(P3) $\bigcap_{\beta \in [0,\infty)} \phi(\beta) = \emptyset$.

Note that (P2) implies the semi-continuity and also the monotonicity of ϕ:
$\beta \leq \alpha \Rightarrow \phi(\beta) \supset \phi(\alpha)$.

The mapping $h \mapsto h^*$, where $h \in F(Y)$ and $h^*(\alpha) = \{y \mid h(y) \geq \alpha\} \forall [0, \infty)$, is a bijection from $F(Y)$ to $\Phi(Y)$, and its inverse mapping is $\phi \mapsto {}^*\phi$, where $\phi \in \Phi(Y)$ and ${}^*\phi(y) = \sup\{\alpha \mid y \in \phi(\alpha)\} = \sup_\alpha \alpha 1_{\Phi(\alpha)}(y) \; \forall y \in Y$. In fact we have $h = {}^*(h^*)$, i.e., $h(y) = \sup\{\alpha \mid y \in h^*(\alpha)\} = \sup_\alpha \alpha 1_{h^*(\alpha)}(y)$. Note that this is a generalization of the representation theorem of a fuzzy set where the range of h as a membership function is restricted to $[0, 1]$.

Accordingly if $g(\theta) = \sup\{\alpha \mid \theta \in \eta \circ f^*(\alpha)\} \forall \theta \notin \Theta$, then the following diagram commutes since $g^* = \eta \circ f^*$:

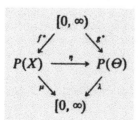

Therefore,

$$(C) \int f \, d\mu = \int_0^\infty \mu(\{x \mid f(x) \geq \alpha\}) \, d\alpha = \int_0^\infty \mu \circ f^*(\alpha) \, d\alpha$$

$$= \int_0^\infty \lambda \circ \eta \circ f^*(\alpha) \, d\alpha = \int_0^\infty \lambda \circ g^*(\alpha) \, d\alpha$$

$$= \int_0^\infty \lambda(\{\theta \mid g(\theta) \geq \alpha\}) \, d\alpha = \int g \, d\lambda,$$

since λ is additive. □

Let θ be an arbitrary element of Θ. Since $g(\theta) = \sup\{\alpha \mid \theta \in \eta \circ f^*(\alpha)\}$,

$$\theta \in \eta(f^*(\alpha)) \quad \forall \alpha \leq g(\theta)$$

and

$$\theta \notin \eta(f^*(\alpha)) \quad \forall \alpha > g(\theta).$$

That is, $f^*(\alpha)$ has the feature θ for $\alpha \leq g(\theta)$ and $f^*(\alpha)$ does not have the feature θ for $\alpha > g(\theta)$. Hence, the value $g(\theta)$ is just the value the function f brings on the feature θ.

Remark 3.1. If

$$f = \bigvee_{i=1}^{n} a_i 1_{A_i} \quad (0 \leq a_1 \leq a_2 \leq \cdots \leq a_n, A_1 \supset A_2 \supset \cdots \supset A_n),$$

then g is expressed as

$$g = \bigvee_{i=1}^{n} a_i 1_{\eta(A_i)}$$

since taking into account that $A_i = \{x \mid f(x) \geq a_i\}$, we have $f^*(a_i) = A_i$ and also $g^*(a_i) = \eta(A_i)$, hence $g^* = \eta \circ f^*$ and $g = \bigvee_{i=1}^{n} a_i 1_{g^*(a_i)}$.

In this section we have first defined the integral (the Choquet integral) from two examples, and second considered the relation between the integral and the representation. In contrast to this, next we show two theorems which show that we can induce the Choquet integral from the representation.

First we define the notation used here as follows: let Y be a non-empty finite set,

$CM(Y)$: the set of all the finite classical measures on $P(Y)$,
$FM(Y)$: the set of all the fuzzy measures on $P(Y)$,
$F(Y)$: the set of all the nonnegative functions on Y.

Let X be a non-empty finite set. We consider a mapping $I: F(X) \times FM(X) \to [0,\infty)$ and deduce that

$$I(f, \mu) = (C) \int f d\mu \quad \forall f \in F(X) \quad \forall \mu \in FM(X).$$

from certain conditions.

The following condition (C0) means that $I(f, \mu)$ is expressed by a representation of μ.

(C0) For every non-empty finite set Θ and for every 0-1 preserving order homomorphism $\eta: P(X) \to P(\Theta)$, there exists a mapping $\overline{\eta}: F(X) \to F(\Theta)$ such that

$$I(f, \lambda \circ \eta) = \int \overline{\eta}(f) d\lambda \quad \forall f \in F(X) \quad \forall \lambda \in CM(\Theta)$$

In order that $I(f, \mu)$ is regarded as an extended Lebesgue integral of f with respect to μ, the following three conditions are necessary.

(C1) For a given $f = a 1_A \in F(X)$, and for given $\mu_1, \mu_2 \in FM(X)$, if $\mu_1(A) = \mu_2(A)$, then $I(f, \mu_1) = I(f, \mu_2)$.

(C2) For a given $\mu \in FM(X)$, if μ is a classical measure, then
$$I(f,\mu) = \int f\, d\mu \quad \forall f \in F(X).$$

(C3) For given $f, g \in F(X)$, if $f(x) \le g(x)\ \forall\, x \in X$, then
$$I(f,\mu) \le I(g,\mu).$$

Condition (C1) means that the value $I(a1_A, \mu)$ depends on $\mu(A)$, and not on $\{\mu(B)|B \ne A\}$. (C2) means that the mapping I is an extension of Lebesgue integral. (C3) means the monotonicity of the mapping I.

Theorem 3.2. *The mapping* $I: F(X) \times FM(X) \to [0, \infty)$ *satisfies the conditions* (C0)-(C3) *iff*
$$I(f,\mu) = (C)\int f\, d\mu \quad \forall f \in F(X)\ \forall \mu \in FM(X).$$

Replacing (C1) with (C1'), we can remove (C3).

(C1') For a given
$$f = \bigvee_{i=1}^{n} a_i 1_{A_i},$$
where $a_1 \le a_2 \le \ldots \le a_n$ and $A_1 \supset A_2 \supset \ldots \supset A_n$, and for given $\mu_1, \mu_2 \in FM(X)$, if $\mu_1(A_i) = \mu_2(A_i)$, $i = 1, 2, \ldots, n$, there holds
$$I(f,\mu_1) = I(f,\mu_2) \quad \forall f \in F(X).$$

Theorem 3.3. *The mapping* $I: F(X) \times FM(X) \to [0,\infty)$ *satisfies the conditions* (C0), (C1'), *and* (C2) *iff*
$$I(f,\mu) = (C)\int f\, d\mu \quad \forall f \in F(X)\ \forall \mu \in FM(X).$$

The proofs of the above theorems are in Appendix 3.

4. Concluding remarks

As we have obtained the interpretation of fuzzy measures, we should develop a theory of fuzzy measure and the Choquet integral on the basis of the interpretation. In a forthcoming paper we will deal with fuzzy measures, defined on an infinite set and discuss some mathematical properties of fuzzy measures and the Choquet integral.

Let us consider the interpretation of fuzzy measures from the viewpoint of measurement theory. "... Measurement may be regarded as the construction of homomorphisms (scales) from empirical relational structures of interest into numerical relational structures that are useful" (Krantz et al. [6]). A homomorphism ϕ into **R**, the set of the real numbers, is called a scale. In terms of permissible transformations, scales are classified into ratio scales, interval scales, ordinal scales, etc. A permissible transformation $\phi \to f(\phi)$ also yields a homomorphism into the same numerical relational structure. A scale whose permissible transformations are the similarity transformations, i.e.,
$$f(x) = ax,\ a > 0,$$

is called a ratio scale, which is a homomorphism into the one-dimensional linear space R^1. A scale whose permissible transformations are the monotonic increasing transformations, i.e., f is strictly increasing,
is called an ordinal scale, which is a homomorphism into the ordered set $(R, \leq)$.

The fuzzy measure we have interpreted is a ratio scale. Our interpretation says that the fuzzy measure $\mu(A)$ of A is the sum of all the measures of the features of A:

$$\mu(A) = \sum_{\theta \in \eta(A)} \lambda_\theta = \lambda(\eta(A)). \tag{14}$$

λ is a classical measure, and a classical measure is a ratio scale. Hence, the fuzzy measure we have interpreted is a ratio scale. The assumption in the beginning of Section 2, "addition is a meaningful operation in measuring an attribute", means that a fuzzy measure is a ratio scale. The Choquet integral is suitable for a fuzzy measure which is a ratio scale. Since various attributes can be measured by ratio scales, the interpretation based on Eq. (14) and the Choquet integral apply to various phenomena besides the examples in this paper.

There are many attributes which cannot be measured by ratio scales. Hence, it is often the case that we measure an attribute to get a fuzzy measure which is not a ratio scale. Let us recall the example of a workshop. This time by a set function μ we do not measure quantity but quality of the products made by each group $A \subset X = \{x_1, x_2\}$. Suppose that the products are graded 1, 2, 3 according to quality and that grade 3 is the best. In addition, suppose that $\mu(\emptyset) = 0$. If each of the workers x_1 and x_2 makes products of grade 2 and if the group $\{x_1, x_2\}$ makes products of grade 3, then we obtain the following fuzzy measure μ:

$$\mu(\{x_1\}) = \mu(\{x_1, x_2\}) = 3.$$

This fuzzy measure μ is not a ratio scale but an ordinal-like scale. In such a case, Eq. (14) loses its meaning. A rational integral with respect to such fuzzy measures, if it can be defined, does not necessarily coincide with the Choquet integral. Moreover, it is not necessarily an extension of the Lebesgue integral. We think that the fuzzy integral [10] is suitable for some such fuzzy measures. We expect the development of the theory of such fuzzy measures.

Appendix 1

Here we show that for every non-empty set X, there exists a set Θ_x and a 0-1 preserving order-homomorphism $\eta_x: P(X) \to P(\Theta_x)$ such that for any fuzzy measure μ on $P(X)$ and for any representation R of μ, a representation equivalent to R is constructed by Θ_x and η_x. This representation is universal in the sense that Θ_x and η_x do not depend on μ.

First of all, we define the concept of equivalence between two representations of a fuzzy measure. Let $(X, P(X), \mu)$ be a fuzzy measure space and let $R = (\Theta, P(\Theta), \lambda, \eta)$ be a representation of μ. Assume that $\mu(X) > 0$. Let

$$\Theta^+ = \{\theta \in \Theta \mid \lambda(\{\theta\}) > 0\}, \quad \eta^+(A) = \eta(A) \cap \Theta^+ \quad \forall A \in P(X),$$

$B(R)$ be the Boolean algebra of subsets of Θ^+ generated by $\{\eta^+(A)|A \in P(X)\}$, i.e., the smallest Boolean algebra containing $\{\eta^+(A)|A \in P(X)\}$, and $\lambda^+ = \lambda|\, B(R)$, the restriction of λ to $B(R)$. Then $(\Theta^+, B(R), \lambda^+)$ is a classical measure space. For every $A \in P(X)$,

$$\mu(A) = \lambda(\eta(A)) = \lambda(\eta(A) \cap \Theta^+) + \lambda(\eta(A) - \Theta^+)$$
$$= \lambda(\eta(A) \cap \Theta^+) = \lambda(\eta^+(A))$$
$$= \lambda^+(\eta^+(A)).$$

Hence, $\mu = \lambda^+ \circ \eta^+$. Therefore, the classical measure space $(\Theta^+, B(R), \lambda^+)$ is sufficient to represent the fuzzy measure μ.

Definition A1.1. Let μ be a fuzzy measure on $P(X)$. If $\mu(X) = 0$, all representations are equivalent. If $\mu(X) > 0$, two representations of μ, $R_1 = (\Theta_1, P(\Theta_1), \lambda_1, \eta_1)$ and $R_2 = (\Theta_2, P(\Theta_2), \lambda_2, \eta_2)$, are said to be equivalent if and only if there is a Boolean algebra isomorphism $T: B(R_1) \to B(R_2)$ such that

$$\lambda_1^+(E) = \lambda_2^+(T(E)) \quad \forall E \in B(R_1)$$

and

$$T(\eta_1^+(A)) = \eta_2^+(A) \quad \forall A \in P(X),$$

that is, T makes the following diagram commute:

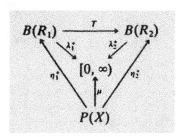

Remark A1.1. A fuzzy measure μ for which $\mu(X) > 0$ generally has infinitely many representations not equivalent to each other. See Remark 2.1 on Proposition 2.2.

Next we show three lemmas on Boolean algebras. Let Θ be a non-empty finite set. It is known that there is a canonical bijection from the class of all Boolean algebras of subsets of Θ to the class of all partitions of Θ; the canonical Boolean algebra corresponding to a partition P is a Boolean algebra $B = \{\bigcup_{i=1}^{n} E_i \mid E_i \in P, n = 0, 1, 2, ...\}$ and the canonical partition corresponding to a Boolean algebra B is a partition $P = \{E \mid E \text{ is an atom of } B\}$, where an atom E is an element of B such that $E \subset F$ or $E \cap F = \emptyset$ for every $F \in B$ and $E = \emptyset$. The following lemma is well known.

Lemma A1.1. Let Θ_1 and Θ_2 be non-empty finite sets, B_1 and B_2 be Boolean algebras of subsets of Θ_1 and Θ_2, respectively, and P_1, and P_2 be the canonical partitions corresponding to B_1 and B_2, respectively. If a mapping $T: B_1 \to P(\Theta_2)$ is a $\cup$ - homomorphism, i.e.,

$$T(E \cup F) = T(E) \cup T(F) \quad \forall E, F \in B_1,$$

and if $T \mid P_1$, the restriction of T to P_1, is a bijection to P_2, then T is a Boolean algebra isomorphism from B_1 to B_2.

It is also known that there is a canonical bisection from the class of all partitions of Θ to the class of all equivalence relations on Θ; the canonical equivalence relation $\sim$ corresponding to a partition P is defined by

$$\theta \sim \theta' \Leftrightarrow \theta \in E \text{ and } \theta' \in E \text{ for some } E \in P$$

and the canonical partition corresponding to an equivalence relation $\sim$ is a partition $P = \{[\theta] \mid \theta \in \Theta\}$, where $[\theta]$ is the equivalence class of θ, i.e., $[\theta] = \{\theta' \in \Theta \mid \theta \sim \theta'\}$. Therefore, there is a canonical bijection from the class of all Boolean algebras of subsets of Θ to the class of all equivalence relations on Θ.

The following lemma is on a Boolean algebra isomorphism.

Lemma A1.2. Let Θ_1 and Θ_2 be non-empty finite sets, B_1 and B_2 be Boolean algebras of subsets of Θ_1 and Θ_2, respectively, $\sim_1$ and $\sim_2$ be the canonical equivalence relations corresponding to B_1 and B_2 respectively, and τ be a mapping from Θ_2 onto Θ_1. If

$$\theta_2 \sim_2 \theta_2' \Leftrightarrow \tau(\theta_2) \sim_1 \tau(\theta_2') \quad \forall \theta_2, \theta_2' \in \Theta_2$$

and if

$$T(E) = \tau^{-1}(E) \quad \forall E \in B_1,$$

then T is a Boolean algebra isomorphism from B_1 to B_2.

Proof. Let P_1 and P_2 be the canonical partitions corresponding to B_1 and B_2, respectively. Since

$$\theta_2 \sim_2 \theta_2' \Leftrightarrow \tau(\theta_2) \sim_1 \tau(\theta_2') \quad \forall \theta_2, \theta_2' \in \theta_2$$

and τ is onto, $T \mid P_1$ is a bijection to P_2. In addition, by the definition of T we have that T is a $\cup$-homomorphism. Then it follows from Lemma A1.1 that T is a Boolean algebra homomorphism from B_1 to B_2.

It is easy to prove the following lemma.

Lemma A1.3. Let $(X, P(X), \mu)$ be a fuzzy measure space and $R = (\Theta, P(\Theta), \lambda, \eta)$ be a representation of μ. We define a relation $\sim$ on Θ^+ by

$$\theta \sim \theta' \Leftrightarrow [\theta \in \eta^+(A) \Leftrightarrow \theta' \in \eta^+(A) \forall A \in P(X)] \quad \forall \theta, \theta' \in \Theta^+.$$

Then $\sim$ is the canonical equivalence relation corresponding to $B(R)$; if $[\theta]$ is the equivalence class of θ, then $[\theta] \in B(R) \ \forall \ \theta \in \Theta^+$ and $E = \bigcup_{\theta \in E} [\theta] \forall E \in B(R)$.

Now we show the main theorem in Appendix 1. Let X be a non-empty finite set and let Θ_x, be the set of all semi-filters in $P(X)$, where a semi-filter in $P(X)$ is a

subset S of $P(X)$ with the properties:
(1) $X \in S$
(2) $\emptyset \in S$,
(3) $A \in S, A \subset B \Rightarrow B \in S$

Let η_x be a mapping from $P(X)$ to $P(\Theta_x)$ defined by
$$\eta_x(A) = \{S \in \Theta_x \mid A \in S\} \quad \forall A \in P(X).$$

Obviously η_x is a 0-1 preserving order-homomorphism, where 0 stands for the empty set and 1 for the whole set. Then, for any fuzzy measure μ on $P(X)$ and for any representation R, a representation equivalent to R can be constructed by Θ_x and η_x.

Theorem A1.4. *For every fuzzy measure μ on $P(X)$ and for every representation $R = (\Theta, P(\Theta), \lambda, \eta)$ of μ, there exists a classical measure λ_x on $P(x)$ such that $R_x = (\Theta_x, P(\Theta_x), \lambda_x, \eta_x)$ is a representation equivalent to R.*

Proof. Since the theorem is valid for $\mu(X) = 0$, we assume that $\mu(X) > 0$. For each $\theta \in \Theta$, we define a subfamily $\tau(\theta)$ of $P(X)$ by
$$\tau(\theta) = \{A \in P(X) \mid \theta \in \eta(A)\}.$$
Then, since η is a 0-1 preserving order-homomorphism, $\tau(\theta)$ is a semi-filter in $P(X)$ for every $\theta \in \Theta$, that is, τ is a mapping from Θ to Θ_x.
By (A1.1) the definitions of η_x and τ, for every $A \in P(X)$ and $\theta \in \Theta$,
$$\theta \in \tau^{-1}(\eta_x(A)) \Leftrightarrow \tau(\theta) \in \eta_x(A) \Leftrightarrow A \in \tau(\theta) \Leftrightarrow \theta \in \eta(A).$$
So we obtain
$$\tau^{-1}(\eta_x(A)) = \eta(A) \quad \forall A \in P(X). \tag{A1.1}$$

We define a classical measure λ_x on $P(\Theta_X)$ by $\lambda_x = \lambda \circ \tau^{-1}$ in an ordinary way, i.e.,
$$\lambda_X(E) = \lambda(\tau^{-1}(E)) \quad \forall E \in P(\Theta_X).$$
By (A1.1) we have that, for every $A \in P(X)$,
$$\lambda_X(\eta_X(A)) = \lambda(\tau^{-1}(\eta_x(A))) = \lambda(\eta(A)) = \mu(A).$$
Then the following diagram commutes:

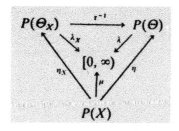

and $R_x = (\Theta_x, P(\Theta_x), \lambda_x, \eta_x)$ is a representation of μ.

Now we prove that R_x is equivalent to R. Let

$$T(E) = (\tau | \Theta^+)^{-1}(E) \quad \forall E \in B(R_x),$$
where $\tau|\Theta^+$ is the restriction of to $\tau\Theta^+$.

First we show that T is a Boolean algebra isomorphism from $B(R_x)$ to $B(R)$.
Let $S \in \Theta_x$. Then
$$S = \tau(\theta) \text{ for some } \theta \in \Theta^+ \Leftrightarrow \lambda(\tau^{-1}(\{S\})) > 0$$
$$\Leftrightarrow \lambda_x(\{S\}) > 0 \Leftrightarrow S \in \Theta_X^+.$$
Hence $\tau|\Theta^+$ is a mapping onto Θ_X^+. Let $\theta \in \Theta^+$ and $A \in P(X)$. It follows from (A1.1) and the definitions of η^+ and η_X^+ that
$$\theta \in \eta^+(A) \Leftrightarrow \theta \in \eta(A) \Leftrightarrow \theta \in \tau^{-1}(\eta_X(A)) \Leftrightarrow \tau(\theta) \in \eta_X(A).$$
Since $\tau|\Theta^+$ is a mapping onto Θ_X^+, we have $\tau(\theta) \in \Theta_X^+$, and hence
$$\tau(\theta) \in \eta_X(A) \Leftrightarrow \tau(\theta) \in \eta_X^+(A).$$
Thus
$$\theta \in \eta^+(A) \Leftrightarrow \tau(\theta) \in \eta_X^+(A) \quad \forall \theta \in \Theta^+ \ \forall A \in P(X). \tag{A1.2}$$
Let $\sim$ and $\sim_X$ be the canonical equivalence relations corresponding to $B(R)$ and $B(R_x)$, respectively. Then it follows from Lemma A1.3 and (A1.2) that, for every $\theta, \theta' \in \Theta^+$
$$\theta \sim \theta' \Leftrightarrow [\theta \in \eta^+(A) \Leftrightarrow \theta' \in \eta^+(A) \forall A \in P(X)]$$
$$\Leftrightarrow [\tau(\theta) \in \eta_X^+(A) \Leftrightarrow \tau(\theta') \in \eta_A^+(A) \ \forall A \in P(X)]$$
$$\Leftrightarrow \tau(\theta) \sim_X \tau(\theta').$$
By Lemma A1.2 we obtain the following fact:

T is a Boolean algebra isomorphism from $B(R_x)$ to $B(R)$. (A1.3)

Let $A \in P(X)$. Then
$$T(\eta_X^+(A)) = (\tau|\Theta^+)^{-1}(\eta_X^+(A)) = \tau^{-1}(\eta_X^+(A)) \cap \Theta^+$$
$$= \tau^{-1}(\eta_X^+(A) \cap \Theta_X^+) = \tau^{-1}(\eta_X^+(A)) \cap \tau^{-1}(\Theta_X^+) \cap \Theta^+.$$
Since $\tau|\Theta^+$ is a mapping onto Θ_X, we have $\Theta^+ \subset \tau^{-1}(\Theta_X^+)$, and hence $\tau^{-1}(\Theta_X^+) \cap \Theta^+ = \Theta^+$. Then
$$\tau^{-1}(\eta_X(A)) \cap \tau^{-1}(\Theta_X^+) \cap \Theta^+ = \tau^{-1}(\eta_X(A)) \cap \Theta^+ = \eta(A) \cap \Theta^+ = \eta^+(A).$$
Therefore we obtain that
$$T(\eta_X^+(A)) = \eta^+(A) \quad \forall A \in P(X). \tag{A1.4}$$
Let $E \in B(R_X)$. Then we have
$$\lambda_X^+(E) = \lambda_X(E) = \lambda(\tau^{-1}(E)) = \lambda(\tau^{-1}(E) \cap \Theta^+) + \lambda(\tau^{-1}(E) - \Theta^+)$$
$$= \lambda(\tau^{-1}(E) \cap \Theta^+) = \lambda((\tau|\Theta^+)^{-1}(E)) = \lambda(T(E)).$$
Since $T(E) \in B(R)$, $\lambda(T(E)) = \lambda^+(T(E))$. Hence we obtain that
$$\lambda_X^+(E) = \lambda^+(T(E)) \quad \forall E \in B(R_X). \tag{A1.5}$$
It follows from (A1.3), (A1.4), and (A1.5) that R_X is equivalent to R. □

By Proposition 2.2 and Theorem A1.4 we have the following corollary.

Corollary A1.5. *For every fuzzy measure μ on $P(X)$, there exists a classical measure λ_X on $P(\Theta_X)$ such that $R_X = (\Theta_X, P(\Theta_X), \lambda_X, \eta_X)$ is a representation of μ.*

Theorem A1.4 and Corollary A1.5 mean that the set Θ_X is sufficient for representation of fuzzy measures on $P(X)$; other sets are unnecessary for representation.

Remark A1.2. Theorem 1.12 in [3] implies Corollary A1.5. The proof in this paper is, however, different from that in [3]. As far as the concept of equivalence between two representations is concerned, we first discuss it in this paper.

Appendix 2

Here we compare the Choquet integral with the other integrals. Since Weber [12] has shown some comparisons between the fuzzy integral, Choquet's and Weber's, we consider some matters he did not mention.

There is a similarity between the fuzzy integral and the Choquet integral. Both definitions use the quantities $\mu(\{x \mid f(x) \geq \alpha\})$. Therefore, the fuzzy integral also reflects the interaction of subsets, because the fuzzy integral of the function $f(3.5)$ in Section 3 is $\sup_i \min[a_i, (A_i)]$ and not $\sup_i \min[a_i, (B_i)]$.

For a measurable function $h: X \to [0, 1]$ and for a probability measure P, the following inequality holds [10]:

$$\left| \oint h \circ P - \int h \, dP \right| \leq \tfrac{1}{4}. \tag{A2.1}$$

As an extension of this inequality we have the next proposition.

Proposition A2.1. *If $h : X \to [0, 1]$ is a measurable function and μ is a fuzzy measure such that $\mu(X) = 1$, then*

$$\left| \oint h \circ \mu - (C)\int h \, dP \right| \leq \tfrac{1}{4}. \tag{A2.2}$$

Proof. For $\alpha \in [0, 1]$, let $h^*(\alpha) = \{x \mid h(x) \geq \alpha\}$. Let $\oint h \circ \mu = M$; then $0 \leq M \leq 1$, $\mu(h^*(\alpha)) \geq M \; \forall \; \alpha < M$, and $\mu(h^*(\alpha)) \leq M \; \forall \; \alpha > M$ (see [10]). Therefore,

$$(C)\int h \, d\mu - \oint h \circ \mu = \int_0^1 \mu(h^*(\alpha)) d\alpha - M$$

$$= \int_0^M \mu(h^*(\alpha)) d\alpha + \int_M^1 \mu(h^*(\alpha)) d\alpha - M$$

$$\leq \int_0^M d\alpha + \int_M^1 M d\alpha - M = M(1-M) \leq \tfrac{1}{4}.$$

Similarly,

$$\oint h \circ d\mu - (C)\int h\, d\mu = M - \int_0^M \mu(h^*(\alpha))d\alpha - \int_M^1 \mu(h^*(\alpha))d\alpha$$

$$\leq M - \int_0^M M\, d\alpha = M - M^2 \leq \tfrac{1}{4}.$$

Ralescu [8] showed that the supremum of the left-hand side in (A2.1) is equal to 1/4. His result applies to our case: the supremum of the left-hand side of (A2.2) is also equal to 1/4.

We cannot directly compare the Choquet integral with an integral with respect to a pseudo-additive measure because, as we have mentioned in Section 1, the latter is based on specific operations rather than the monotonicity of the fuzzy measure. We can define another integral

$$\sum_{i=1}^{n} (a_i - a_{i-1}) \hat{\cdot}\, \mu(A_i)$$

by changing Σ to $\hat{\Sigma}$ and $\cdot$ (multiplication) to $\hat{\cdot}$ in Definition 3.1. Weber [12] discussed such integrals.

Höhle's integral is the 'dual' of Choquet's: if we denote Höhle's integral of f by $(H)\int f d\mu$ and if $\mu^*(A) = \mu(X) - \mu(A^c)\ \forall A \in P(X)$, then we have

$$(H)\int f d\mu = (C)\int f d\mu^*.$$

Therefore, Höhle's integral has unacceptable properties:
(1) $(H)\int 1_A\, d\mu = \mu(X) - \mu(A^c) \neq \mu(A)$,
(2) $\mu(X) = v(X), \mu \leq v \Rightarrow (H)\int f\, d\mu \geq (H)\int f\, dv$.

Appendix 3

Here we prove Theorem 3.2 and Theorem 3.3.
It is obvious that if

$$I(f,\mu) = (C)\int f d\mu \quad \forall f \in F(X) \forall \mu \in FM(X),$$

then the mapping I satisfies (C0), (C2), (C3), and (C1'); we have shown in Section 3 that the Choquet integral satisfies these conditions. We will prove the converse.

Corollary A1.5 in Appendix 1 implies that there exists a non-empty finite set and a 0-1 preserving order-homomorphism $\eta:P(X)\ P(\Theta)$ with the property (P).

(P) For every $\mu \in FM(X)$, there is $\lambda \in CM(\Theta)$ such that $\mu = \lambda \circ \eta$.

We use these Θ and η. That is, we can use the above property (P). (C0) implies that there exists a mapping $\overline{\eta} : F(X) \to F(\Theta)$ such that

$$I(f, \lambda \circ \eta) = \int \overline{\eta}(f) d\lambda \quad \forall f \in F(X)\ \forall \lambda \in CM(\Theta).$$

By the property (P), in order to show

$$I(f,\mu) = (C)\int f\mathrm{d}\mu \quad \forall\, f \in F(X)\,\forall\, \mu \in FM(X).$$

it is sufficient to show that

$$\int \overline{\eta}(f)\mathrm{d}\lambda = (C)\int f\mathrm{d}(\lambda \circ \eta) \quad \forall\, f \in F(X)\forall\, \lambda \in CM(\Theta).$$

The conditions (C1), (C2), (C3), and (C1') can be rewritten as (D1), D2) (D3), and (D1'), respectively:

(D1) For a given $f = a1_A \in F(X)$ and for given $\lambda_1, \lambda_2 \in CM(\Theta)$, if $\lambda_1(A) = \lambda_2(A)$, then

$$\int \overline{\eta}(f)\mathrm{d}\lambda_1 = \int \overline{\eta}(f)\mathrm{d}\lambda_2.$$

(D2) For a given $\hat{\lambda} \in CM(\Theta)$, if $\hat{\lambda} \circ \eta$ is a classical measure on $P(X)$, then

$$\int \overline{\eta}(f)\mathrm{d}\hat{\lambda} = \int f\,\mathrm{d}(\hat{\lambda} \circ \eta) \quad \forall f \in F(X).$$

(D3) For a given $f, g \in F(X)$, if $f(x) \leq g(x)\ \forall\, x \in X$, then

$$\int \overline{\eta}(f)\,\mathrm{d}\lambda \leq \int \overline{\eta}(g)\,\mathrm{d}\lambda \quad \forall\, \lambda \in CM(\Theta).$$

(D1') For a given $f = \vee_{i=1}^{n} a_i 1_{A_i} \in F(X)$,, where $a_1 \leq a_2 \leq \ldots \leq a_n$ and $A_1 \supset A_2 \supset \ldots \supset A_n$, and for given $\lambda_1, \lambda_2 \in CM(\Theta)$, if

$$\lambda_1(\eta(A_i)) = \lambda_2(\eta(A_i)), \quad i = 1, 2, \ldots, n,$$

then

$$\int \overline{\eta}(f)\mathrm{d}\lambda_1 = \int \overline{\eta}(f)\mathrm{d}\lambda_2.$$

By Proposition 3.1 and Remark 3.1, in order to prove Theorem 3.2 [resp. Theorem 3.3], it is sufficient to deduce from (D1) - (D3) [resp. (D1') and (D2)] that

$$\overline{\eta}\left(\bigvee_{i=1}^{n} a_i 1_{A_i}\right) = \bigvee_{i=1}^{n} a_i 1_{\eta(A_i)},$$

where $0 \leq a_1 \leq a_2 \leq \ldots \leq a_n$ and $A_1 \supset A_2 \supset \ldots \supset A_n$. We do it by proving a sequence of lemmas.

The most important tool we use here is the Dirac measure. Let Y be a non-empty finite set and let y_0 be an arbitrary element of Y. The Dirac measure focused on y_0 is a classical measure on $P(Y)$ defined by, for every $A \in P(Y)$,

$$v_{y_0}(A) = \begin{cases} 1, & y_0 \in A, \\ 0, & y_0 \notin A. \end{cases}$$

v_{y_0} has the following important property:

$$\int g\,\mathrm{d}v_{y_0} = g(y_0) \quad \forall g \in F(Y).$$

Lemma A3.1. *Let us assume* (D2). *If* $f = \vee_{i=1}^{n} a_i 1_{A_i} \in F(X)$, *where* $a_1 \leq a_2 \leq \ldots \leq a_n$ *and* $A_1 \supset A_2 \supset \ldots \supset A_n$, *if* $\hat{\lambda} \in CM(\Theta)$, *and if* $\hat{\lambda} \circ \eta$ *is a classical measure on* $P(X)$, *then*

$$\int \overline{\eta}(f) d\hat{\lambda} = \int \bigvee_{i=1}^{n} a_i 1_{\eta(A_i)} d\hat{\lambda},$$

Proof. We easily obtain that

$$\int \bigvee_{i=1}^{n} a_i 1_{\eta(A_i)} d\hat{\lambda} = \sum_{i=1}^{n} (a_i - a_{i-1}) \hat{\lambda}(\eta(A_i)),$$

and

$$\int f d(\hat{\lambda} \circ \eta) = \sum_{i=1}^{n} (a_i - a_{i-1}) \hat{\lambda}(\eta(A_i))$$

where $a_0 = 0$. It follows from (D2) that $\int \overline{\eta}(f) d\hat{\lambda} = \int f d(\hat{\lambda} \circ \eta)$. Then the lemma holds.

Lemma A3.2. *For every $\theta \in \Theta$ and for every family $\{A_i \mid i = 1, 2, ..., n\}$ of subsets of X for which $A_1 \supset A_2 \supset ... \supset A_n$ there exists $\hat{\lambda} \in CM(\Theta)$ such that $\hat{\lambda} \circ \eta$ is a classical measure on $P(X)$ and*

$$\hat{\lambda}(\eta(A_i)) = \lambda_\theta(\eta(A_i)), \quad i = 1, 2, ..., n,$$

where λ_θ is the Dirac measure focused on θ.

Proof. Let $X = A_0 \supset A_1 \supset A_2 ... A_n \supset A_{n+1} = \emptyset$ and let θ be an arbitrary element of Θ. Let $N = \max\{i \mid \theta \in \eta(A_i)\}$. Since $A_N \neq A_{N+1}$, there is $x_0 \in A_N - A_{N+1}$.

Let μ_{x_0} be the Dirac measure focused on x_0. Then

$$\mu_{x_0}(A_i) = \lambda_\theta(\eta(A_i)) = \begin{cases} 1, & i \leq N, \\ 0, & i > N. \end{cases}$$

By the property (P), there is $\hat{\lambda} \in CM(\Theta)$ such that $\mu_{x_0} = \hat{\lambda} \circ \eta$. It is obvious that $\hat{\lambda}$ satisfies the requirements.

The following lemma gives the proof of Theorem 3.3.

Lemma A3.3. *Let us assume (D1') and (D2). If $f = \bigvee_{i=1}^{n} a_i 1_{A_i} \in F(X)$, where $a_1 \leq a_2 \leq ... \leq a_n$ and $A_1 \supset A_2 \supset ... \supset A_n$, then*

$$\overline{\eta}(f) = \bigvee_{i=1}^{n} a_i 1_{\eta(A_i)},$$

Proof. Let θ be an arbitrary element of Θ. By the previous lemma, there is $\hat{\lambda} \in CM(\Theta)$ such that $\hat{\lambda} \circ \eta$ that is a classical measure on $P(X)$ and

$$\hat{\lambda}(\eta(A_i)) = \lambda_\theta(\eta(A_i)), \quad i = 1, 2, ..., n. \tag{A3.1}$$

Then it follows from (D1') that

$$\int \overline{\eta}(f) d\lambda_\theta = \int \overline{\eta}(f) d\hat{\lambda}. \tag{A3.2}$$

By Lemma A3.1, we have
$$\int \overline{\eta}(f)d\hat{\lambda} = \int \bigvee_{i=1}^{n} a_i 1_{\eta(A_i)} d\hat{\lambda}. \tag{A3.3}$$

(A3.1) implies that
$$\int \bigvee_{i=1}^{n} a_i 1_{\eta(A_i)} d\hat{\lambda} = \int \bigvee_{i=1}^{n} a_i 1_{\eta(A_i)} d\lambda_\theta. \tag{A3.4}$$

By (A3.2), (A3.3), and (A3.4) we obtain that $\int \overline{\eta}(f)d\lambda_\theta = \int \bigvee_{i=1}^{n} a_i 1_{\eta(A_i)} d\lambda_\theta$ and hence
$$\overline{\eta}(f)(\theta) = \int \bigvee_{i=1}^{n} a_i 1_{\eta(A_i)} \theta.$$

Then the proof of Theorem 3.3 is complete. We will prove Theorem 3.2.

If we replace (D1') with (D1) and set $n = 1$ in the proof of Lemma A3.3, then we obtain the following lemma.

Lemma A3.4. *Let us assume* (D1) *and* (D2). *If* $f = a1_A \in F(X)$, *then*
$$\overline{\eta}(f)(\theta) = a1_{\eta(A_i)}.$$

Lemma A3.5. *The condition* (D3) *is equivalent to the following condition:*
if $f(x) \leq g(x) \ \forall x \in X$, *then* $\overline{\eta}(f)(\theta) \leq \overline{\eta}(g)(\theta) \ \forall \theta \in \Theta$.

Proof. If $\theta \in \Theta$ and if λ_θ is the Dirac measure focused on θ, then $\int \overline{\eta}(f) d\lambda_\theta = \overline{\eta}(f)(\theta)$. Therefore the condition
$$\int \overline{\eta}(f) d\lambda \leq \int \overline{\eta}(g) d\lambda \quad \forall \lambda \in CM(\Theta)$$
is equivalent to the condition
$$\overline{\eta}(f)(\theta) \leq \overline{\eta}(g)(\theta) \quad \forall \theta \in \Theta.$$
Immediately, the lemma holds.

In the rest of the paper we assume (D1)-(D3).

Lemma A3.6. *If* $f = \bigvee_{i=1}^{n} a_i 1_{A_i} \in F(X)$, , *where* $a_1 \leq a_2 \leq \ldots \leq a_n$ *and* $A_1 \supset A_2 \supset \ldots \supset A_n$, *then* $\overline{\eta}(f) \geq \bigvee_{i=1}^{n} a_i 1_{\eta(A_i)}$.

Proof. Since $f \geq a_i 1_{Ai}$, $i = 1, 2, \ldots, n$, it follows from Lemma A3.4 and Lemma A3.5 that
$$\overline{\eta}(f) \geq \overline{\eta}(a_i 1_{A_i}) = a_i 1_{\eta(A_i)}, \quad i = 1, 2, \ldots, n.$$

Thus $\overline{\eta}(f) \geq \bigvee_{i=1}^{n} a_i 1_{\eta(A_i)}$.

For any two set functions v_1 and v_2 with the same domain, we define $v_1 + v_2$ and $v_1 - v_2$ by
$$(v_1 + v_2)(A) = v_1(A) + v_2(A) \quad \text{and} \quad (v_1 - v_2)(A) = v_1(A) - v_2(A)$$
in an ordinary way.

Lemma A3.7. *For every,* $\theta \in \Theta$ *there exists* $\lambda \in CM(\Theta)$ *such that* $(\lambda + \lambda_\theta) \circ \eta$ *is a classical measure on* $P(X)$, *where* λ_θ *is the Dirac measure focused on* θ.

Proof. Let μ_c be the counting measure on $P(X)$, that is, $\mu_c(A)$ is the number of the elements of the subset A. Note that μ_c is a classical measure on $P(X)$. Let $\theta \in \Theta$, λ_θ be the Dirac measure focused on θ, and $\mu = \mu_c - (\lambda_\theta \circ \eta)$.

μ is a fuzzy measure on $P(X)$. Since

$$\mu(A) = \begin{cases} \mu_c(A) - 1, & \theta \in \eta(A), \\ \mu_c(A), & \text{otherwise,} \end{cases}$$

we have

$$\mu_c(A) - 1 \leq \mu(A) \leq \mu_c(A) \quad \forall A \in P(X).$$

Since μ_c is the counting measure, $\mu_c(A) \leq \mu_c(B) - 1$ whenever $A \subsetneq B$. Therefore $\mu(A) \leq \mu(B)$ whenever $A \subsetneq B$. In addition, $\mu(\emptyset) = \mu_c(\emptyset) - \lambda_\theta(\eta(\emptyset)) = 0 - 0 = 0$. By the property (P), there is $\lambda \in CM(\Theta)$ such that $\mu = \lambda \circ \eta$. Since

$$\mu_c = \mu + (\lambda_\theta \circ \eta) = (\lambda \circ \eta) + (\lambda_\theta \circ \eta) = (\lambda + \lambda_\theta) \circ \eta,$$

λ satisfies the requirement. □

The following lemma gives the proof of Theorem 3.2.

Lemma A3.8. *If* $f = \vee_{i=1}^n a_i 1_{A_i} \in F(X)$, *where* $a_1 \leq a_2 \leq \ldots \leq a_n$ *and* $A_1 \supset A_2 \supset \ldots \supset A_n$ *then* $\overline{\eta}(f) \geq \vee_{i=1}^n a_i 1_{\eta(A_i)}$.

Proof. Let θ be an arbitrary element of Θ and let λ_θ be the Dirac measure focused on θ. By the previous lemma, there is $\lambda \in CM(\Theta)$ such that $(\lambda + \lambda_\theta) \circ \eta$ is a classical measure on $P(X)$. From Lemma A3.1 it follows that

$$\int \overline{\eta}(f) \, d(\lambda + \lambda_\theta) = \int \vee_{i=1}^n a_i 1_{\eta(A_i)} \, d(\lambda + \lambda_\theta).$$

Hence,

$$\int \overline{\eta}(f) \, d\lambda + \int \overline{\eta}(f) \, d\lambda_\theta = \int \vee_{i=1}^n a_i 1_{\eta(A_i)} \, d\lambda + \int \vee_{i=1}^n a_i 1_{\eta(A_i)} \, d\lambda_\theta. \quad (A3.5)$$

On the other hand, it follows from Lemma A3.6 that

$$\int \overline{\eta}(f) \, d\lambda \geq \int \vee_{i=1}^n a_i 1_{\eta(A_i)} \, d\lambda \quad (A3.6)$$

and

$$\int \overline{\eta}(f) \, d\lambda_\theta \geq \int \vee_{i=1}^n a_i 1_{\eta(A_i)} \, d\lambda_\theta. \quad (A3.7)$$

By (A3.5), (A3.6), and (A3.7) we obtain

$$\int \overline{\eta}(f) \, d\lambda_\theta = \int \vee_{i=1}^n a_i 1_{\eta(A_i)} \, d\lambda_\theta.$$

Therefore

$$\overline{\eta}(f)(\theta) = \vee_{i=1}^n a_i 1_{\eta(A_i)}(\theta).$$

References

[1] G. Choquet, Theory of capacities, *Ann. Inst. Fourier* **5** (1953) 131-295.
[2] A. P. Dempster, Upper and lower probabilities induced by a multivalued mapping, *Ann. Math Statist.* **38** (1967) 325-339.
[3] U. Höhle, A mathematical theory of uncertainty, in: R.R. Yager, Ed., *Fuzzy Sets and Possibility Theory* (Pergamon Press, New York, 1982) 344-355.
[4] U. Höhle, Integration with respect to fuzzy measures, in: *Proc. IFAC Symposium on Theory and Application of Digital Control*, New Delhi (January 1982) 35-37.
[5] K. Ishii and M. Sugeno, A model of human evaluation process using fuzzy measure, *Internat. J. Man-Machine Stud.* **22** (1985) 19-38.
[6] D. H. Krantz. R. D. Luce, P. Suppes and A. Tversky, *Foundations of Measurement*, Vol. I (Academic Press, New York, 1971).
[7] R. Kruse, Fuzzy integrals and conditional fuzzy measures, *Fuzzy Sets and Systems* **10** (1983) 309-313.
[8] D. Ralescu, Toward a general theory of fuzzy variables, *J. Math. Anal. Appl.* **86** (1982) 176-193.
[9] A. Seif and J. Aguilar-Martin, Multi-group classification using fuzzy correlation, *Fuzzy Sets and Systems* **3** (1980) 109-122.
[10] M. Sugeno, Theory of fuzzy integrals and its applications, Doctoral Thesis, Tokyo Institute of Technology (1974).
[11] M. Sugeno and T. Murofushi, Pseudo-additive measures and integrals, *J. Math. Anal. Appl.* **122** (1987) 197-200.
[12] S. Weber, $\perp$-decomposable measures and integrals for Archimedean t-conorms , *J. Math. Anal. Appl.* **101** (1984) 114-138.

Fuzzy Measure of Fuzzy Events Defined By Fuzzy Integrals*

Michel Grabisch
*Thomson-Sintra ASM,
1 Avenue Aristide Briand,
94117 Arcueil Cedex, France*

Toshiaki Murofushi and Michio Sugeno
*Tokyo Institute of Technology,
Dept. of Systems Science,
4259 Nagatsuta, Midori-ku,
Yokohama 227, Japan*

*Received April 1991
Revised August 1991*

Abstract

We define the concept of fuzzy measure of a fuzzy event by using a general form of fuzzy integral proposed by Murofushi, called fuzzy t-conorm integral, encompassing previous definitions. Zadeh defined the probability measure of a fuzzy event, and later the possibility measure of a fuzzy event. Using a duality property of fuzzy t-conorm integral, we propose a general definition of fuzzy measure of fuzzy events, which is compatible with previous definitions of Zadeh, and possesses all properties of a fuzzy measure, in particular the duality property. Using our definition, we examine the case of decomposable measures and belief functions. A comparison with previous works is provided.

Keywords: fuzzy measure, fuzzy integral, fuzzy event, decomposable measure, belief function, duality.

---

*© 1992 North-Holland. Retyped with written permission from
Fuzzy Sets and Systems, **50** (1992) 293-313.*

1. Introduction

The concept of fuzzy integral with respect to a fuzzy measure proposed in the early 1970s by Sugeno [18] using *min* and *max* operators. Later, several authors tried to generalize or modify the original definition, aiming at different purposes. Kruse [12] and later Weber [21] defined fuzzy integrals with respect to decomposable measures, replacing *min* and *max* operators by product-like operators and Archimedean t-conorms. Similar attempts were also made by Ichihashi et al. [8], Sugeno and Murofushi [19], using pseudo-addition and multiplication. These definitions can be considered as generalizations of the Sugeno integral, and are all distinct from the usual Lebesgue integral. In order to be coherent with the Lebesgue integral, Murofushi and Sugeno [13] proposed the so-called Choquet integral, referring to a functional defined by Choquet in a different context [1]. The main property of the Choquet integral is that it reduces to the Lebesgue integral when the fuzzy measure is additive. This was not the only attempt to generalize the Lebesgue integral, and Höhle [6] proposed a similar integral (in fact a dual expression) without referring to the work of Choquet.

Recently, Murofushi and Sugeno [14] have introduced the general concept of fuzzy t-conorm integral, which can encompass the original definition and those of Kruse, Weber and Choquet. Moreover, this class of integrals is defined for any fuzzy measure, and is not restricted to decomposable measures.

As early as 1974, Sugeno suggested to define the fuzzy measure of a fuzzy event $\tilde{A}$ by the Sugeno integral of its membership function [18]. Independently, four years later Zadeh defined the concept of possibility measure of a fuzzy (or non-fuzzy) event [26]. Interestingly enough, the definition proposed by Zadeh turned out to be a Sugeno integral with respect to a possibility measure, thus coinciding with the previous definition of Sugeno (see, e.g., [2]). Besides this, necessity of fuzzy events have been defined by Dubois and Prade [2] in a dual manner: $N(\tilde{A}) := 1 - \Pi(\tilde{A}^c)$. Although the fact that the possibility of fuzzy events is a Sugeno integral is well known, strangely enough very few researchers (as far as we know, only Inuiguchi et al. [9]) have verified if this definition was coherent with that of Sugeno, i.e., if $N(\tilde{A})$ is effectively the Sugeno integral of the membership function of $\tilde{A}$ with respect to the necessity measure N. In [5], the authors investigated a more general problem, namely for what class of fuzzy t-conorm integrals the following kind of duality relation holds: $1 - (\mathcal{F})\int (1-f) \Diamond d\mu^* = (\mathcal{F})\int f \Diamond d\mu$ where (μ, μ^*) indicates a pair of dual measures. It was found in particular that the Sugeno integral always verifies this relation, thus leading to the conclusion that necessity of fuzzy events are actually a Sugeno integral with respect to a necessity measure.

Unfortunately, in the very beginning of the history of fuzzy sets theory, Zadeh defined the probability (which can be viewed as a special case of fuzzy measure) of a fuzzy event by a Lebesgue integral [25], and as is stated above, the Sugeno integral and its generalizations by Kruse, Weber or Ichihashi are mathematically quite different from the Lebesgue integral. As a consequence, it becomes necessary to consider the general class of fuzzy t-conorm integrals in order to encompass all the cases. The results in [5] permit us to define the fuzzy event $\tilde{A}$ by the fuzzy t-conorm integral of its membership function in such a way which is consistent with the original definition and properties of fuzzy measures, that is monotonicity, continuity and duality (i.e., every measure has a dual measure), and consistent with previous definitions of Zadeh. In the 1980s, several researchers, such as Weber [22], Ichihashi et al. [7], and Wang [20], have proposed definitions of fuzzy measure of fuzzy events based on different kinds of fuzzy integral, but only in some limited cases (decomposable measures for Weber, Ichihashi, L-valued possibility measures for Wang) and without examining the duality property, which we consider an important one, nor being consistent with the probability of a fuzzy event.

The paper is organized as follows. This section gives the general framework and motivations, and the organization of the paper. Section 2 provides basic definitions of the necessary concepts (t-conorm, fuzzy measure, simple function, . . .). Section 3 gives the definition of fuzzy t-conorm integrals and some particular cases (Sugeno integral, Choquet integral, . . .). Section 4 enlarges results presented in [5] about the above cited duality relation between integrals, and Section 5 defines fuzzy measure of fuzzy events. Section 6 considers decomposable measures as well as belief and plausibility functions, which are particular cases of fuzzy measures, defined on fuzzy events. A comparison with previous works is provided. Section 7 concludes the paper.

In the following, *min* and *max* operators are denoted by $\wedge$ and $\vee$, respectively. We will often use the following notation:

Notation 1. Let X be a non-empty set and $\mathcal{X}$ a σ-algebra on X. The set of all measurable mappings from X to $[0, 1]$ is denoted by $\mathcal{F}(X)$.

2. Preliminaries

2.1. *t-conorms*

Definition 1. A *triangular conorm* (or t-*conorm* for short) $\perp$ is a binary operation from $[0,1]^2$ to $[0, 1]$ fulfilling:
 (i) $x \perp 0 = x$,
 (ii) $x \perp y \leq u \perp v$ whenever $x \leq u$ and $y \leq v$,
 (iii) $x \perp y = y \perp x$,
 (iv) $(x \perp y) \perp z = x \perp (y \perp z)$.

The associativity property permits us to define unambiguously t-conorms of any number of arguments $\perp (x_1, \cdots, x_n)$. Commonly used t-conorms are the *max* operator and the 'bounded sum' defined by $\hat{+}(x_1, \cdots, x_n) := \min(1, \sum_{i=1}^{n} x_i)$.

A t-conorm is said to be strict if it is continuous on $[0, 1]^2$ and strictly increasing in each place, i.e.,
$$x \perp y < x \perp v \text{ whenever } x > 0 \text{ and } y < v$$
and similarly for the other argument.

A t-conorm is said to be Archimedean if it verifies $\forall x \in]0,1[$, $x \perp x > x$.

Note that *max* is not Archimedean but the bounded sum is.

For every continuous Archimedean t-conorm there exists a continuous and strictly increasing function $k: [0, 1] \to [0, \infty]$, with $k(0) = 0$ such that,

$$x \perp y = k^{-1}\left[k(1) \wedge (k(x) + k(y))\right]. \qquad (1)$$

Remark that k is not uniquely defined, and every αk, $\alpha > 0$, will give the same t-conorm. T-conorms such that $k(1)$ is finite are said to be nilpotent, otherwise they are strict. This function is called the generator of the t-conorm. The generator of the bounded sum is the identity function, so it is clearly nilpotent.

2.2. *Fuzzy measures*

Definition 2. Let X be a non-empty set and $\mathscr{X}$ a σ-algebra defined on X. A fuzzy measure μ defined on the measurable space $(X, \mathscr{X})$ is a set function $\mu: \mathscr{X} \to [0, 1]$ verifying the following axioms:

(i) $\mu(\emptyset) = 0$, $\mu(X) = 1$.
(ii) $A \subseteq B \Rightarrow \mu(A) \leq \mu(B)$.

$(X, \mathscr{X}, \mu)$ is said to be a fuzzy measure space.

Fuzzy measures include as particular cases probability measures, possibility and necessity measures, etc.

The condition $\mu(X) = 1$ is only a matter of convention and can be dropped. However in the sequel we will consider $\mu(X) = 1$ unless otherwise indicated. In [18] and related papers, a third axiom is added, concerning continuity:

(*) for every increasing (or decreasing) sequence $\{A_n\}$ of measurable sets,

$$\mu\left(\lim_{n \to \infty} A_n\right) = \lim_{n \to \infty} \mu(A_n).$$

However, this axiom is not necessary for defining integrals, and will not be considered here.

Definition 3 [21]. Let $\perp$ be a t-conorm and μ a fuzzy measure verifying the continuity axiom for every increasing sequence, μ is said to be $\perp$-decomposable if it has the following property:

$$\mu(A \cup B) = \mu(A) \perp \mu(B) \quad \text{whenever} \quad A \cap B = \emptyset.$$

A possibility measure is a $\vee$-decomposable measure, and a probability measure is a $\hat{+}$-decomposable measure.

When $\perp$ is Archimedean with generator g, Weber [21] distinguishes between three types of decomposable measures, namely:

- (S): $\perp$ is a strict t-conorm, thus $g \circ \mu : \mathscr{X} \to [0, \infty]$ is an infinite (i.e., $g \circ \mu(X)$ is infinite) (σ-) additive measure whenever μ is a (σ-) decomposable one.
- (NSA): $\perp$ is a non-strict (nilpotent) t-conorm and $g \circ \mu : \mathscr{X} \to [0, g(1)]$ is a finite (i.e., $g \circ \mu(X)$ is finite) (σ-) additive measure.
- (NSP): $\perp$ is a non-strict t-conorm and $g \circ \mu : \mathscr{X} \to [0, g(1)]$ is a finite *pseudo-additive measure in the sense that* $(g \circ \mu)(\bigcup_{i \in I} A_i)$

$= g(1) < \sum_{i \in I}(g \circ \mu)(A_i)$ can occur for a family $\{A_i\}_{i \in I}$ of disjoint subsets.

2.3. Representation of a fuzzy measure

Murofushi and Sugeno [15] have proposed the concept of representation of a fuzzy measure by a usual additive measure.

Definition 4. Let $(X, \mathscr{X})$ and $(Y, \mathscr{Y})$ be two measurable spaces. A mapping $H: \mathscr{X} \to \mathscr{Y}$ is said to be an *interpreter for measurable sets* iff it verifies:
  (i) $H(\varnothing) = \varnothing, H(X) = Y$.
  (ii) $A \subseteq B \Rightarrow H(A) \subseteq H(B)$.

Definition 5. Let $(X, \mathscr{X}, \mu)$ be a fuzzy measure space. A quadruplet $(Y, \mathscr{Y}, m, H)$ is said to be a *representation* of $(X, \mathscr{X}, \mu)$ iff it verifies:
  (i) H is an interpreter from $\mathscr{X}$ to $\mathscr{Y}$.
  (ii) m is an additive measure on $(Y, \mathscr{Y})$.
  (iii) $\mu = m \circ H$.

It is proved in [15] that every fuzzy measure has its representation, which is non-unique.

2 4. Simple functions

Definition 6. A simple function $f : X \to [0, a]$ is such that

$$\forall x \in X, \quad f(x) = \sum_{i=1}^{n} a_i 1_{D_i}(x) \tag{2}$$

where the D_i's are a disjoint family of subsets of X, 1_{D_i} denotes the characteristic function of D_i, and we assume without loss of generality that $0 \leq a_1 \leq a_2 \leq \ldots \leq a_n \leq a$.

Two equivalent forms of Eq. (2) are given by

$$\forall x \in X, \quad f(x) = \bigvee_{i=1}^{n} a_i 1_{A_i}(x), \tag{3}$$

$$\forall x \in X, \quad f(x) = \sum_{i=1}^{n} (a_i - a_{i-1}) 1_{A_i}(x), \tag{4}$$

with $A_i = \bigcup_{j=i}^{n} D_j$ and $a_0 = 0$ by convention.

This notation will be used throughout the paper.

3. Fuzzy t-conorm integrals

(The reader is referred to [14] for all proofs and details.) We begin this section by defining the Sugeno integral and the Choquet integral.

Definition 7. The *Sugeno integral* of a measurable function $f: X \to [0, 1]$ is defined by

$$\mathcal{f} f \circ \mu \sup_{\alpha \in [0,1]} \left[\alpha \wedge \mu(\{x \mid f(x) > \alpha\}) \right]. \tag{5}$$

For simple functions, the expression reduces to

$$\mathcal{f} f \circ \mu \bigvee_{i=1}^{n} (a_i \wedge \mu(A_i)). \tag{6}$$

Definition 8. The *Choquet integral* of a measurable function $f: X \to [0, 1]$ is defined by

$$(C)\int f \, d\mu := \int_0^1 \mu(\{x \mid f(x) > \alpha\}) \, d\alpha. \tag{7}$$

For simple functions the expression reduces to:

$$(C)\int f \, d\mu := \sum_{i=1}^{n} (a_i - a_{i-1}) \mu(A_i). \tag{8}$$

Note the similarity of Eq. (2) with the Lebesgue integral, of Eq. (3) with the Sugeno integral, and of Eq.(4) with the Choquet integral.

Using the concept of representation of a fuzzy measure, one can also represent the Choquet integral by a Lebesgue integral [15]. Let $(Y, \mathcal{Y}, m, H)$ be a representation of $(X, \mathcal{X}, \mu)$ and let f be a measurable mapping from X to $[0, 1]$, i.e., an element of $\mathcal{F}(X)$. H, which is a mapping from $\mathcal{X}$ to $\mathcal{Y}$, can be extended to a mapping from $\mathcal{F}(X)$ to $\mathcal{F}(Y)$ in the following way:

$$H(f)(y) = \sup\{r \in [0, 1] \mid y \in H(\{x \mid f(x) > r\})\}, \quad \forall y \in Y. \tag{9}$$

It is easy to see that $H(1_A) = 1_{H(A)}$, thus the above definition is indeed an extension of H. An equivalent expression of Eq. (9) is

$$H(f)(y) = \sup_{\alpha \in [0,1]} \alpha \cdot 1_{H(\{x \mid f(x) > \alpha\})}(y) \tag{10}$$

where $>$ can be replaced by $\geq$ as well. If f is a simple function on X, then $H(f)$ is a simple function on Y given by

$$H(f)(y) = \bigvee_{i=1}^{n} a_i 1_{H(A_i)}(y) \tag{11}$$

with usual notations of simple functions. Murofushi and Sugeno [15] have shown that for every measurable function f on X, we have

$$(C)\int f \, d\mu = \int H(f) \, dm \tag{12}$$

where the right side is a usual Lebesgue integral, and H, m are defined as above.

We now proceed to define fuzzy t-conorm integrals. First we need the following definitions:

Definition 9. $(\Delta, \bot, \underline{\bot}, \Diamond)$ is a *t-conorm system for integration* iff
    (a) $\Delta, \bot, \underline{\bot}$, are continuous t-conorms, which are $\vee$ or Archimedean.
    (b) $\Diamond: ([0, 1], \Delta) \times ([0, 1], \bot) \to ([0, 1], \underline{\bot})$ is a product-like operation

fulfilling
(i) ◊ is continuous on $[0, 1]^2$;
(ii) $a \lozenge x = 0 \Leftrightarrow a = 0$ or $x = 0$;
(iii) when $x \perp y < 1$, $a \lozenge (x \perp y) = (a \lozenge x) \perp (a \lozenge y)$, $\forall a \in [0, 1]$;
(iv) when $a \Delta b < 1$, $(a \Delta b) \lozenge x = (a \lozenge x) \underline{\perp} (b \lozenge x)$, $\forall x \in [0, 1]$.

$([0, 1), \Delta)$, $([0, 1], \perp)$ and $([0, 1], \underline{\perp})$ are, respectively, the spaces of values of integrand, measure and integral. An interesting case arises when all three t-conorms are Archimedean (the system is then called Archimedean): let us denote by k, g and h the generators of Δ, $\perp$, and $\underline{\perp}$ respectively. The following property holds for the product-like operation:
$$\forall a > 0, \quad \forall x > 0, \quad a \lozenge x = 1$$
or
$$\forall a \in [0, 1], \quad \forall x \in [0, 1], \quad a \lozenge x = h^{-1}[h(1) \wedge (k(a) \cdot g(x))]. \tag{13}$$

Definition 10. For a given t-conorm Δ, we define an operation $-_\Delta$ on $[0, 1]^2$ by:
$a -_\Delta b := \inf\{c \mid b \Delta c \geq a\}$.
Note that when $\Delta = \vee$, we have
$$a -_\Delta b = \begin{cases} a & \text{if } a \geq b, \\ 0 & \text{if } a < b. \end{cases} \tag{14}$$
If Δ has a generator k, then
$$a -_\Delta b = k^{-1}\left[0 \vee (k(a) - k(b))\right]. \tag{15}$$

In particular when $\Delta = \hat{+}$ we have $k = \text{Id}$, thus $a -_\Delta b = 0 \vee (a - b)$. When Δ is continuous, the following property is useful:
$$a \geq b \Rightarrow (a -_\Delta b) \Delta b = a, \tag{16}$$
i.e., $-_\Delta$ and Δ act as the ordinary arithmetical operators $-$ and $+$. Thus $-_\Delta$ will be called a pseudo-difference with respect to the t-conorm Δ. Using this definition we can give an equivalent form of Eq. (2) and Eq. (4) for simple functions:
$$\forall x \in X, \quad f(x) = \bigtriangleup_{i=1}^{n} a_i 1_{D_i}(x), \tag{17}$$
$$\forall x \in X, \quad f(x) = \bigtriangleup_{i=1}^{n} (a_i -_\Delta a_{i-1}) 1_{A_i}(x), \tag{18}$$
with the same notations as above. Note that Eq. (18) reduces to Eq. (3) when $\Delta = \vee$. We are now ready to define fuzzy t-conorm integrals.

Definition 11. Let $(X, \mathcal{X}, \mu)$ be a fuzzy measure space and $\mathcal{F} = (\Delta, \perp, \underline{\perp}, \lozenge)$ a t-conorm system. For a measurable simple function $f : X \to [0, 1]$, the *fuzzy t-conorm integral* of f based on $\mathcal{F}$ with respect to μ is defined by
$$(\mathcal{F}) \int f \lozenge d\mu := \underline{\perp}_{i=1}^{n}\left((a_i -_\Delta a_{i-1}) \lozenge \mu(A_i)\right). \tag{19}$$

In particular, we have:

- With the t-conorm system $(\hat{+},\hat{+},\hat{+},\cdot)$ where $\cdot$ stands for the ordinary product, we recover the Choquet integral.
- With $(\vee, \vee, \vee, \wedge)$ we have the Sugeno integral.
- Let μ be a normal $\bot$-decomposable measure, i.e., a $\bot$-decomposable measure with $\bot = \vee$ or belonging to the (S) or the (NSA) type. Then Eq. (19) reduces to

$$(\mathscr{F})\int f \Diamond d\mu = \underset{i=1}{\overset{n}{\bot}} \left(a_i \Diamond \mu(D_i)\right). \tag{20}$$

When $\underline{\bot} = \bot$ and the t-conorm system is Archimedean, Eq. (20) coincides with the Weber integral. Kruse integrals are a special case of Weber integrals.

The above results show the expressive power of fuzzy t-conorm integrals.

If the t-conorm system is Archimedean, then Eq. (19) can be expressed using the generators of $\mathscr{F}$ and the Choquet integral:

$$(\mathscr{F})\int f \Diamond d\mu = h^{-1}\left[h(1) \wedge (C)\int k \circ f \, d(g \circ \mu)\right] \tag{21}$$

with k, g, h denoting the generators of $\Delta, \bot, \underline{\bot}$, respectively. The above equation is defined for any function and is not restricted to simple measurable functions. It is worth noting that in the case of an Archimedean t-conorm integral can be expressed by the Choquet integral, so the properties of the two integrals are nearly the same.

Now we turn to integrals with respect to dual fuzzy measures.

4. Integrals with respect to dual measures

In this section we will study what class of fuzzy t-conorm integrals verify the duality relation $1 - (\mathscr{F})\int(1-f)\Diamond d\mu^* = (\mathscr{F})\int f \Diamond d\mu$ for a pair of dual measures (μ, μ^*). We will show that the Sugeno integral verifies this relation for every pair of dual measures. On the other hand, it is not difficult to show that the Choquet integral also possesses this property, but this result can be extended to a larger class of integrals we call Choquet-like integrals, if we consider a more general duality relation:

4.1. $\bot$-dual measures

Usually the notion of dual measures refers to the relation holding between possibility and necessity, or belief and plausibility functions, i.e., a pair of dual measures (μ, μ^*) $\mu^*(\cdot) = 1 - \mu(\cdot^c)$ (or $\mu^*(\cdot) = \mu(X) - \mu(\cdot^c)$ for generality), where ^{c} denotes the complement. Note that probability measures are self-dual measures.

In fact, this definition can be generalized using the previously defined pseudo-difference operator:

Definition 12. Let $\bot$ be an Archimedean nilpotent t-conorm. $(\mu, \mu^\bot)$ *is a pair of* $\bot$-*dual measures* iff $\forall A \in \mathscr{X}$, $\mu^\bot(A) = 1 -_\bot \mu(A^c)$, with $-_\bot$ defined as in Definition 8.

It can be easily verified that the above defined dual measure is indeed a fuzzy measure (use Eq. (15) and the monotonicity of k). We note that the usual definition is recovered with $\bot = \hat{+}$. It is important to stress that $\bot$ must be a nilpotent t-conorm because the expression $1 -_\bot a$ does not make sense for strict t-conorms or $\vee$ (see Eqs. (14) and (15)). In fact $1 - {}_\bot(\cdot)$ is strictly equivalent to the general negation operator defined by Trillas (see, e.g., [3]), and possesses desirable properties (involution, continuity, decreasingness).

We now define the Choquet-like integral:

4.2 Choquet-like integrals

Definition 13. A fuzzy t-conorm integral is a *Choquet-like integral* iff $(\Delta, \bot, \underline{\bot}, \Diamond)$ verifies:

(α) $a \Diamond x = 1 \Leftrightarrow a = 1$ and $x = 1$,
(β) $\Delta, \bot$, and $\underline{\bot}$ are Archimedean nilpotent t-conorms.

It is easy to check that the Choquet integral is a Choquet-like integral. As seen before, the properties of the Choquet integral are fundamental when Archimedean systems are considered. Thus we will first show the following lemma concerning Choquet integral:

Lemma 1. *Let f be a measurable function such that $\forall x \in X, 0 \leq f(x) \leq a < \infty$, and $\mu : \mathscr{X} \to [0, \mu(X)]$ a fuzzy measure, whose dual measure μ^* is defined by $\mu^*(\cdot) = \mu(X) - \mu(\cdot^c)$. Then the following property holds:*

$$(C) \int f[a - f(x)] \, d\mu = a\mu(X) - (C) \int f \, d\mu^* . \qquad (22)$$

Proof. We use directly the expression of the Choquet integral in the continuous case. Expanding the left side of Eq. (22) gives:

$$(C) \int f[a - f(x)] \, d\mu = \int_0^a \mu(\{x \mid a - f(x) \geq r\}) \, dr = \int_0^a \mu(\{x \mid f(x) \leq s\}) \, ds$$

with $s := a - r$. We expand now the right side:

$$a\mu(X) - (C) \int f \, d\mu^* = a\mu(X) - \int_0^a \mu^*(\{x \mid f(x) > s\}) \, ds$$

$$= a\mu(X) - \int_0^a [\mu(X) - \mu(\{x \mid f(x) > s\}^c)]$$

$$= a\mu(X) - a\mu(X) + \int_0^a \mu(\{x \mid f(x) \leq s\}) \, ds$$

$$= \int_0^a \mu(\{x \mid f(x) \leq s\}) \, ds$$

where we have used the fact that $(C) \int f \, d\mu = \int_0^\infty \mu(\{f \geq r\}) \, dr = \int_0^\infty \mu(\{f > r\}) \, dr$

(see [15]). □

Choquet-like integrals can be expressed using the generator functions in a rather simple way; this is stated in the following lemma:

Lemma 2. *Let $\mathscr{F} = (\Delta, \perp, \underline{\perp}, \Diamond)$ define a Choquet-like integral, and $f: X \to [0, 1]$ be a measurable function. Then the following holds:*

(i) *condition (α) of Definition 13 is equivalent to:*
$$h(1) = k(1) \cdot g(1) \tag{23}$$

(ii) $(\mathscr{F}) \int f \Diamond d\mu = h^{-1}\left[(C) \int k \circ f \, d(g \circ \mu)\right]. \tag{24}$

Proof. (i) (α) $\Rightarrow$ Eq. (23): By $1 \Diamond 1 = 1$ and Eq. (13) we get $h(1) \leq k(1) \cdot g(1)$. Suppose that $h(1) < k(1) \cdot g(1)$. Then $h(1)/g(1)$ lies in the interval $]0 = k(0), k(1)[$, and thus using the continuity of k and the intermediate value theorem, there exists $0 < a < 1$ such that $k(a) = h(1)/g(1)$, i.e., $a \Diamond 1 = 1$, which contradicts the hypothesis. Thus, $h(1) = k(1) \cdot g(1)$.

Eq. (23) $\Rightarrow$ (α): Suppose $a = x = 1$. Then $a \Diamond x = h^{-1}(h(1) \wedge k(1) \cdot g(1)) = h^{-1}(h(1)) = 1$ by hypothesis. Suppose now $a \Diamond x = 1$. Then $h(1) = h(1) \wedge k(a) \cdot g(x)$, and because k, g are strictly increasing, it follows from the hypothesis that $a = 1$ and $x = 1$.

(ii) Observe that because of the monotonicity of k, $k \circ 1 \geq k \circ f$. Also from the monotonicity of the Choquet integral [1, 13, 15] it follows that:

$$(C) \int k \circ f \, d(g \circ \mu) \leq (C) \int k \circ 1 \, d(g \circ \mu) = k(1) g(\mu(X)) = k(1) g(1) = h(1),$$

using (i) above. Thus, Eq. (21) reduces to Eq. (24) as expected.

Using the two lemmas we can show the following result:

Theorem 1. *Let $\mathscr{F} = (\Delta, \perp, \underline{\perp}, \Diamond)$ define a Choquet-like integral, and $f: X \to [0, 1$ be a measurable function. Then for every pair $(\mu, \mu^{\perp})$ of $\perp$-dual measures, the following holds:*

$$(\mathscr{F}) \int f \Diamond d\mu = 1 -_{\underline{\perp}} (\mathscr{F}) \int (1 -_{\Delta} f) \Diamond d\mu^{\perp} \tag{25}$$

Proof. We denote by $-_\Delta, -_\perp, -_{\underline{\perp}}$, the pseudo-difference operators associated to the t-conorms $\Delta, \perp, \underline{\perp}$, respectively, whose generators are k, g, h, respectively. We first need the following result:

$(g \circ \mu^{\perp})^*(A) = g \circ \mu$, with $\mu^*(\cdot) \underline{\Delta} \mu(X) - \mu(\cdot^c)$

Proof: For every A in $\mathscr{X}$ we have

$$(g \circ \mu^{\perp})^*(A) = g \circ \mu^{\perp}(X) - g \circ \mu^{\perp}(A^c) = g(1) - g[1 -_\perp \mu(A)]$$
$$= g(1) - [0 \vee (g(1) - g \circ \mu(A))] = g(1) - g(1) + g \circ \mu(A)$$
$$= g \circ \mu(A).$$

We are now able to prove the theorem. Expanding the right side of Eq. (25) and using Lemmas 1, 2(ii), and the above results gives:

$$1-_{\perp}(\mathscr{F})\int(1-_{\Delta} f) \Diamond \, d\mu^{\perp} = h^{-1}\left[0 \vee (h(1) - (C)\int k \circ 1 - k \circ f) \, d(g \circ \mu^{\perp}))\right]$$
$$= h^{-1}\left[h(1) - k(1)(g \circ \mu^{\perp})(X)\right.$$
$$\left. + (C)\int k \circ f \, d(g \circ \mu^{\perp})^*\right]$$
$$= h^{-1}\left[h(1) - k(1)g(1) + (C)\int k \circ f \, d(g \circ \mu)\right]$$
$$= h^{-1}\left[(C)\int k \circ f \, d(g \circ \mu)\right]$$
$$= (\mathscr{F})\int f \Diamond \, d\mu. \quad \Box$$

The reader can ask himself if the class of Choquet-like integrals does not reduce to a singleton, namely the Choquet integral itself. Fortunately this is not the case and we provide here an example. Let us introduce the Sugeno family of nilpotent t-conorms defined by

$$a \perp_{\lambda} b := 1 \wedge (a + b - \lambda ab) \tag{26}$$

where $\lambda \in]-1, +\infty[$. The generator of $\perp_{\lambda}$ is $k(x) = \alpha \ln(1 + \lambda x)$, where α is a positive arbitrary constant. Note that when $\lambda = 0$ we recover $\hat{+}$, the bounded sum operator. We propose for $\mathscr{F}$ the following system: $\mathscr{F} \underline{\Delta} (\perp_{\lambda_1}, \perp_{\lambda_2}, \perp_{\lambda_3}, \Diamond)$ with $\lambda_1, \lambda_2, \lambda_3 \neq 0$, and $\Diamond$ defined below. If we denote as above by k, g, h the generators of $\perp_{\lambda_1}, \perp_{\lambda_2}, \perp_{\lambda_3}$, the constants $\alpha_1, \alpha_2, \alpha_3$ are chosen so as to have the property $h(1) = k(1) \cdot g(1)$. It is easy to see that they must verify

$$\frac{\alpha_1 \alpha_2}{\alpha_3} = \frac{\ln(1 + \lambda_3)}{\ln(1 + \lambda_1) \ln(1 + \lambda_2)}.$$

Substituting into the equation $h(a \Diamond x) = h(1) \wedge (k(a) \cdot g(x))$ and rearranging terms leads to the following definition of $\Diamond$:

$$a \Diamond x = 1 \wedge \frac{1}{\lambda_3}\left[(1 + \lambda_3)^{\log_{1+\lambda_1}(1+\lambda_1 a) \cdot \log_{1+\lambda_2}(1+\lambda_2 x)} - 1\right].$$

Now we turn to Sugeno-like integrals:

4.3. Sugeno-like integrals

As stated above, the t-conorm system underlying the Sugeno integral is $(\vee, \vee, \vee, \wedge)$. Murofushi and Sugeno [14] have shown that any t-conorm system combining both $\vee$ operator and Archimedean t-conorms has poor mathematical properties. Moreover, Archimedean t-conorms are essentially different from the $\vee$ operator, so the only meaningful generalization of Sugeno integral appears to be done through the use of t-conorm systems as $(\vee, \vee, \vee, T)$, where T denotes a t-norm. In fact, this is precisely what was suggested by Weber [21].

Definition 14. A *Sugeno-like integral*, denoted by $(\vee) \int f \, T \, \mu$, is a fuzzy t-conorm integral whose t-conorm system is $(\vee, \vee, \vee, T)$, with T being a t-norm.

Using Definition 11 and Eq. (14), Sugeno-like integrals of simple functions are expressed as

$$(\vee) \int f \top \mu = \bigvee_{i=1}^{n} (a_i \top \mu(A_i)) \qquad (27)$$

(compare with Eq. (6)).

In [5], the authors have found that, in the case of simple functions the duality relation between integrals holds only for $\top = \wedge$. The reason is the following: it is known that any pair of mutually distributive operators $(\oplus, \otimes)$ verify the relation

$$\bigotimes_{i \in I} \left(\bigoplus_{j \in J_i} c_{ij} \right) = \bigoplus_{\{j_i\} \in K} \left(\bigotimes_{i \in I} c_{ij_i} \right)$$

with $K := \times_{i \in I} J_i$, $\times$ denoting the Cartesian product, and I, J_i being two finite index sets (this is the most general form of distributivity relations as $a \wedge (b \vee c) = (a \wedge b) \vee (a \wedge c)$). This property is necessary in the proof of the theorem, and unfortunately the only pair of mutually distributive t-norm, t-conorm is $(\wedge, \vee)$, so that only the ordinary Sugeno integral possesses the property. We restate here the result in the continuous case (remark that the proof uses a similar result established by Kandel [11]).

Theorem 2. *Let f be a measurable function and (μ, μ^*) a pair of dual measures. Then the following relation holds for the (ordinary) Sugeno integral:*

$$1 - (\vee) \int (1 - f) \wedge \mu^* = (\vee) \int f \wedge \mu. \qquad (28)$$

Proof. We will use the following property of Sugeno integral:

$$(\vee) \int f \wedge \mu = \sup_{\alpha} [\alpha \wedge \mu(\{x \mid f(x) \geq \alpha\})] = \sup_{\alpha} [\alpha \wedge \mu(\{x \mid f(x) > \alpha\})]$$

(see [18], p. 29 for a proof using continuity of fuzzy measure, but this property is in fact not necessary). We expand the left side of Eq. (28):

$$1 - (\vee) \int (1 - f) \wedge \mu^* = 1 - \sup_{\alpha} [\alpha \wedge \mu^*(\{x \mid 1 - f(x) > \alpha\})]$$

$$= 1 - \sup_{\alpha} [\alpha \wedge (1 - \mu(\{x \mid 1 - f(x) \leq \alpha\}))]$$

$$= 1 - \sup_{\beta} [(1 - \beta) \wedge (1 - \mu(\{x \mid f(x) \geq \beta\}))] \quad (\alpha = 1 - \beta)$$

$$= 1 - \sup_{\beta} [1 - (\beta \vee \mu(\{x \mid f(x) \geq \beta\}))]$$

$$= \inf_{\beta} [\beta \vee \mu(\{x \mid f(x) \geq \beta\})]$$

$$= (\vee) \int f \wedge \mu.$$

The last equality comes from a result established by Kandel [11] (Kandel provides two proofs of this result, of which the second one does not use continuity of fuzzy measures). □

Note that a partial proof (for possibility and necessity measures only) has been established by Inuiguchi et al. (see [9], p. 125).

The above results permit us to define fuzzy measure of fuzzy events by fuzzy t-conorm integrals.

5. Fuzzy measure of fuzzy events

5.1. *Motivations and definition*

In 1968 Zadeh defined the concept of probability of a fuzzy event $\widetilde{A}$, in order to extend to fuzzy sets the fundamental concept of probability of an event in classical probability theory. The definition is the following [25]:

$$P(\widetilde{A}) := \int \phi_{\widetilde{A}}(x)\, dP \tag{29}$$

where $\phi_{\widetilde{A}}$ denotes the membership function of $\widetilde{A}$, and P a probability. Remark that this expression is the Lebesgue integral of $\phi_{\widetilde{A}}$ with respect to the probability measure P. Later Zadeh defined possibility measures as $\vee$-decomposable measures, and extended the definition to fuzzy events [26]:

$$\Pi(\widetilde{A}) := \sup_{x}\bigl(\phi_{\widetilde{A}}(x) \wedge \pi(x)\bigr) \tag{30}$$

where π is the possibility distribution of the possibility measure Π. As stated in the introduction, it can be shown that Eq. (30) is the Sugeno integral of $\phi_{\widetilde{A}}$ with respect to the possibility measure Π. A direct consequence of Theorem 2 is that the necessity of a fuzzy event $\widetilde{A}$ is the Sugeno integral of $\phi_{\widetilde{A}}$, with respect to the necessity measure.

We stress here that despite a similarity between the two definitions of Zadeh (they are both defined by integrals of the membership function), they are intrinsically different on a mathematical point of view because the Sugeno integral is not an extension of the Lebesgue integral, and the Choquet integral (which is one) is not an extension of the Sugeno integral, so that the only solution is to use fuzzy t-conorm integrals.

Let us denote by $\widetilde{\mu}$ an extension of the fuzzy measure μ for measurable fuzzy sets (i.e., whose membership function is measurable), i.e., a fuzzy set function $\mathscr{F}(X) \to [0, 1]$. Intuitively, the definition of $\widetilde{\mu}$ must satisfy the following properties:

(i) $\widetilde{\mu}$ must be an extension of μ on $\mathscr{F}(X)$, i.e., $\widetilde{\mu}(A) = \mu(A)$, whenever $A \in \mathscr{X}$.

(ii) Monotonicity: $\widetilde{A} \subseteq \widetilde{B} \Rightarrow \widetilde{\mu}(\widetilde{A}) \leq \widetilde{\mu}(\widetilde{B})$.

(iii) Continuity: if μ is continuous for every increasing (resp. decreasing) sequence of measurable subsets, then for every increasing (resp. decreasing) sequence of measurable fuzzy subsets $\{\widetilde{A}_n\}$, we have $\widetilde{\mu}(\lim_{n\to\infty}\widetilde{A}_n) = \lim_{n\to\infty}\widetilde{\mu}(\widetilde{A}_n)$.

(iv) $\perp$-duality: for every fuzzy measure on $\mathscr{F}(X)$, a dual measure exists. Using a system of Archimedean nilpotent t-conorms, the most general definition is given by: $\widetilde{\mu}^{\perp}(\widetilde{A}) = 1 -_{\perp} \widetilde{\mu}(\widetilde{A}^{c\Delta})$ where $\widetilde{A}^{c\Delta}$ is the Δ-negation of $\widetilde{A}$ (in the sense of Trillas, as explained in the previous section) defined by $\phi_{\tilde{A}^{\circ}\Delta}(x) \triangleq 1 -_\Delta \phi_{\tilde{A}}(x)$, and $\tilde{\mu}^\perp$ is an extension of $\mu^\perp$, the $\perp$-dual measure of μ.

(v) Consistency with the definitions of Zadeh: when $\mu = P$, we recover Eq. (29); when $\mu = \Pi$ we recover Eq. (30).

Condition (i) ensures us that we do not need a special symbol ($\tilde{\mu}$) for denoting fuzzy measures of fuzzy events. Thus in the sequel we will use only the symbol μ for denoting fuzzy measures on $\mathscr{X}$ or $\mathscr{F}(X)$. Conditions (ii) and (iii), together with condition (i), ensure that the basic axioms defining fuzzy measures still hold for the extended definition. Condition (iv) ensures that a dual measure exists in any case, as is the case for fuzzy measures defined on $\mathscr{X}$. The last condition is necessary in order to be coherent with previous definitions.

Based on these considerations, we propose the following definition:

Definition 15. Let μ be a fuzzy measure and $\tilde{A}$ be a fuzzy event with membership function $\phi_{\tilde{A}}$. The fuzzy measure of $\tilde{A}$ based on μ is defined as:

$$\mu(\tilde{A}) := (\mathscr{F}) \int \phi_{\tilde{A}} \lozenge \, d\mu . \tag{31}$$

However, the t-conorm system $\mathscr{F}$ must belong to one of the two following kinds:

(a) $\mathscr{F} = (\vee, \vee, \vee, \wedge)$ (ordinary Sugeno integral).

(b) $\mathscr{F} = (\Delta, \perp, \perp, \lozenge)$, such that $\mathscr{F}$ defines a Choquet-like integral, i.e., verifying conditions (α) and (β) of Definition 13.

If $\mathscr{F}$ is of the first kind, we recover the original definition of Sugeno and we say that it is a fuzzy measure defined by the Sugeno integral. In the other case, we say that we have a fuzzy measure defined by a Choquet-like integral. Remark that the second and third t-conorms in $\mathscr{F}$ must be identical, because here the result of the integral is a measure value and thus the space of integral values must coincide with the space of measure values. Denoting as before by k, g, h the generators of $\mathscr{F}$, we see that we must have $h = \alpha g$, $\alpha > 0$. Moreover, the condition $1 \lozenge 1 = 1$ entails $\alpha = k(1)$ as is easy to see from Lemma 2. Thus, its expression is given by

$$\mu(\tilde{A}) = g^{-1}\left(\frac{1}{k(1)}\left[(C)\int k \circ \phi_{\tilde{A}} d(g \circ \mu)\right]\right). \tag{32}$$

Usually, for the sake of simplicity we should consider $k(1) = 1$ so that $g \equiv h$.

The following two theorems show us that these definitions are meaningful:

Theorem 3. *Let $\mathscr{F}$ be of the first kind (Sugeno integral). Then for every fuzzy measure μ, the fuzzy measure extended on $\mathscr{F}(X)$ defined by $\mathscr{F}$ verifies all conditions (i) to (v). However, concerning (iv), only ordinary dual measures (for $\Delta = \perp = \underline{\perp} = \hat{+}$) exist, and concerning (v), only the case of the possibility is recovered.*

Proof. (i)-(iii) have already been proved in [18, Section 3.4].

(iv) is clear from Theorem 2.
(v) is clear from the definition. □

Theorem 4. *Let $\mathscr{F} = (\Delta, \bot, \underline{\bot}, \Diamond)$ be of the second kind (Choquet-like integral). Then for every fuzzy measure μ, the fuzzy measure extended on $\mathscr{F}(X)$ defined by $\mathscr{F}$ verifies all conditions (i) to (v). However, concerning (v), only the case of the probability is recovered.*

Proof. Let us denote by k, g, h the generators of Δ, $\bot$, $\underline{\bot}$, respectively, with $h = k(1) \cdot g$.

(i) Let A be an ordinary subset of X. Using the definition we have:

$$\tilde{\mu}(A) = (\mathscr{F}) \int \phi_A \Diamond \, d\mu = h^{-1}\left[(C) \int k \circ 1_A \, d(g \circ \mu)\right]$$

$$= h^{-1}[k(1) \cdot g \circ \mu(A)] = g^{-1}\left[\frac{1}{k(1)} \cdot k(1) \cdot g \circ \mu(A)\right] = \mu(A).$$

(ii) Let $\tilde{A} \subseteq \tilde{B}$, i.e. $\phi_{\tilde{A}} \leq \phi_{\tilde{B}}$. From the monotonicity of the Choquet integral [1, 15, 19] and of k, h we obtain the desired result.

(iii) Let $\{\tilde{A}_n\}$ be an increasing (resp. decreasing) sequence of fuzzy events and μ be a fuzzy measure continuous for every increasing (resp. decreasing) sequence of measurable subsets. The Choquet integral possesses the following property for every increasing (resp. decreasing) sequence of functions f_n, on [0, 1], [15]:

$$\lim_{n \to \infty} (C) \int f_n \, d\mu = (C) \int \lim_{n \to \infty} (f_n) \, d\mu.$$

Then in particular we have:

$$\lim_{n \to \infty} (C) \int \phi_{\tilde{A}_n} \, d\mu = (C) \int \lim_{n \to \infty} (\phi_{\tilde{A}_n}) \, d\mu.$$

Thus, using the continuity of h, k, we can write:

$$\lim_{n \to \infty} \mu(\tilde{A}_n) = \lim_{n \to \infty} h^{-1}\left[(C) \int k(\phi_{\tilde{A}_n}) \, d(g \circ \mu)\right]$$

$$= h^{-1}\left[(C) \int k\left(\lim_{n \to \infty} \phi_{\tilde{A}_n}\right) d(g \circ \mu)\right] = \mu\left(\lim_{n \to \infty} \tilde{A}_n\right).$$

(iv) Clear from Theorem 1.
(v) Clear from the definition. □

We give here an example of a t-conorms system of the second kind, using the Sugeno family of t-conorms as before. The additional constraint $\bot = \underline{\bot}$ leads to:

$$\lambda_2 = \lambda_3 = \lambda, \quad \frac{\alpha_1 \alpha_2}{\alpha_3} = \frac{1}{\ln(1 + \lambda_1)}$$

Taking for example $\alpha_2 = \alpha_3 = \alpha$ leads to the following definition of $\Diamond$:

$$a \Diamond x = 1 \wedge \frac{1}{\lambda}\left[(1 + \lambda x)^{\log_{1 + \lambda_1}(1 + \lambda_1 a)} - 1\right].$$

5.2. Some properties

We state here some properties related to the measure of the union of two fuzzy sets, which will be useful in the sequel. Let $\tilde{A}$ be a fuzzy event whose membership function is $\phi_{\tilde{A}}$. We introduce the following notation: the subset $X_{\tilde{A}}(\alpha)$ of X is defined by

$$X_{\tilde{A}}(\alpha) := \{x \in X \mid k \circ \phi_{\tilde{A}}(x) > \alpha\} \tag{33}$$

where k is a generator of Δ as before. It is easy to see that the following properties hold for every finite family $\{\tilde{A}_i\}_{i \in I}$ of fuzzy events:

$$X_{\cup_{i \in I} \tilde{A}_i}(\alpha) = \bigcup_{i \in I} X_{\tilde{A}_i}(\alpha), \quad \forall \alpha \in [0, k(1)], \tag{34}$$

$$X_{\cap_{i \in I} \tilde{A}_i}(\alpha) = \bigcap_{i \in I} X_{\tilde{A}_i}(\alpha), \quad \forall \alpha \in [0, k(1)], \tag{35}$$

where $\bigcup \tilde{A}_i$ and $\bigcap \tilde{A}_i$ denote the usual union (intersection) of fuzzy sets. We have the following properties:

Theorem 5. *Let μ be a fuzzy measure on $(X, \mathcal{Z})$ and consider its extension on $\mathcal{F}(X)$ defined by a Choquet-like integral $\mathcal{F} = (\Delta, \bot, \bot, \Diamond)$ with g being the generator of $\bot$. Let us denote by $\mathrm{LC}_{i \in I}(x_i)$ a linear combination of the x_i, $i \in I$, with real coefficients. Then the following holds:*
(i) *If μ is such that for every finite family $\{A_i\}_{i \in I}$ in $\mathcal{Z}$ we have*

$$g \circ \mu\left(\bigcup_{i \in I} A_i\right) = \mathrm{LC}_{K \subseteq I, K \neq \emptyset}\left(g \circ \mu\left(\bigcap_{k \in K} A_k\right)\right),$$

then the same formula can be extended to all finite family of fuzzy events $\{\tilde{A}_i\}_{i \in I}$:

$$g \circ \mu\left(\bigcup_{i \in I} \tilde{A}_i\right) = \mathrm{LC}_{K \subseteq I, K \neq \emptyset}\left(g \circ \mu\left(\bigcap_{k \in K} \tilde{A}_k\right)\right).$$

Moreover, the same property holds if $\bigcap$ and $\bigcup$ are interchanged.
(ii) *If μ is such that for every finite family $\{A_i\}_{i \in I}$ in $\mathcal{Z}$ we have*

$$g \circ \mu\left(\bigcup_{i \in I} A_i\right) \geq \mathrm{LC}_{K \subseteq I, K \neq \emptyset}\left(g \circ \mu\left(\bigcap_{k \in K} A_k\right)\right),$$

then the same formula can be extended to all finite family of fuzzy events $\{\tilde{A}_i\}_{i \in I}$:

$$g \circ \mu\left(\bigcup_{i \in I} \tilde{A}_i\right) \geq \mathrm{LC}_{K \subseteq I, K \neq \emptyset}\left(g \circ \mu\left(\bigcap_{k \in K} \tilde{A}_k\right)\right),$$

(iii) *If μ is such that for every finite family $\{A_i\}_{i \in I}$ in $\mathcal{Z}$ we have*

$$g \circ \mu\left(\bigcap_{i \in I} A_i\right) \leq \mathrm{LC}_{K \subseteq I, K \neq \emptyset}\left(g \circ \mu\left(\bigcup_{k \in K} A_k\right)\right),$$

then the same formula can be extended to all finite family of fuzzy events $\{\tilde{A}_i\}_{i \in I}$:

$$g \circ \mu\left(\bigcap_{i \in I} \tilde{A}_i\right) \leq \mathrm{LC}_{K \subseteq I, K \neq \emptyset}\left(g \circ \mu\left(\bigcup_{k \in K} \tilde{A}_k\right)\right),$$

Proof. For (i),

$$g \circ \mu\left(\bigcup_{i \in I} \tilde{A}_i\right) = \frac{1}{k(1)} (C) \int k \circ \phi_{\bigcup_{i \in I} \tilde{A}_i} \, d(g \circ \mu)$$

$$= \frac{1}{k(1)} \int_0^{k(1)} g \circ \mu\left(X_{\bigcup_{i \in I} \tilde{A}_i}(\alpha)\right) d\alpha$$

$$= \frac{1}{k(1)} \int_0^{k(1)} g \circ \mu\left(\bigcup_{i \in I} X_{\tilde{A}_i}(\alpha)\right) d\alpha$$

$$= \frac{1}{k(1)} \int_0^{k(1)} \mathrm{LC}_{K \subseteq I, K \neq \emptyset}\left(g \circ \mu\left(\bigcap_{k \in K} X_{\tilde{A}_k}(\alpha)\right)\right) d\alpha$$

$$= \mathrm{LC}_{K \subseteq I, K \neq \emptyset}\left(\frac{1}{k(1)} \int_0^{k(1)} g \circ \mu\left(\bigcap_{k \in K} X_{\tilde{A}_k}(\alpha)\right) d\alpha\right)$$

$$= \mathrm{LC}_{K \subseteq I, K \neq \emptyset}\left(\frac{1}{k(1)} \int_0^{k(1)} g \circ \mu\left(X_{\bigcap_{k \in K} \tilde{A}_k}(\alpha)\right) d\alpha\right)$$

$$= \mathrm{LC}_{K \subseteq I, K \neq \emptyset}\left(g \circ \mu\left(\bigcap_{k \in K} \tilde{A}_k\right)\right)$$

where we have used properties of Eq. (34) and (35). Of course, the proof is similar when $\cup$ and $\cap$ are interverted.

(ii) is similar to (i): use the monotonicity of g and the fact that usual integrals verify

$$f \leq g \Rightarrow \int f(x) \, dx \leq \int g(x) \, dx.$$

(iii) is similar to (ii). □

Concerning fuzzy measures whose extension is defined through the Sugeno integral, as possibility and necessity, Dubois and Prade [4] have shown the following result:

Theorem 6 (Dubois and Prade [4]). *Zadeh's possibility and necessity of fuzzy events are, respectively, order n subadditive and superadditive, i.e.,*

$$\Pi\left(\bigcap_{i \in I} \tilde{A}_i\right) \leq \sum_{K \subseteq I, K \neq \emptyset} (-1)^{\mathrm{card}\, K + 1} \Pi\left(\bigcup_{k \in K} \tilde{A}_k\right), \tag{36}$$

$$N\left(\bigcup_{i \in I} \tilde{A}_i\right) \geq \sum_{K \subseteq I, K \neq \emptyset} (-1)^{\mathrm{card}\, K + 1} N\left(\bigcap_{k \in K} \tilde{A}_k\right), \tag{37}$$

where $\{\tilde{A}_i\}_{i \in I}$ *is any finite family of fuzzy events on X.*

5.3. Representation of extended fuzzy measures

Let $(X, \mathscr{X}, \mu)$ be a fuzzy measure space, and $(Y, \mathscr{Y}, m, H)$ be its representation by a given interpreter H. In Subsection 5.1 we have extended the definition of μ to measurable fuzzy sets, that is to $(X, \mathscr{F}(\mathscr{X}))$. A question arises here as to how to extend the representation of $(X, \mathscr{X}, \mu)$ to $(X, \mathscr{F}(\mathscr{X}), \mu)$ which will be denoted by $(Y, \mathscr{F}(\mathscr{Y}), m, \tilde{H})$, in a natural and consistent way. (Note that there is no necessity to put a ~ on μ and m because their extensions are already defined and consistent with the crisp case.)

We propose the following definition of $\tilde{H}$, restricted to the case of fuzzy measures extended by the usual Choquet integral:

Definition 16. Let $(X, \mathscr{X}, \mu)$ be a fuzzy measure space, and $(Y, \mathscr{Y}, m, H)$ be its representation by H. Let us consider the extension of μ on $(X, \mathscr{F}(\mathscr{X}))$ by the Choquet integral, i.e., $\mathscr{F} = (\hat{+}, \hat{+}, \hat{+}, \cdot)$. Then the extension of the representation of $(X, \mathscr{F}(\mathscr{X}), \mu)$, denoted by $(Y, \mathscr{F}(\mathscr{Y}), m, \tilde{H})$, is defined by

$$[\tilde{H}(\tilde{A})]_\alpha := H([\tilde{A}]_\alpha), \quad \forall \alpha \in [0,1], \forall \tilde{A} \in \mathscr{F}(\mathscr{X}), \tag{38}$$

where $[\cdot]_\alpha$, denotes the α-cut. or equivalently by:

$$\tilde{H}(\tilde{A}) := H(\phi_{\tilde{A}}) \tag{39}$$

where $\phi_{\tilde{A}}$ denotes the membership of $\tilde{A}$. The additive measure m is extended on $\mathscr{F}(\mathscr{Y})$ by the Lebesgue integral (Zadeh's definition).

The above definition of $\tilde{H}: \mathscr{F}(\mathscr{X}) \to \mathscr{F}(\mathscr{Y})$ is clearly an extension of $H: \mathscr{X} \to \mathscr{Y}$ because we have $\tilde{H}(A) = H(A)$ for every crisp set A. The two expressions shown in Eqs. (38) and (39) are equivalent because if we express $\tilde{A}$ by its α-cuts, i.e., $\phi_{\tilde{A}}(x) = \bigvee_{\alpha \in [0,1]} \alpha \cdot 1_{[\tilde{A}]_\alpha}$, then the image of the mapping $\phi_{\tilde{A}}$ by H is

$$H(\phi_{\tilde{A}}) = \bigvee_{\alpha \in [0,1]} \alpha \cdot 1_{H(\tilde{A})_\alpha}, \tag{40}$$

i.e., H and $\tilde{H}$ are essentially the same.

Let us verify now that this definition is indeed a representation in the usual sense:
- $\tilde{H}(\emptyset) = \emptyset$, $\tilde{H}(X) = Y$: clear.
- $\tilde{A} \subseteq \tilde{B} \Rightarrow \tilde{H}(\tilde{A}) \subseteq \tilde{H}(\tilde{B})$: clear from the fact that $[\tilde{A}]_\alpha \subseteq [\tilde{B}]_\alpha$, $\forall \alpha \in [0,1]$ and H is an interpreter.
- $\mu = m \bullet \tilde{H}$: first we expand the left side. For a given fuzzy set $\tilde{A}$, we have $\mu(\tilde{A}) = (C)\int \phi_{\tilde{A}} d\mu = \int H(\phi_{\tilde{A}}) \, dm$ using the representation of Choquet integral. The right side writes $m(\tilde{H}(\tilde{A})) = \int \phi_{\tilde{H}(\tilde{A})} dm$ (probability of a fuzzy event). Using the

above mentioned equivalence, we get immediately $m(\widetilde{H}(\widetilde{A})) = \int H(\phi_{\widetilde{A}}) dm$ and thus the right and left sides are equivalent.

This shows that again we do not need a special symbol $\widetilde{H}$ for the extension of H on fuzzy sets.

6. Some particular cases of fuzzy measures

In this section we examine the extension of the most usual fuzzy measures to fuzzy events, and compare our approach to previous ones.

6.1. Decomposable measures

Let μ be a $\perp$-decomposable measure, with $\perp$ being $\vee$ or Archimedean with generator g. The case $\perp = \vee$ leads to possibility measures, and we have already seen that their extension to fuzzy events is defined through the Sugeno integral. We consider now the general Archimedean case. If $\perp$ is nilpotent, it seems natural to extend such measures on $\mathcal{F}(X)$ by Choquet-like integrals with $\mathscr{F} = (\Delta, \perp, \perp, \Diamond)$, i.e., the second t-conorm coinciding with the t-conorm defining μ (note that it was also the case with possibility measures). If $\perp$ is a strict t-conorm, there is no preferential way to extend μ, and the extended measure will not possess good properties. As a consequence, in this section we restrict ourself to nilpotent t-conorms, and the extension of μ will be always defined by Eq. (32) (Choquet-like integral).

We have the following property:

Theorem 7. *Let μ be a $\perp$-decomposable measure, with a nilpotent Archimedean t-conorm. Then the extension of μ by a Choquet-like integral defined by $(\Delta, \perp, \perp, \Diamond)$ verifies*

$$\mu(\widetilde{A} \cup \widetilde{B}) = \mu(\widetilde{A}) \perp \mu(\widetilde{B}) \tag{41}$$

whenever $\widetilde{A} \cap \widetilde{B} = \emptyset$ and $g \circ \mu(\text{supp } \widetilde{A}) + g \circ \mu(\text{supp } \widetilde{B}) \le g(1)$, where supp $\widetilde{A}$ indicates the support of $\widetilde{A}$.

Proof. We have

$$\mu(\widetilde{A} \cup \widetilde{B}) = g^{-1}\left(\frac{1}{k(1)}[(C)\int k \circ \phi_{\widetilde{A} \cup \widetilde{B}} \, d(g \circ \mu)]\right).$$

Using the fact that the supports are disjoint, it follows that

$$(C)\int k \circ \phi_{\widetilde{A} \cup \widetilde{B}} \, d(g \circ \mu) = \int_0^{k(1)} g \circ \mu\big(X_{\widetilde{A}}(\alpha) \cup X_{\widetilde{B}}(\alpha)\big) d\alpha$$

$$= \int_0^{k(1)} g\big(\mu(X_{\widetilde{A}}(\alpha)) \perp \mu(X_{\widetilde{B}}(\alpha))\big) d\alpha.$$

Using the condition $g \circ \mu(\text{supp } \widetilde{A}) + g \circ \mu(\text{supp } \widetilde{B}) \le g(1)$ and the monotonicity of the set function $g \circ \mu$, we have

(C) $\int k \circ \phi_{\tilde{A}\cup\tilde{B}} \, d(g \circ \mu) = \int_0^{k(1)} g \circ \mu\big(X_{\tilde{A}}(\alpha)\big) d\alpha + \int_0^{k(1)} g \circ \mu\big(X_{\tilde{B}}(\alpha)\big) d\alpha$.

Now observe that in general $\mu(\tilde{A}) \le \mu(\text{supp } A)$ (monotonicity of the Choquet integral), thus the condition $g \circ \mu(\text{supp } \tilde{A}) + g \circ \mu(\text{supp } \tilde{B}) \le g(1)$ entails:
$$g \circ \mu(\tilde{A}) + g \circ \mu(\tilde{B}) \le g(1).$$

Thus we have:
$$\mu(\tilde{A} \cup \tilde{B}) = g^{-1}\left(\frac{1}{k(1)}\int_0^{k(1)} g \circ \mu\big(X_{\tilde{A}}(\alpha)\big) d\alpha + \frac{1}{k(1)}\int_0^{k(1)} g \circ \mu\big(X_{\tilde{B}}(\alpha)\big) d\alpha\right)$$
$$= g^{-1}\big(g \circ \mu(\tilde{A}) + g \circ \mu(\tilde{B})\big) = \mu(\tilde{A}) \perp \mu(\tilde{B}). \quad \square$$

This result shows that the decomposability property holds for the extended measure.

By analogy with additive measures, which verify
$$\mu(A \cup B) = \mu(A) + \mu(B) - \mu(A \cap B), \forall A, B,$$
we now turn to the study of $\perp$-decomposable measures verifying
$$\mu(A \cup B) = \mu(A) \perp \mu(B) -_{\perp} \mu(A \cap B) = \mu(B) \perp (\mu(A) -_{\perp} \mu(A \cap B)).$$

Note that because $\mu(A)$ and $\mu(B) \ge \mu(A \cap B)$, the right equality always holds. First remark that we have the following equivalence:

Theorem 8. *Let μ be a fuzzy measure on $(X, \mathcal{X})$ and $\perp$ a Archimedean nilpotent t-conorm with the generator g. The following four statements are equivalent:*

(i) μ is $\perp$-decomposable, and $g \circ \mu(\tilde{A}) + g \circ \mu(\tilde{B}) \le g(1)$, for all disjoints A, B in $\mathcal{X}$.

(ii) *For all A, B in $\mathcal{X}$, the two following relations hold:*
$$\mu(A \cup B) = \mu(A) \perp (\mu(B) -_{\perp} \mu(A \cap B)),$$
$$g \circ \mu(\tilde{A}) + g \circ \mu(\tilde{B}) - g \circ \mu(A \cap B) \le g(1).$$

(iii) μ *is of the* NSA *type, i.e.* $g \circ \mu$ *is an additive measure.*

(iv) *The extension of μ on $\mathcal{F}(X)$ by a Choquet-like integral defined by $(\Delta, \perp, \perp, \Diamond)$ is such that $g \circ \mu$ is additive, i.e.*
$$g \circ \mu\left(\bigcup_{i \in I} A_i\right) = \sum_{K \subseteq I, K \ne \emptyset}(-1)^{\text{card } K + 1} g \circ \mu\left(\bigcap_{k \in K} \tilde{A}_k\right)$$
for every finite family $\{\tilde{A}_i\}_{i \in I}$ of fuzzy sets.

Proof. (ii) $\Rightarrow$ (i): Clear.

(i) $\Rightarrow$ (iii): Consider two disjoint A, B in $\mathcal{X}$. Then we obtain by the hypothesis
$$\mu(A \cup B) = \mu(A) \perp \mu(B),$$
which implies
$$g \circ \mu(A \cup B) = g(1) \wedge \big(g \circ \mu(A) + g \circ \mu(B)\big) = g \circ \mu(A) + g \circ \mu(B)$$

using the hypothesis.
(iii) $\Rightarrow$ (ii): Clearly the additivity of $g \circ \mu$ implies
$$g \circ \mu(A) + g \circ \mu(B) - g \circ \mu(A \cap B) = g \circ \mu(A \cup B) \leq g(1)$$
Also using the previous inequality leads to:
$$\mu(A \cup B) = g^{-1}(g \circ \mu(A) + g \circ \mu(B) - g \circ \mu(A \cap B))$$
$$= g^{-1}(g(1) \wedge (g \circ \mu(A) + (0 \vee (g \circ \mu(B) - g \circ \mu(A \cap B)))))$$
$$= \mu(A) \perp (\mu(B) -_\perp \mu(A \cap B)).$$

(iv) $\Rightarrow$ (iii): Clear since (iii) is a restricted case of (iv).

(iii) $\Rightarrow$ (iv): Clear from Theorem 5(i) and the fact that every additive measure p verifies:
$$p\left(\bigcup_{i \in I} A_i\right) = \sum_{K \subseteq I, K \neq \emptyset} (-1)^{\operatorname{card} K + 1} p\left(\bigcap_{k \in K} A_k\right). \qquad \square$$

Finally we have the following result:

Theorem 9. *Let μ be a fuzzy measure and $\perp$ be a Archimedean nilpotent t-conorm verifying conditions (i) or (ii) or (iii) of Theorem 8. Then the following property holds for the extension of μ by $\mathscr{F} = (\Delta, \perp, \perp, \Diamond)$:*
$$\mu(\tilde{A} \cup \tilde{B}) = \mu(\tilde{A}) \perp (\mu(\tilde{B}) -_\perp \mu(\tilde{A} \cap \tilde{B})), \quad \forall \tilde{A}, \tilde{B} \in \mathscr{F}(X). \qquad (43)$$

Proof. From the hypothesis and Theorem 8(iv) we infer that for every $\tilde{A}$, $\tilde{B}$:
$$g \circ \mu(\tilde{A} \cup \tilde{B}) = g \circ \mu(\tilde{A}) + g \circ \mu(\tilde{B}) - g \circ \mu(\tilde{A} \cap \tilde{B})$$
$$= g(1) \wedge (g \circ \mu(\tilde{A}) + (0 \vee (g \circ \mu(\tilde{B}) - g \circ \mu(\tilde{A} \cap \tilde{B}))))$$
$$= g(\mu(\tilde{A}) \perp (\mu(\tilde{B}) -_\perp \mu(\tilde{A} \cap \tilde{B}))) \qquad \square$$

The case of decomposable measures has been extensively studied by Weber [21, 22]. In particular, Weber showed the following theorem, which is similar to our Theorem 7:

Theorem 10 (Weber [21, 22]). *Let μ be a $\perp$-decomposable measure, and two fuzzy sets $\tilde{A}$, $\tilde{B}$ with membership functions $\phi_{\tilde{A}}$, $\phi_{\tilde{B}}$, respectively. Then*
$$\forall x \in X, \; \phi_{\tilde{A}}(x) + \phi_{\tilde{B}}(x) \leq 1 \Rightarrow \int (\phi_{\tilde{A}} + \phi_{\tilde{B}}) \perp \mu = \int \phi_{\tilde{A}} \perp \mu \perp \int \phi_{\tilde{B}} \perp \mu$$

where $\int \phi \perp \mu$ denotes the Weber integral.

This result seems to contain Theorem 7 since if the supports of $\tilde{A}$ and $\tilde{B}$ are disjoint, then Weber's theorem applies and $\phi_{\tilde{A}}(x) + \phi_{\tilde{B}}(x) = \phi_{\tilde{A}}(x) \vee \phi_{\tilde{B}}(x)$. However, if μ is of the NSP type (nilpotent, $g \circ \mu$ non-additive), Weber's integral is not always equivalent to the fuzzy t-conorm integral. In fact, in the NSP case, Weber's integral can be defined only if X is μ-achievable, that is, iff there exists a countable partition $\{X_i\}_{i \in I}$ of X such that $X_i \in \mathscr{X}$ and $\mu(X_i) < 1$ for all $i \in I$. In this case, the Weber integral is defined by

$$\int \phi \perp \mu = g^{-1}\left(\sum_{i \in I} \int_{X_i} \phi \, \mathrm{d}(g \circ \mu)\right).$$

Murofushi and Sugeno [14] have investigated the relation between the two types of integral in detail, and have shown the following result:

Theorem 11 (Murofushi and Sugeno [14]). *Let μ be a $\perp$-decomposable measure on $(X, \mathscr{X})$ of the NSP type, and suppose X is μ-achievable. Then the following relation exists between the Weber integral defined by Eq. (45) and the fuzzy t-conorm integral defined by*

$$(\mathscr{F})\int \phi \lozenge \, \mathrm{d}\mu \le \int \phi \perp \mu$$

with equality only in one of the two following cases:

(a) $(\mathscr{F})\int \phi \lozenge \, \mathrm{d}\mu = 1$;

(b) *there exists a sequence of measurable disjoint sets $\{D_i\}_{i \in I}$ such that*
- $\mu(D_i) < 1, \forall i \in I$,
- $\{x \mid \phi(x) > 0\} = \bigcup_{i \in i} D_i$,
- $g \circ \mu(\{x \mid \phi(x) > 0\}) = \sum_{i \in I} g \circ \mu(D_i)$.

The above leads to the following remarks:
- If $\Delta \ne \hat{+}$, then the result of Weber does not recover Theorem 7.
- Consider $\Delta = \hat{+}$ and X is μ-achievable. In Theorem 7 we have supposed the support of $\tilde{A}$ and $\tilde{B}$ are disjoint. Thus, in the case where $\mu(\mathrm{supp}\,\tilde{A}) < 1$ and $\mu(\mathrm{supp}\,\tilde{B}) < 1$ we can deduce immediately from condition (b) of Theorem 11 that $\mu(\tilde{A} \cup \tilde{B}) = \int \phi_{\tilde{A} \cup \tilde{B}} \perp \mu$, and $\mu(\tilde{A}) = \int \phi_{\tilde{A}} \perp \mu$ and $\mu(\tilde{B}) = \int \phi_{\tilde{B}} \perp \mu$.

    Consequently, our result is recovered by Weber's result.
- Consider again $\Delta = \hat{+}$ and X μ-achievable, but this time we suppose $\mu(\mathrm{supp}\,\tilde{A}) = 1$. But from the hypothesis of Theorem 7, this entails $\mu(\mathrm{supp}\,\tilde{B}) = 0$ and thus $\mu(\tilde{B}) = 0$, leading to a trivial case.
- If X is not μ-achievable, then Weber's integral is not defined, but Theorem 7 still holds.

6.2. Belief functions

Considered as particular cases of fuzzy measures, belief and plausibility functions of Shafer on a non-empty set X are characterized by the following property:

$$\mathrm{Bel}\left(\bigcup_{i \in I} A_i\right) \ge \sum_{K \subseteq I, K \ne \emptyset} (-1)^{\mathrm{card}\,K+1} \mathrm{Bel}\left(\bigcap_{k \in K} A_k\right),$$

$$\mathrm{Pl}\left(\bigcap_{i\in I} A_i\right) \leq \sum_{K\subseteq I, K\neq \emptyset}(-1)^{\mathrm{card}\,K+1}\mathrm{Pl}\left(\bigcup_{k\in K} A_k\right),$$

for every finite family $\{A_i\}_{i\in I}$ of subsets of X.

Considering results of Subsection 5.2, the only possible extension of belief and plausibility functions on fuzzy events which can be made is through the t-conorm system $(\Delta, \hat{+}, \hat{+}, \Diamond)$, i.e., with g being the identity function. Thus we can state:

Definition 17. Let Bel$(\cdot)$ and Pl$(\cdot)$ be belief and plausibility functions defined on a non-empty set X. Their extension on $\mathscr{F}(X)$ is defined by a Choquet-like integral whose t-conorm system is $(\Delta, \hat{+}, \hat{+}, \Diamond)$, leading to the following expression:

$$\mathrm{Bel}(\widetilde{A}) = \frac{1}{k(1)}(C)\int k \circ \phi_{\widetilde{A}}\mathrm{d\,Bel},$$

$$\mathrm{Pl}(\widetilde{A}) = \frac{1}{k(1)}(C)\int k \circ \phi_{\widetilde{A}}\mathrm{d\,Pl},$$

with $\widetilde{A}$ being any fuzzy set on X.

Some remarks are noteworthy here:
- In the above definition, the basic case, and the most significant one, is with $k = \mathrm{Id}$, i.e., with the usual Choquet integral. The general case with $k \neq \mathrm{Id}$ is merely the basic definition applied to a fuzzy set $\widetilde{A}$ which is deformed by k. Consequently, in the sequel we will always assume $k = \mathrm{Id}$, unless otherwise indicated.
- This is not the first attempt to extend the definition of belief functions on fuzzy events (see a recent survey along with a new proposal by Yen [24]), and several authors, the first of them being perhaps Smets [19], proposed their own definitions, with different assumptions and purposes. We can roughly distinguish two groups among them. The first group considers belief functions defined by a basic probability assignment m on a finite set whose focal elements are itself fuzzy: Yager [23], Ishizuka et al. [10], Pal and Gupta [16] among others. As pointed out by Yen [24], their definitions, all based on $\mathrm{Bel}(\widetilde{A}) := \sum_{\widetilde{B}} I(\widetilde{B} \subseteq \widetilde{A}) m(\widetilde{B})$ with $I(\widetilde{B} \subseteq \widetilde{A})$ being a degree of inclusion of $\widetilde{B}$ in $\widetilde{A}$, suffer from a lack of theoretical foundations and justifications (particularly, how to define the degree of inclusion?). Anyway, as our definition is based on the extension of a usual belief function (thus assuming crisp focal elements), we are not concerned with this kind of approach. The second group (Smets [19]) considers crisp focal elements, and define belief and plausibility functions of a fuzzy event by the lower and upper expectation of its membership function (recall that belief and plausibility are lower and upper probabilities), thus being consistent with the definition of the probability of a fuzzy event by Zadeh. Interestingly enough, the definition of Smets is identical to our definition (with the usual Choquet integral). Moreover, Yen [24] reaches the same conclusion

by a completely different way: he considers belief functions as the solution of an optimization problem (minimize $\sum_{x\in \tilde{A}}\sum_{B} m(x; B)\phi_{\tilde{A}}(x)$ where $m(x; B)$ is the probability mass allocated to x from the basic probability of a focal element B, subject to the constraint that m has the properties of a mass function). This shows that our definition, although on a (fuzzy) measure theoretical point of view, and not referring explicitly to the formalism of Shafer, is well rooted in the theory of Shafer.
- Höhle [6], who has proposed an integral similar to the Choquet integral, has also shown that integrals with respect to a belief (plausibility) measure exhibit order n superadditivity (subadditivity).

Finally, we want to give another interpretation of our definition using the concept of representation of a fuzzy measure (see Subsections 2.3 and 5.3). We consider a finite space X on which a probability mass function m_0 is defined, generating a belief function $\mathrm{Bel}: \mathcal{P}(X) \to [0, 1]$, where $\mathcal{P}$ is the power set of X. Let us define a mapping $H: \mathcal{P}(X) \to \mathcal{P}(\mathcal{P}(X))$ by

$$H(A) := \mathcal{P}(A), \quad \forall A \subseteq X \tag{50}$$

and a mapping $m: \mathcal{P}(\mathcal{P}(X)) \to [0, 1]$ by

$$m(\mathcal{A}) := \sum_{B \in \mathcal{A}} m_o(B), \quad \forall \mathcal{A} \subseteq \mathcal{P}(X). \tag{51}$$

Clearly, we have $\mathrm{Bel} = m \circ H$, and $(\mathcal{P}(X), \mathcal{P}(\mathcal{P}(X)), m, H)$ is a representation of $(X, \mathcal{P}(X), \mathrm{Bel})$. In other words,

$$\mathrm{Bel}(A) = m(H(A)) = m(\mathcal{P}(A)), \tag{52}$$

i.e., $\mathrm{Bel}(A)$ can be viewed as a probability measure of the crisp set $\mathcal{P}(A)$.

By analogy, for a fuzzy event $\tilde{A}$ we may rewrite Eq. (52) as

$$\mathrm{Bel}(\tilde{A}) = m(H(\tilde{A})). \tag{53}$$

$H(\tilde{A})$ can be understood as $\mathcal{P}(\tilde{A})$, i.e., the power set of a fuzzy set $\tilde{A}$. However, how can we define $H(\tilde{A})$, which is a fuzzy subset of $\mathcal{P}(X)$, i.e., a fuzzy family of subsets of X? It seems reasonable to define $H(\tilde{A})$ by

$$[H(\tilde{A})]_\alpha = H([\tilde{A}]_\alpha) \tag{54}$$

From this, the membership function of $H(\tilde{A})$ is given by

$$\phi_{H(\tilde{A})}(x) = \bigvee_{\alpha \in [0,1]} \alpha \cdot \mathbf{1}_{[H(\tilde{A})]_\alpha}(x) = \bigvee_{\alpha \in [0,1]} \alpha \cdot \mathbf{1}_{H([\tilde{A}]_\alpha)}(x) \tag{55}$$

where $H([\tilde{A}]_\alpha) = \mathcal{P}([\tilde{A}]_\alpha)$ the power set of the crisp set $[\tilde{A}]_\alpha$.

Now we show the adequacy of the definition of $H(\tilde{A})$ by applying Eq. (12). Using the above defined extension of Bel, we have

$$\mathrm{Bel}(\tilde{A}) = (C) \int \phi_{\tilde{A}} \mathrm{d}\, \mathrm{Bel}$$

and using Eq. (12) we get

$$\text{Bel}(\widetilde{A}) = \int H(\phi_{\widetilde{A}}) \, dm \tag{56}$$

with H and m defined above in Eq. (52). By Eq. (10) we have

$$H(\phi_{\widetilde{A}}) = \bigvee_{\alpha \in [0,1]} \alpha \cdot \mathbf{1}_{H((x|\phi_{\widetilde{A}}(x) \geq \alpha))} = \bigvee_{\alpha \in [0,1]} \alpha \cdot \mathbf{1}_{H([\widetilde{A}]_\alpha)}. \tag{57}$$

Therefore, $\text{Bel}(\widetilde{A}) = \int \phi_{H(\widetilde{A})} dm$. This shows that the definition in Eq. (54) is consistent. On the other hand, we have $m(H(\widetilde{A})) = \int \phi_{H(\widetilde{A})} dm$, so we can write

$$\text{Bel}(\widetilde{A}) = m(H(\widetilde{A})),$$

thus defining the belief of a fuzzy event as the expectation (in the sense of Zadeh) of its (interpreted) membership function.

7. Concluding remarks

In this paper, we have tried to find out the most general class of fuzzy integrals which possess a duality property in a broad sense. Two distinct classes were found, the first one we called Choquet-like integrals, and the second one which is in fact a singleton, namely the Sugeno integral.

Starting from these two classes, we defined the concept of fuzzy measure of fuzzy event, in a way that is consistent with previous definitions. The proposed definition ensures the existence for each (extended) fuzzy measure of a dual measure, an intuitively desirable feature if we think of all the theories of uncertainty which have reached some degree of maturity: probability theory, evidence theory, possibility theory, ...

Also we have shown that the characteristic properties of some classes of fuzzy measures, as decomposable measures, belief functions, are preserved in the extension procedure. In particular, for belief and plausibility functions, their extension by fuzzy integral has been revealed to be meaningful in the framework of the Dempster-Shafer theory.

References

[1] G. Choquet, Theory of capacities, *Ann. Inst. Fourier* **5** (1953) 131-295.
[2] D. Dubois and H. Prade, *Fuzzy Sets and Systems: Theory and Applications* (Academic Press, New York, 1980).
[3] D. Dubois and H. Prade, A class of fuzzy measures based on triangular norms, *Internat. J. General systems* **8** (1982) 43-61.
[4] D. Dubois and H. Prade, Evidence measures based on fuzzy information, *Automatica* **21** (1985) 547-562.
[5] M. Grabisch and M. Sugeno, Fuzzy integrals and dual measures-application to pattern classification, *Proc. Sino-Japan Joint Meeting on Fuzzy Sets and Systems*, Beijing (1990).
[6] U. Höhle, Integration with respect to fuzzy measures, *Proc. IFAC Symp. on Theory and Appl. of Digital Control*, New Delhi (1982) 35-37.
[7] H. Ichihashi and H. Tanaka, A fuzzy fault tree formulated by a class of fuzzy measures, *Bull. Univ. Osaka Prefecture Ser.* A **35** (2) (1986).
[8] H. Ichihashi, H. Tanaka and K. Asai, Fuzzy integrals based on pseudo-addition and multiplication, *J. Math. Anal. Appl* **130** (1988) 354-364.

[9] M. Inuiguchi, H. Ichihashi and H. Tanaka, Fuzzy linear programming using multiattribute value function, *J. Oper. Res. Soc. Japan* **31** (1) (1988) 121-141 (in Japanese).

[10] M. Ishizuka, K.S. Fu and J.T.P. Yao, A rule-based inference with fuzzy set for structural damage assessment, in: M.M. Gupta, E. Sanchez, Eds., *Approximate Reasoning in Decision Analysis* (North-Holland, Amsterdam, 1982) 261-268.

[11] A. Kandel and W.J. Byatt, Fuzzy sets, fuzzy algebra, and fuzzy statistics, *Proc. of IEEE* **66** (1978) 1619-1639.

[12] R. Kruse, Fuzzy integrals and conditional fuzzy measures, *Fuzzy Sets and System* **10** (1983) 309-313.

[13] T. Murofushi and M. Sugeno, An interpretation of fuzzy measure and the Choquet integral as an integral with respect to a fuzzy measure, *Fuzzy Sets and Systems* **29** (1989) 201-227.

[14] T. Murofushi and M. Sugeno, Fuzzy t-conorm integral with respect to fuzzy measures: Generalization of Sugeno integral and Choquet integral, *Fuzzy Sets and System* **42** (1991) 57-71.

[15] T. Murofushi and M. Sugeno, A theory of fuzzy measures. Representation, the Choquet integral and null sets, *J. Math. Anal. Appl.* **159** (2) (1991) 532-549.

[16] S. K. Pal and A. Das Gupta, A way to handle subjective uncertainties and a quantified measure of the same, *Proc. Int. Conf. on Fuzzy Logic and Neural Networks*, Iizuka, Japan (1990) 299-302.

[17] P. Smets, The degree of belief in a fuzzy event, *Inform. Sci.* **25** (1981) 1-19.

[18] M. Sugeno, - *Theory of fuzzy integrals and its applications*, Doct. Thesis, Tokyo Institute of Technology (1974).

[19] M. Sugeno and T. Murofushi, Pseudo-additive measures and integrals, *J. Math. Anal Appl* **122** (1987) 197-222.

[20] Zi Xiao Wang, On the fuzzy measures and the measures of fuzziness for L-fuzzy sets, *Proc. of the IFAC Fuzzy Information Symposium*, Marseille, France (1983) 341-346.

[21] S. Weber, $\perp$-decomposable measures and integrals for Archimedean t-conorms, *J. Math. Anal. Appl.* **101** (1984) 114-138.

[22] S. Weber, Measures of fuzzy sets and measures of fuzziness, *Fuzzy Sets and Systems* **13** (1984) 247-271.-

[23] R.R. Yager, Generalized probabilities of fuzzy events from fuzzy belief structures, *Inform. Sci.* **28** (1982) 45-62.

[24] J. Yen, Generalizing the Dempster-Shafer theory to fuzzy sets, IEEE *Trans. Systems. Man Cybernet.* **20** (1990) 559-570.

[25] L.A. Zadeh, Probability measures of fuzzy events, *J. Math. Anal. Appl.* **23** (1968) 421-427.

[26] L.A. Zadeh, Fuzzy sets as a basis for a theory of possibility, *Fuzzy Sets and Systems* **1** (1978) 3-28.

Fuzzy Measure Analysis of Public Attitude Towards The Use of Nuclear Energy[*]

T. Onisawa, M. Sugeno
Department of Systems Science,
Tokyo Institute of Technology
4259 Nagatsuta. Midori-ku.
Yokohama, 227,Japan

Y. Nishiwaki
International Atomic Energy Agency, Vienna, Austria

H. Kawai
Kinki University, Osaka, Japan

Y. Harima
Tokyo Institute of Technology, Tokyo, Japan

Received January 1985
Revised June 1985

Abstract

This paper is concerned with applying fuzzy measures and fuzzy integrals to analyze public attitude towards the use of nuclear energy. To this end, a questionnaire on the use of nuclear energy is set up and data are collected in Japan, the Philippines and the FRG. Factor analysis is performed to get the primary structure of public attitude. It is shown that the attitude of the responders to the questionnaire in each country is well explained with its hierarchical structure obtained by fuzzy measure analysis.

Keywords: fuzzy measure, fuzzy integral, fuzzy measure analysis, structure identification, overlapping coefficient, necessity coefficient, public attitude.

[*] © 1986 North-Holland. Retyped with written permission from *Fuzzy Sets and Systems*, **20** (1986) 259-289.

1. Introduction

It is important to identify the structure of public acceptance or rejection when new technologies are developed and implemented. The structure of attitudes should have the essential attributes and their interrelation. In such a structural analysis the attitudes need to be decomposed into meaningful attributes by a suitable model. Fishbein et al. have proposed an attitude model as follows [2, 3]. This model has been used at the Joint IAEA/IIASA Risk Assessment Project of International Atomic Energy Agency [5, 8]. The attitude A_0 towards an object or an event is assumed to be expressed by the summation of $(e_i \times b_i)$:

$$A_0 = \sum_{i=1}^{n} e_i \times b_i, \tag{1}$$

where e_i is the subject's evaluation of the attribute i, b_i is the strength of his belief of the attribute i about the object, and n is the number of his salient beliefs about the object.

However, the data obtained in this type of study may be more or less subjective, i.e., fuzzy, and the following problems may be pointed out:
(1) A man does not always have an additive measure such as probability to evaluate fuzzy objects.
(2) The attributes of an object in his evaluation process are not always independent of each other.

In either case a linear model such as Eq. (1) may not be applicable since it is based on the assumptions of additivity and independency.

Sugeno has proposed the concepts of fuzzy measures and fuzzy integrals and applied these concepts to model human subjectivity [6, 7]. The concept 'fuzzy measure' is considered as a generalized probability measure, which is interpreted as a subjective measure to evaluate fuzzy objects. It is not necessary to assume subjective and/or objective independency and additivity in a fuzzy integral model. Using fuzzy integrals we can build a subjective evaluation model of a fuzzy object associated with various attributes.

This paper aims to apply fuzzy measures and fuzzy integrals to the analysis of public attitude towards the use of nuclear energy. To build a model of human evaluation process, it is of crucial importance to identify its structure, i.e., to select the relevant attributes which may play an important role at the evaluation. The problem of structure identification is also discussed in applying a fuzzy integral model. The public attitude is well explained with a hierarchical structure by using factor analysis and fuzzy measure analysis.

2. Fuzzy measures and integrals [6, 7]

2.1. *Fuzzy measures*

A fuzzy measure is an extended probability measure in one sense, which has, in general, only monotonicity without additivity.

Let X be a universal set and $\mathcal{B}$ be a Borel field. Then a set function g defined on $\mathcal{B}$ with the following properties is called a fuzzy measure.
(i) $g(\emptyset) = 0, g(X) = 1$.
(ii) If $A, B \in \mathcal{B}$ and $A \subset B$, then $g(A) \leq g(B)$.
(iii) If $F_n \in \mathcal{B}$ for $1 \leq n < \infty$ and the sequence $\{F_n\}$ is monotone (in the sense of inclusion), then $\lim_{n \to \infty} g(F_n) = g(\lim_{n \to \infty} F_n)$.

The triplet $(X, \mathcal{B}, g)$ is called a fuzzy measure space, and g is called a fuzzy measure of $(X, \mathcal{B})$.

For applications, it is enough to consider the finite case. Let K be a finite set, $K = \{s_1, s_2, \cdots, s_n\}$ and $P(K)$ be a class of all the subsets of K. Then a fuzzy measure g of $(K, P(K))$ is characterized by the first two properties since the third one implies continuity. In particular, $g(\{s_i\})$ for a subset with a single element s_i is called a fuzzy density like a probability density. We denote $g^i = g(\{s_i\})$.

The fuzzy measure in this paper may be interpreted as the grade of subjective importance, which is non-additive and one attaches to an attribute or attributes an object in his evaluation process. For example suppose that the evaluation process of the *niceness* of a house with attributes such as s_1: *space*, s_2: *look*, s_3: *price*, etc. Then $g(\{s_1\})$ is the grade of subjective importance of '*space*' and $g(\{s_1, s_2\})$ is that of '*space*' and '*look*'. It is very rare that one's grade of subjective importance of '*space*' and '*look*' equals that of '*space*' added by that of '*look*'. However, it satisfies at least monotonicity: the grade of subjective importance of '*space*' and '*look*' is greater than that of '*space*'.

The concept of comparative probability exemplified by the statement "A is at least as probable as B" has been proposed [1]. In [1] the events are represented as elements of a field $\mathcal{F}$ of subsets of a set Ω. The comparative probability relation $\succsim$ is characterized by the following five axioms.

C0 (*Nontriviality*): $\Omega \succ \emptyset$, where $\emptyset$ is the null or empty set.

C1 (*Comparability*): $A \succsim B$ or $B \succsim A$.

C2 (*Transitivity*): $A \succsim B, B \succsim C \Rightarrow A \succsim C$.

C3 (*Improbability of impossibility*): $A \succsim \emptyset$.

C4 (*Disjoint unions*): $A \cap (B \cup C) = \emptyset \Leftrightarrow (B \succsim C \Leftrightarrow A \cup B \succsim A \cup C)$.

From the viewpoint of the subjective evaluation C4 is strict. The subjective evaluation of '$A \cup C$' may be often greater than that of '$A \cup B$' even if the subjective evaluation of 'B' is greater than that of 'C'. Fuzzy measure does not necessarily require C4.

2.2. Fuzzy integrals

Let h be a measurable function from X to $[0, 1]$. Then the fuzzy integral of h over A with respect to g is defined as

$$\oint_A h(x) \circ g = \sup_{\alpha \in [0,1]} \left[\alpha \wedge g(A \cap F_\alpha)\right], \tag{2}$$

where $F_\alpha = \{x \mid h(x) \geq \alpha\}$ and $\wedge$ stands for minimum. A is the domain of the fuzzy integral which is omitted if A is X.

Now let us see how to calculate a fuzzy integral. For simplicity, consider a fuzzy measure g of $(K, P(K))$ where K is a finite set previously defined.

Let h: K [0, 1] and assume without loss of generality that $h(s_1) \geq h(s_2) \geq \cdots \geq h(s_n)$. Renumber the elements of K, if not. Then we have

$$\int h(s) \circ g = \bigvee_{i=1}^{n} \left[h(s_i) \wedge g(K_i)\right], \tag{3}$$

where $K_i \triangleq \{s_1, s_2, \cdots, s_n\}$ and $\vee$ stands for maximum.

A fuzzy integral can be used as a model of subjective evaluation of fuzzy objects [7] where the attributes of an object are measured by a fuzzy measure and the characteristic function of an object is integrated with respect to a fuzzy measure. In the previous example about a house, let h: $K \to [0, 1]$ be the characteristic function of a house, i.e., the function expressing the characteristics of a house. For example, we set $h('space') = 0.9$ for a big house and $h('price') = 0.3$ for a cheap house, etc. Then the overall evaluation of a house is given by the fuzzy integral of h with respect to g, i.e., the grade of subjective importance of each attribute. As is clear from the definition of a fuzzy measure, a fuzzy measure is not only a subjective scale for guessing whether an *a priori* non-located element in X, the universe, belongs to a subset A of X, but also is concerned with such cases as the grade of subjective importance of an attribute referred to in the above from a practical point of view. The fuzzy integral model is applicable to non-linear cases, where one does not have to assume independency of an attribute from another. In the above example it is highly possible that there is a certain dependency between *'space'* and *'price'*. If it is the case, we cannot use a linear model as far as we regard *'space'* and *'price'* as the attributes of a house. We have to consider the dependency of attributes from two points of view. One is objective dependency such as between *'space'* and *'price'*, and the other is subjective dependency. Even if an attribute seems physically independent of another, one may consider that they are subjectively dependent.

3. Fuzzy measure analysis

3.1. *Fuzzy integral model*

In this paper the fuzzy measure analysis means to build a fuzzy integral model for the evaluation process of an object with various attributes. The problem concerned with the identification of the structure of a model is included in the fuzzy measure analysis. That is, we have to select relevant attributes to evaluate an object and estimate interrelation between those attributes and the object. The sort of attributes used in a model depends on both the nature of an object and the subjectivity of a human evaluator.

Let A be an object and $K = \{s_1, \cdots, s_n\}$ be a set of its attributes. By h_A: $\to K$ [0

,1], we denote the characteristic function of the object A which can be given in two ways:

(1) objectively from the physical properties of the attributes like the characteristic function of a geometrical pattern, and
(2) subjectively by an evaluator, according to his judgement.

Let g be a fuzzy measure to express the grade of subjective importance of the attributes in the overall evaluation of the object. Then the overall evaluation of A is given by

$$E_A = \int h_A(s) \circ g. \tag{4}$$

There are two cases in applying a fuzzy integral to evaluation problems.

(i) We have r objects similar to each other with the same attributes and model person's subjective evaluation process. So we have

$$E_i = \int h_i(s) \circ g, \quad 1 \leq i \leq r, \tag{5}$$

where i is the number of an object, h_i is the characteristic function of the i-th object and g is a person's subjective measure. This evaluation problem has been studied in [4, 6, 7].

(ii) We have only one object and model the evaluation process, on an average, of m persons. In this case we also have

$$E_i = \int h_i(s) \circ g, \quad 1 \leq j \leq m, \tag{6}$$

where j is the number of a person, h_j is the subjective characteristic function of the object given by the j-th person, and g is the mean fuzzy measure among persons in the sense that g minimizes a performance index, e.g., Eq. (7). This is the case considered in this paper.

In short we have many objects or many evaluators. In either case, let E_j^* be a subjective evaluation of the j-th object or that of the unique object by the j-th person. The parameter g in the model is identified so as to minimize the performance index

$$J = \left[\frac{1}{m} \sum_{j=1}^{m} (E_j^* - E_j)^2 \right]^{1/2} \tag{7}$$

Different methods have been proposed to identify g [4, 9]. For example, if a fuzzy measure is g_λ [6, 7], the identification is not so difficult since the value of a fuzzy measure of any set is calculated by fuzzy densities. However, for a general fuzzy measure, the algorithm of its identification is complicated. Suppose that we have n attributes. Then the number of the subsets of K is 2^n. We have to determine the values of a fuzzy measure for all the subsets so that they satisfy the condition of monotonicity. In this paper we use a general fuzzy measure and identify it according to the procedure given in [4]. We omit the identification algorithm in this paper.

3.2. Structure identification

Now let us discuss how to identify the structure of a model. There is no general method for structure identification as we know. Suppose that we can list all the possible attributes of an object.

Define

$$\mu_{ij} = \begin{cases} \dfrac{g(\{s_i, s_j\}) - (g^i + g^j)}{g^i \wedge g^j}, & i \neq j, \\ 0, & i = j, \end{cases} \quad (8)$$

$$m_{ij} = \begin{cases} \mu_{ij}, & \mu_{ij} \leq 0, \\ \mu_{ij}/(\mu_{ij} + 1) & \mu_{ij} > 0, \end{cases} \quad (9)$$

$$\eta_j = \sum_{i=1}^{n} m_{ij}^3 / (n-1). \quad (10)$$

$\mu_{ij} \geq -1$ is the degree of overlapping of g between the i-th and the j-th attributes. In the above if $\mu_{ij} > 0$, then g is super-additive for s_i and s_j, i.e., $g(\{s_i, s_j\}) > g^i + g^j$. This means that s_i and s_j are important since the grade of subjective importance becomes very big if s_i and s_j are joined together. It can be considered that s_i supports s_j. If $\mu_{ij} < 0$, then g is sub-additive. It can be considered that s_i casts a doubt on s_j in the evaluation of an object when both are evaluated. As an extreme case, suppose that s_i and s_j are the same attribute. In such a case we would have

$$g(\{s_i, s_j\}) > g^i \vee g^j.$$

The grade of subjective importance of $\{s_i, s_j\}$ does not increase. From this follows $\mu_{ij} = -1$.

m_{ij} is the normalized μ_{ij}. Then η_j is the value related to the mean degree of overlapping between the j-th attribute and other attributes. In order to indicate greater contribution of m_{ij} to η_j when $|m_{ij}|$ is large and to reflect the signs of m_{ij} on η_j, Eq. (10) is assumed. We call $\eta_j \in [-1, 1)$ an overlapping coefficient. If $0 \leq \eta_j < 1$, then the j-th attribute does not overlap with others on an average, i.e., this attribute is relevant to evaluate an object.

Now define

$$\xi_j = 1 + \eta_j(1 - g^j), \quad -1 \leq \eta_j < 0. \quad (11)$$

We may eliminate those attributes which may be considered irrelevant in a model by referring to the value of ξ_j, i.e., by taking account of both importance g^j and overlapping η_j. ξ_j is called a necessity coefficient expressing the degree of necessity of the j-th attribute in the structure of a model. Note that $\xi_j = 0$ if $\eta_j = -1$, i.e., complete overlapping, and if $g^j = 0$, i.e., no importance. The attribute s_j may be eliminated if the following inequality is satisfied:

$$\xi_j / \max_{j, \eta_j < 0} \xi_j < 0.7, \quad (12)$$

where s_j is not eliminated if $\eta_j \geq 0$.

In other words in case $\eta_j \geq 0$, s_j may be regarded as no overlapping on an average and therefore this attribute is considered relevant and will not be eliminated. In case $\eta_j < 0$, then Eq. (11), which takes g^j into consideration, will be used. However, a threshold 0.7 is introduced in this case so that the necessity coefficient of the relevant attribute may not have too widely different values. And, if the grade of subjective importance for s_k, a fuzzy density g^k, is very small among others, then we may also drop the k-th attribute s_k.

4. Analysis of public attitude

4.1. *Questionnaire*

We apply the method presented in this paper to the analysis of public attitude towards the use of nuclear energy. To this aim we use the data obtained from the questionnaire prepared by the International Atomic Energy Agency. The questionnaire concerned the use of nuclear energy and is shown in Appendix I where the object 'the use of nuclear energy' consists of 30 attributes such as "the use of nuclear energy" (1) "improves our standard of living", (2) "restricts personal freedom through rigorous security measures", etc. The subjects are asked to give their judgements on these attributes in a 7 bipolar scale from -3 to 3 under 3 categories '*Evaluation*', '*Belief*' and '*Importance*' where, '*Evaluation*' is labeled as bad to good, '*Belief*' as unlikely to likely, and '*Importance*' as unimportant to important. The questionnaire also includes '*Favorability*', i.e., the overall evaluation of the use of nuclear energy, which is labeled as unfavorable to favorable.

4.2. *Data for modeling*

We denote the data obtained from the questionnaire as follows.

e_{ij}: evaluation of the i-th attribute by subject j,
b_{ij}: belief of the i-th attribute by subject j,
w_{ij}: importance of the i-th attribute by subject j,
F_j: favorability of the use of nuclear energy by subject j.

Note that all the data take integers in the interval [-3, 3] as those values. Responders are divided into three groups on the basis of whether they give positive, negative or zero F: PRO GROUP, CON GROUP or NEUTRAL. Table 1 shows the details of three countries, Japan, the Philippines, and German Federal Republic. JAPAN CON GROUP and NEUTRAL GROUPS of each country are eliminated from the analysis because of their small sizes.

Cross-national surveys always run into the difficulty of finding appropriate samples which allow comparison between countries. Since attitudes on energy systems are partly determined by the level of knowledge - which differs from country to country - and there is no indication that nuclear energy is perceived in terms of identical values, it is necessary to restrict the scope of the social positions in order to create a homogeneous background. In order to avoid the creation of artifacts, students of technical and national sciences are selected as responders. Students in engineering and natural sciences all over the world have at least a basic understanding of the functions and purposes of nuclear energy [8].

From e_{ij} and b_{ij} we define the effective evaluation of attributes $e_{ij}*$ by

$$e_{ij}^* = \left[e_{ij}(b_{ij} + 0.75) + 11.25\right]/22.5, \quad 0 \le e_{ij}^* \le 1. \quad (13)$$

Table 1. Details of three countries

		Japan	Philippines	FRG
PRO GROUP	(+1)	21	27	19
	(+2)	23	20	39
	(+3)	55	8	27
(Total)		99	55	85
CON GROUP	(−1)	2	37	12
	(−2)	3	26	14
	(−3)	6	18	24
(Total)		11	81	50
NEUTRAL	(0)	6	10	14
Defect		4	28	1
Sum total		120	174	150

Eq. (13) is based on the following:
(1) If $e_{ij} = 0$ (*neutral*), then $e_{ij}^* = 0.5$ (*neutral*) in spite of b_{ij}.
(2) If $e_{ij} = -3$ (*very bad*) and $b_{ij} = +3$ (*very likely*), then $e_{ij}^* = 0$ (*worst evaluation*).
(3) If $e_{ij} = +3$ (*very good*) and $b_{ij} = +3$ (*very likely*), then $e_{ij}^* = 1$ (*best evaluation*).
(4) If $b_{ij} < 0$, then $e_{ij}^*(e_{ij} < 0) > e_{ij}^*(e_{ij} > 0)$ where $e_{ij}^*(e_{ij} \gtrless 0)$ is the effective evaluation when ($e_{ij} \gtrless 0$), respectively.
(5) If $b_{ij} = 0$, then $e_{ij}^*(e_{ij} > 0) > e_{ij}^*(e_{ij} < 0)$.
(6) $e_{ij}^* (e_{ij} = n, b_{ij} = m) \geq e_{ij}^*(e_{ij} = -n, b_{ij} = -m)$ and $e_{ij}^* (e_{ij} = n, b_{ij} = -m) \geq e_{ij}^* (e_{ij} = -n, b_{ij} = m)$, where $n, m = 1, 2, 3$ and $e_{ij}^*(e_{ij} = n, b_{ij} = m)$ is the effective evaluation when $e_{ij} = n, b_{ij} = m$.

At the moment we do not have a good idea to determine the size of the shift. In Appendix II we shall show that another size of the shift, which satisfies the above (1)-(6), results in almost the same structure. Now e_{ij}^* takes its value in [0, 1] by the normalization w_{ij} and F_j are normalized in [0, 1] as follows; +3 → 1, +2 → 5/6 , +1 → 2/3, 0 →1/2, -1 → 1/3, -2 → 1/6 and -3 → 0.

4.3. Factor analysis

Factor analysis is performed for the five groups of responders where the data e_{ij}^* are used in order to find a small number of salient factors which are latent and to estimate the global structure of the model by reducing the large number of original variables. Factor analysis is used as a pretreatment of the data. The Varimax method with orthogonal rotation is employed; the software package used is MSL009 (MELCOM COSMO-700, Mitsubishi Electric Corp.). The criterion of rotating factors is whether eigenvalues are greater than one or not.

Factor loadings and cumulative contribution rates are shown in Tables 2 through 6. Tables 7 through 11 show the aspects, their attributes and their labels. Factors with low values of factor loadings are eliminated. Less than 6 of the attributes with factor loadings greater than 0.4 are selected.

Table 2. Factor loadings and cumulative contribution rates (CCR): JAPAN PRO GROUP

	F1	F2	F3	F4	F5
1	-0.122	0.559	0.315	-0.059	0.005
2	0.078	0.216	-0.300	-0.050	0.261
3	-0.090	0.200	0.214	0.064	0.628
4	0.496	0.036	0.005	0.175	0.173
5	-0.183	0.196	-0.081	0.165	0.604
6	0.046	0.048	0.330	0.281	-0.067
7	0.182	0.597	0.177	-0.035	0.052
8	0.336	0.134	0.390	0.049	0.016
9	0.220	0.349	0.032	0.101	0.419
10	0.163	-0.032	0.055	-0.059	0.269
11	0.586	0.340	0.000	-0.368	0.212
12	0.099	0.345	-0.050	-0.138	0.507
13	0.339	-0.019	0.326	-0.439	0.304
14	0.053	-0.050	-0.116	0.026	0.371
15	0.197	0.096	0.546	-0.077	-0.183
16	0.528	-0.110	0.108	0.170	-0.065
17	-0.021	0.561	-0.072	-0.040	0.102
18	0.572	0.243	0.124	-0.378	0.100
19	0.206	0.247	-0.183	0.027	0.076
20	0.634	0.099	0.166	-0.302	0.164
21	0.472	0.099	0.014	0.006	-0.078
22	-0.058	0.327	0.178	0.392	0.204
23	-0.041	-0.062	0.461	-0.057	0.208
24	0.227	-0.005	0.418	-0.208	0.309
25	0.115	0.257	-0.050	0.249	-0.011
26	0.110	0.007	-0.044	0.527	0.084
27	0.163	0.092	0.273	0.063	-0.038
28	0.084	0.057	0.321	-0.255	-0.006
29	0.591	-0.092	0.206	0.078	0.036
30	0.021	0.450	0.035	0.249	0.115
CCR	0.135	0.211	0.255	0.297	0.331

4.4. Hierarchical structure

Starting from the results of factor analysis, let us build a fuzzy integral model. Using fuzzy densities and overlapping and/or necessity coefficients, the relevant attributes or aspects are selected and the interpretations of the structures are tried.

First we express the relation between an aspect and its attributes. An aspect is regarded as an object to be evaluated. Unfortunately, we do not have the data concerned with the overall evaluation of the aspects in the questionnaire. Using importance w_{ij}, define

$$E^*_{Xj} = \sum_{i \in X} w_{ij} e^*_{ij} \Big/ \sum_{i \in X} w_{ij}, \qquad (14)$$

where X is an aspect and the summation is taken over the attributes belonging to X. E_{Xj}^* is the weighted average of e_{ij}^* by the importance w_{ij} and thus can be regarded as the overall evaluation of an aspect X by the j-th person. The characteristic function of an aspect X is given by

$$h_{Xj}(i) = e^*_{ij}, \qquad i \in X. \qquad (15)$$

Now we can set a fuzzy integral model shown in Eq. (6) which gives the overall evaluation of an aspect. Let,

$$E_{Xj} = \int h_{Xj} \circ g_X, \qquad (16)$$

Table 3. *Factor loadings and cumulative contribution rates* (CCR): PHILIPPINE PRO GROUP

	F1	F2	F3	F4	F5	F6
1	0.045	0.156	0.171	-0.011	0.664	-0.123
2	0.069	-0.114	0.151	0.604	0.039	-0.031
3	-0.019	0.357	0.379	0.400	0.352	0.068
4	-0.349	0.012	0.340	-0.236	0.287	0.235
5	-0.392	0.628	0.235	0.066	0.163	0.032
6	0.318	-0.103	0.067	0.045	0.023	0.611
7	0.111	0.624	-0.055	0.088	0.146	-0.068
8	-0.085	-0.007	-0.118	0.425	0.568	0.058
9	0.214	0.141	-0.020	0.477	0.203	-0.044
10	0.373	0.158	-0.005	0.221	0.564	0.175
11	0.779	0.131	0.121	0.170	0.046	0.303
12	0.700	0.046	0.223	0.257	0.138	0.079
13	0.426	0.341	-0.130	0.460	0.037	0.151
14	0.482	0.263	0.005	0.246	0.220	-0.319
15	0.214	0.101	0.217	0.532	0.018	0.297
16	0.531	0.028	-0.166	0.421	-0.070	-0.193
17	0.264	0.430	0.136	-0.272	0.232	-0.113
18	0.815	-0.000	-0.023	0.010	0.066	-0.002
19	0.073	-0.044	0.761	0.013	0.038	-0.184
20	0.633	-0.153	0.144	0.014	0.284	-0.037
21	0.045	0.102	0.832	0.161	0.036	-0.016
22	0.020	0.335	0.344	0.09	0.004	0.160
23	0.314	0.406	0.015	0.023	-0.086	0.205
24	0.738	0.280	-0.032	0.007	0.044	0.190
25	0.055	0.339	0.454	-0.026	0.301	0.339
26	-0.359	0.509	0.434	0.079	0.058	-0.003
27	0.093	-0.058	0.131	-0.002	-0.054	-0.347
28	0.319	0.109	0.181	0.145	-0.300	0.408
29	0.779	-0.124	0.022	0.148	-0.068	0.016
30	0.267	0.208	0.254	0.060	0.380	0.205
CCR	0.212	0.325	0.383	0.429	0.474	0.509

Table 4. Factor loadings and cumulative contribution rates
(CCR): PHILIPPINE CON GROUP

	F1	F2	F3	F4
1	·0.030*	0.485	0.116	0.089
2	·0.240	−0.035	0.158	0.274
3	−0.108	0.758	0.235	−0.010
4	0.671	0.002	−0.159	0.015
5	−0.018	0.652	0.022	0.170
6	0.406	0.086	−0.025	0.165
7	0.075	0.186	0.395	0.064
8	0.291	0.035	0.026	−0.026
9	0.134	0.085	−0.029	0.516
10	0.511	−0.074	0.117	0.332
11	0.818	−0.133	−0.008	0.240
12	0.415	0.199	0.257	0.360
13	0.124	−0.127	−0.083	0.472
14	0.153	−0.317	0.393	0.134
15	0.134	0.168	−0.093	0.583
16	0.132	−0.045	−0.006	0.581
17	0.066	−0.004	0.671	0.006
18	0.558	−0.025	0.097	0.309
19	−0.042	0.175	0.678	−0.010
20	0.487	−0.227	0.086	0.185
21	0.039	0.438	0.028	−0.154
22	−0.087	0.555	0.141	−0.070
23	0.092	−0.020	0.380	0.550
24	0.103	−0.037	0.070	0.591
25	−0.074	0.435	−0.023	−0.014
26	0.051	0.654	−0.312	0.125
27	−0.160	−0.015	0.223	−0.041
28	0.443	−0.378	−0.020	0.201
29	0.678	−0.294	0.206	0.193
30	0.322	0.052	0.414	−0.015
CCR	0.161	0.259	0.318	0.360

and identify a fuzzy measure g_X to minimize Eq. (7) according to the procedure given in [4]. The identified g_X is interpreted as the mean grade of subjective importance of the attributes belonging to X in a group. If irrelevant attributes are selected by referring overlapping coefficients, necessity coefficients and/or fuzzy densities, we can eliminate them. Fuzzy measure analysis is then repeated for a new set of attributes. The fuzzy integral model of the evaluation of the aspect from its attributes is improved according to the above procedure.

Next the relation between the aspects and the final object is found as the following. The characteristic function of the object in a PRO GROUP is given by

$$H_j(X) = E_{Xj} \Big/ \max_X E_{Xj}, \qquad (17)$$

and that in a CON GROUP is given by

$$H_j(X) = \left(E_{Xj} - \min_X E_{Xj}\right) \Big/ \left(1 - \min_X E_{Xj}\right), \quad (18)$$

where $E_{Xj} = \int h_{Xj} \circ g_X$, g_X is the identified fuzzy measure, X is an aspect and j is a person.

Here an aspect is regarded as an attribute of the object, i.e., the use of nuclear energy.
As the overall evaluation, we have favorability F_j. Set

$$F_j = \int H_j(X) \circ g, \quad (19)$$

and identify a fuzzy measure g which is the grade of subjective importance of the aspects. Then the fuzzy integral model of the evaluation of the final object from its aspects is also improved by referring to η, ξ and/or g.

Table 5. Factor loadings and cumulative contribution rates
(CCR): FGR PRO GROUP

	F1	F2	F3	F4	F5
1	0.149	0.550	0.177	0.051	0.179
2	-0.095	-0.055	0.421	0.318	-0.060
3	0.020	0.677	-0.060	-0.044	0.262
4	0.317	0.184	0.653	-0.010	-0.092
5	0.202	0.665	0.012	0.283	0.156
6	0.278	0.034	-0.187	0.042	0.301
7	0.049	0.164	-0.005	0.626	-0.065
8	-0.023	0.019	0.715	0.176	0.093
9	0.261	-0.117	0.195	0.490	0.056
10	0.015	0.032	0.407	-0.064	0.030
11	0.180	0.118	0.597	-0.053	0.269
12	-0.104	0.665	0.242	0.054	0.068
13	0.548	-0.046	0.404	0.219	0.072
14	0.253	0.273	0.164	-0.171	0.408
15	0.076	0.135	0.372	0.281	0.076
16	0.431	0.106	0.522	-0.092	0.075
17	-0.051	0.503	-0.159	-0.038	-0.082
18	0.672	-0.093	0.239	0.170	0.259
19	-0.064	0.399	0.280	0.156	0.169
20	0.638	0.078	0.294	0.063	0.096
21	0.421	-0.153	0.135	-0.026	0.123
22	0.006	0.535	0.180	-0.140	-0.187
23	0.562	-0.002	-0.022	0.084	0.058
24	-0.029	0.159	0.090	-0.078	0.498
25	0.387	0.288	0.215	0.215	-0.153
26	0.481	0.362	-0.143	0.222	-0.200
27	-0.165	-0.109	0.057	0.045	0.032
28	0.169	-0.121	0.238	0.358	0.535
29	0.244	0.029	0.434	0.346	0.383
30	0.089	0.007	0.488	0.149	0.048
CCR	0.181	0.265	0.320	0.358	0.393

Table 6. Factor loadings and cumulative contribution rates (CCR): FGR CON GROUP

	F1	F2	F3	F4	F5
1	0.149	0.339	0.209	0.209	0.255
2	−0.048	−0.017	0.029	0.642	0.143
3	0.049	0.305	−0.012	0.187	0.644
4	0.754	0.074	−0.041	0.332	0.063
5	−0.012	0.531	0.167	0.167	0.578
6	−0.101	0.073	0.385	0.189	0.416
7	0.194	0.500	0.124	−0.165	0.109
8	0.530	−0.002	−0.244	−0.077	−0.317
9	0.415	0.244	0.149	0.207	0.298
10	0.307	0.245	0.183	0.719	0.056
11	0.650	0.035	0.005	0.410	−0.116
12	0.008	0.649	−0.102	0.006	0.192
13	0.352	0.039	0.369	0.625	0.143
14	0.313	0.040	−0.124	0.676	−0.116
15	0.343	−0.061	0.125	0.584	0.253
16	0.028	−0.018	0.699	0.037	−0.037
17	−0.207	0.374	−0.230	0.127	0.447
18	0.737	−0.012	0.264	0.088	0.029
19	−0.001	0.731	−0.063	0.162	−0.133
20	0.690	−0.061	0.023	0.067	0.192
21	0.322	0.255	0.160	0.584	−0.244
22	0.306	−0.054	0.026	0.020	0.519
23	0.233	0.176	0.177	0.506	0.194
24	0.506	0.068	0.211	0.545	0.235
25	0.008	0.469	0.171	0.443	0.083
26	−0.221	0.479	0.475	0.159	0.335
27	−0.053	0.062	−0.166	0.610	0.194
28	0.599	0.203	−0.214	0.156	−0.105
29	0.484	0.053	0.010	0.405	0.193
30	0.298	0.374	−0.036	0.349	0.167
CCR	0.259	0.369	0.425	0.473	0.510

(i) JAPAN PRO GROUP

Fuzzy -integral model of the evaluation of the aspect from its attribute

Table 12 shows the results of fuzzy measure analysis for the JAPAN PRO GROUP. We can see that the performance indices of the evaluation models of the aspects are satisfactory. All attributes except 18 concerned with the aspect A have negative overlapping coefficients and so the fuzzy measures of those attributes are sub-additive. That is, there is mutual dependency among those attributes except the attribute 18. Now let us improve the structure obtained by factor analysis. We drop the attributes 11 and 20 with the necessity coefficients 0.63 and 0.60, respectively, in the aspect A and also the attribute 12 with necessity 0.53 in the aspect D. The results for the new aspects A* and D* are shown in Table 13. We can see that for the aspect A* the performance is almost the same as before though the attributes 11 and 20 are eliminated. The overlapping coefficients of the attributes 16, 18 and 29 in A* are positive. These attributes are absolutely relevant for the aspect A*. For the aspect D*, the performance is also the same as before. Finally, we conclude that the aspects A*, B, C and D* are explained best by the attributes in Tables 12 and 13. It can be considered that the attributes in A* support one another in the evaluation of A* by the students in this group because of their positive overlapping coefficients. On the other hand, the attributes in B, C and D* do not necessarily support one another in their evaluation. These interpretations cannot be presented in factor analysis only.

Table 7. Results of factor analysis: JAPAN PRO GROUP

Aspects	Attributes	Factor loadings
Negative Impacts of Large Scale Technology A	4. is harmful to future generation	0.496
	11. exposes people to hazards which they cannot influence by any actions of their own	0.586
	16. has a long-term impact on climate	0.528
	18. leads to accidents which affect large numbers of people at the same time	0.572
	20. leads to environmental pollution	0.634
	29. involves hazardous agents which cannot be detected by man's senses	0.591
Fringe Benefits B	1. improves our standard of living	0.559
	7. helps to conserve natural resources	0.597
	17. provides a cheap energy source	0.561
	30. leads to a more even distribution of income among nations	0.450
Impact on Society C	15. involves a technology which is usable as a tool in international politics	0.546
	23. leads to consumption-oriented society	0.461
	24. concentrates power in big industrial enterprises	0.418
Economic Progress D	3. promotes my nation's industrial development	0.628
	5. leads to technological progress	0.604
	9. uses up valuable land	0.419
	12. assures the economic independence of my country	0.507

Fuzzy integral model of the evaluation of the final object from its aspects

Table 14 shows the results of fuzzy measure analysis for the evaluation of the final object. Though the performance 0.14 may seem reasonable, let us improve the model as we have just done in the evaluation of the aspects. We may drop the aspect A* with the necessity coefficient 0.41 as is seen from Table 14. Table 15 shows the new results. We can see that the model is a little bit improved since the performance is the same but the aspect A* is eliminated. The remaining aspects B, C and D* have approximately the same necessity. Figure 1 shows the structure of the attitude of the JAPAN PRO GROUP.

The grades of subjective importance of the aspects '*fringe benefits*', '*impact on society*' and '*economic progress*' are high, but their overlapping coefficients are nearly equal to -1. The Japanese students in this group seem to evaluate those aspects subjectively dependently. These interpretations cannot be obtained only through factor analysis. The aspect '*negative impacts of large scale technology*' is dropped. They seem to consider that the use of nuclear energy gives them great benefits. These results seem to reflect that 55 subjects out of 99 answer +3, i.e., the use of nuclear energy is '*very favorable*'. Hereafter only final results are shown in tables.

Table 8. Results of factor analysis: PHILIPPINE PRO GROUP

Aspects	Attributes	Factor loadings
Negative Impacts of Large Scale Technology E	11. exposes people to hazards which they cannot influence by any actions of their own	0.779
	12. assures the economic independence of my country	0.700
	18. leads to accidents which affect large numbers of people at the same time	0.815
	20. leads to environmental pollution	0.633
	24. concentrates power in big industrial enterprises	0.738
	29. involves hazardous agents which cannot be detected by mans' senses	0.779
Technical & Economic Progress F	5. leads to technological progress	0.628
	7. helps to conserve natural resources	0.624
	17. provides a cheap energy source	0.430
	23. leads to consumption-oriented society	0.406
	26. stimulates scientific and technical research	0.509
Impact on Society G	19. is a long-term solution to energy needs	0.761
	21. restricts options for future societal development	0.832
	25. leads to increased employment	0.454
Potential for Threat H.	2. restricts personal freedom through rigorous security measures	0.604
	8. provides a source of threats from terrorists	0.425
	9. uses up valuable land	0.477
	13. has an impact on people's health	0.460
	15. involves a technology which is usable as a tool in international politics	0.532

Table 9. Results of factor analysis: PHILIPPINE CON GROUP

Aspects	Attributes	Factor loadings
Negative Impacts of Large Scale Technology K	4. is harmful to future generations	0.671
	10. leads to dependency on small groups of specialists	0.511
	11. exposes people to hazards which they cannot influence by any actions of their own	0.818
	18. leads to accidents which affect large numbers of people at the same time	0.558
	29. involves hazardous agents which cannot be detected by man's senses	0.678
Progress in National Development L	3. promotes my nation's industrial development	0.755
	5. leads to technological progress	0.652
	22. increases my nation's prestige	0.535
	26. stimulates scientific and technical research	0.654
Fringe benefits M	17. provides a cheap energy source	0.671
	19. is a long-term solution to energy needs	0.678
	30. leads to a more even distribution of income among nations	0.414
Potential for Threat N	9. uses up valuable land	0.516
	13. has an impact on people's health	0.472
	15. involves a technology which is usable as a tool in international politics	0.583
	16. has a long-term impact on climate	0.581
	23. leads to consumption-oriented society	0.550
	24. concentrates power in big industrial enterprises	0.591

Table 10. Results of factor analysis: FRG PRO GROUP

Aspects	Attributes	Factor loadings
Negative Impacts of Large Scale Technology P	13. has an impact on people's health	0.548
	18. leads to accidents which affect large numbers of people at the same time	0.672
	20. leads to environmental pollution	0.638
	23. leads to consumption-oriented society	0.562
Economic Progress Q	1. improves our standard of living	0.550
	3. promotes my nation's industrial development	0.677
	5. leads to technological progress	0.665
	12. assures the economic independence of my country	0.665
	17. provides a cheap energy source	0.503
	22. increases my nation's prestige	0.535
Potential for Threat R	4. is harmful to future generations	0.653
	8. provides a source of threats from terrorists	0.715
	11. exposes people to hazards which they cannot influence by any actions of their own	0.597
	16. has a long-term impact on climate	0.522
Impact on Society S	14. postpones the development of alternative energy sources	0.408
	24. concentrates power in big industrial enterprises	0.498
	28. leads to diffusion of knowledge for construction of weapons	0.535

Table 11. Results of factor analysis: FRG CON GROUP

Aspects	Attributes	Factor loadings
Negative Impacts of Large Scale Technology T	4. is harmful to future generations	0.754
	8. provides a source of threats from terrorists	0.530
	11. exposes people to hazards which they cannot influence by any actions of their own	0.650
	18. leads to accidents which affect large numbers of people at the same time	0.737
	20. leads to environmental pollution	0.690
	28. leads to diffusion of knowledge for construction of weapons	0.599
Fringe Benefits U	7. helps to conserve natural resources	0.500
	12. assures the economic independence of my country	0.649
	19. is a long-term solution to energy needs	0.731
	25. leads to increased employment	0.469
	26. stimulates scientific and technical research	0.479
Potential for Threat V	2. restricts personal freedom through rigorous security measures	0.642
	10. leads to dependency on small groups of specialists	0.719
	13. has an impact on people's health	0.625
	14. postpones the development of alternative energy sources	0.676
	27. reduces the need to conserve energy	0.610
Progress in National Development W	3. promotes my nation's industrial development	0.644
	5. leads to technological progress	0.578
	17. provides a cheap energy source	0.447
	22. increases my nation's prestige	0.519

Table 12. Results of fuzzy measure analysis: JAPAN PRO GROUP

Aspects	Attributes	Fuzzy densities	Overlapping coefficients	Necessity coefficients	Performance indices
A	4	0.11	−0.34	0.70	0.04
	11	0.19	−0.46	0.63	
	16	0.09	−0.37	0.66	
	18	0.05	+0.27	−	
	20	0.33	−0.61	0.60	
	29	0.17	−0.09	0.93	
B	1	0.55	−0.60	0.73	0.06
	7	0.56	−0.46	0.80	
	17	0.58	−0.28	0.88	
	30	0.45	−0.53	0.71	
C	15	0.43	−0.29	0.83	0.08
	23	0.19	−0.26	0.79	
	24	0.28	−0.01	0.99	
D	3	0.40	−0.12	0.93	0.07
	5	0.50	−0.49	0.75	
	9	0.25	−0.45	0.66	
	12	0.28	−0.66	0.53	

Table 13. New aspects A^* and D^*: JAPAN PRO GROUP

Aspects	Attributes	Fuzzy densities	Overlapping coefficients	Necessity coefficients	Performance indices
A^*	4	0.18	−0.01	0.99	0.06
	16	0.16	+0.00	−	
	18	0.12	+0.09	−	
	29	0.22	+0.07	−	
D^*	3	0.55	−0.09	0.96	0.07
	5	0.50	−0.14	0.93	
	9	0.22	−0.18	0.86	

Table 14. Primary structure of evaluation of the object: JAPAN PRO GROUP

Aspects	Fuzzy densities	Overlapping coefficients	Necessity coefficients	Performance index
A^*	0.25	−0.78	0.41	0.14
B	0.86	−0.66	0.91	
C	0.50	−0.61	0.70	
D^*	0.83	−0.73	0.88	

Table 15. Final structure of evaluation of the object: JAPAN PRO GROUP

Aspects	Fuzzy densities	Overlapping coefficients	Necessity coefficients	Performance index
B	0.86	−0.98	0.86	0.14
C	0.99	−0.97	0.99	
D*	0.86	−0.98	0.86	

Table 16. Results of fuzzy measure analysis: PHILIPPINE- PRO GROUP

Aspects	Attributes	Fuzzy densities	Overlapping coefficients	Necessity coefficients	Performance indices
E	11	0.21	−0.42	0.67	0.05
	12	0.13	−0.13	0.89	
	18	0.17	+0.05	−	
	20	0.13	−0.13	0.89	
	24	0.06	−0.35	0.67	
	29	0.19	−0.24	0.81	
F	5	0.40	−0.67	0.60	0.09
	7	0.44	−0.58	0.67	
	17	0.55	−0.68	0.69	
	23	0.40	−0.68	0.59	
	26	0.28	−0.42	0.70	
G	19	0.53	−0.00	1.00	0.10
	21	0.24	+0.02	−	
	25	0.34	+0.02	−	
H*	8	0.24	+0.29	−	0.10
	13	0.18	+0.53	−	
	15	0.08	+0.38	−	

(ii) PHILIPPINE PRO GROUP

Fuzzy integral model of the evaluation of the aspect from its attributes

Table 16 shows that the relevant attributes have almost the same necessity coefficients in each aspect. The students in this group seem to evaluate the attributes in G and H* subjectively independently while they seem to evaluate the attributes in E and F subjectively dependently. The positive or negative overlapping coefficients show these interpretations.

Fuzzy integral model of the evaluation of the final object from its aspects

The grade of subjective importance of 'technical & economic progress' is high. The Philippine students in this group seem to consider that the use of nuclear energy gives them 'technical & economic progress'. However, the grade of subjective importance of 'potential for threat' is as high as that of 'technical & economic *progress*'. The necessity coefficients of both aspects are almost the same. They also seem to be rather anxious about nuclear energy. These results may reflect the following: only 8 students out of 55 answer that the use of nuclear energy is '*very favorable*'.

The structure of their attitudes shows not only their support of the use of nuclear energy but also their anxiety. They seem to evaluate the use of nuclear energy in a different way than the Japanese.

(iii) PHILIPPINE CON GROUP

Fuzzy integral model of the evaluation of the aspect from its attributes

Table 18 shows that some of the attributes obtained by factor analysis are eliminated by the fuzzy measure analysis but that the overlapping coefficients are positive. The attributes in K*, M* and N* support one another in the evaluation of these aspects by the responders in this group. These attributes are relevant for them to evaluate K*, M* and N*. However, the attributes in L do not necessarily support one another.

Table 17. Structure of evaluation of the object; PHILIPPINE PRO GROUP

Aspects	Fuzzy densities	Overlapping coefficients	Necessity coefficients	Performance index
F	0.78	−0.30	0.93	0.14
H*	0.67	−0.30	0.89	

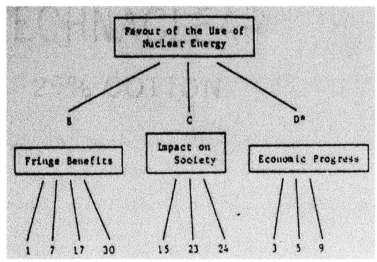

Fig. 1. Improved model of public attitude: JAPAN PRO GROUP.

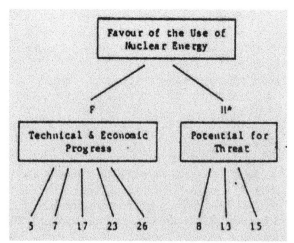

Fig. 2. Model of public attitude: PHILIPPINE PRO GROUP.

Table 18. Results of fuzzy measure analysis: PHILIPPINE CON GROUP

Aspects	Attributes	Fuzzy densities	Overlapping coefficients	Necessity coefficients	Performance indices
K*	11	0.05	+0.60	–	0.08
	18	0.05	+0.56	–	
	29	0.14	+0.55	–	
L	3	0.53	−0.53	0.61	0.08
	5	0.29	−0.00	1.00	
	22	0.37	−0.16	0.92	
	26	0.51	−0.28	0.80	
M*	17	0.40	+0.01	–	0.08
	19	0.51	+0.01	–	
N*	9	0.15	+0.04	–	0.10
	13	0.18	+0.00	–	
	15	0.09	+0.22	–	
	24	0.07	+0.24	–	

Fuzzy integral model of the evaluation of the final object from its aspects

The students in this group seem to consider that '*progress in national development*' and '*fringe benefits*' are relevant aspects for the use of nuclear energy but that these are not so important. On the other hand, '*negative impact of large scale technology*' is a relevant and important aspect for their evaluation of the object. They may consider that it is relevant to evaluate benefits such as L and M* for the use of nuclear energy. But they answer that the use of nuclear energy is unfavorable attaching greater importance to the aspect K*. These interpretations cannot be obtained only through factor analysis. The results may reflect that almost half of the

81 subjects answer -3, i.e., 'very unfavorable'. Figure 3 shows the structure of the attitude of the PHILIPPINE CON GROUP. Dotted lines mean that the fuzzy densities are low.

Table 19. Structure of the evaluation of the object: PHILIPPINE PRO GROUP

Aspects	Fuzzy densities	Overlapping coefficients	Necessity coefficients	Performance index
K*	0.17	-0.08	0.93	0.14
L	0.03	-0.34	0.67	
M*	0.03	+0.58	-	

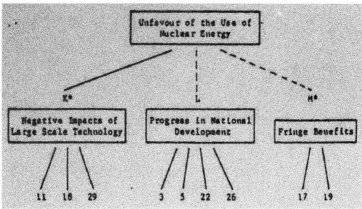

Fig. 3. Model of public attitude: PHILIPPINE CON GROUP.

(iv) FRG PRO GROUP

Fuzzy integral model of the evaluation of the aspect from its attributes

P*, R and S* are explained best by the attributes in Table 20 because the necessity coefficients of these attributes are high and/or their overlapping coefficients are positive. The responders in this group seem to evaluate the attributes in P* and S* subjectively independently. On the other hand, they seem to evaluate the attributes in Q* subjectively dependently.

Fuzzy integral model of the evaluation of the final object from its aspects

The primary results are shown in Table 21. We can see that the fuzzy density and the necessity coefficient of the aspect Q* are highest among those of all the aspects and that the necessity coefficients of the other aspects are less than 70% of that of Q*. The subjects in this group seem to consider that the use of nuclear energy is favorable since it gives them benefits such as '*economic progress*'. But the grades of subjective importance of risk considerations such as '*negative impacts of large scale* *technology*' and '*potential for threat*' are not low. That is, they may be more or less anxious about nuclear energy. These results may reflect that only 27 out of 87 responders answer +3, i.e., '*very favorable*'. This ratio is less than that in the Japanese group, 55/99. Figure 4 shows the structure of the attitude of the FRG PRO GROUP.

Table 20. *Results of fuzzy measure analysis:* FRG PRO GROUP

Aspects	Attributes	Fuzzy densities	Overlapping coefficients	Necessity coefficients	Performance indices
P*	13	0.35	−0.04	0.97	0.16
	18	0.30	+0.04	−	
	20	0.23	+0.08	−	
Q*	1	0.60	−0.52	0.79	0.06
	3	0.40	−0.66	0.61	
	5	0.57	−0.50	0.78	
	12	0.40	−0.74	0.55	
	17	0.69	−0.72	0.78	
R	4	0.10	+0.02	−	0.12
	8	0.21	−0.25	0.80	
	11	0.15	+0.19	−	
	16	0.10	−0.29	0.74	
S*	14	0.11	+0.31	−	0.14
	24	0.11	+0.31	−	

(v) FRG CON GROUP

Fuzzy integral model of the evaluation of the aspect from its attributes

Table 22 shows that some of the attributes obtained by factor analysis are eliminated by the fuzzy measure analysis but that the necessity coefficients are nearly equal to one or the overlapping coefficients are positive. T*, U*, V* and W* can be explained by the attributes in Table 22. It can be considered that the attributes in T*, U*, V* and W* support one another in the evaluation of these aspects by the students in this group.

Table 21. *Structure of evaluation of the object:* FRG PRO GROUP

Aspects	Fuzzy densities	Overlapping coefficients	Necessity coefficients	Performance index
P*	0.50	−1.00	0.50	0.12
Q*	0.83	−0.98	0.83	
R	0.50	−1.00	0.50	
S*	0.25	−0.98	0.26	

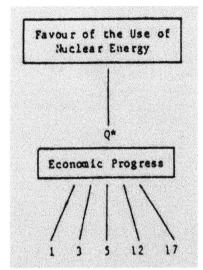

Fig. 4. Model of public attitude: FRG PRO GROUP.

Table 22. Results of fuzzy measure analysis: FPG CON GROUP

Aspects	Attributes	Fuzzy densities	Overlapping coefficients	Necessity coefficients	Performance indices
T*	11	0.06	+0.02	-	0.08
	18	0.12	+0.28	-	
	20	0.07	+0.26	-	
U*	7	0.50	−0.07	0.97	0.08
	19	0.49	−0.07	0.96	
	25	0.33	−0.02	0.99	
V*	2	0.05	+0.36	-	0.16
	10	0.05	+0.66	-	
	14	0.05	+0.32	-	
W*	3	0.16	+0.02	-	0.14
	5	0.54	−0.01	1.00	
	17	0.51	−0.01	0.99	

Fuzzy integral model of the evaluation of the final object from its aspects

Table 23 shows the results of fuzzy measure analysis of this group. It is found that the fuzzy density of T* is highest among those of the aspects in this group and its necessity coefficient is very high. The fuzzy density of V* is high but its necessity coefficient is low. On the contrary the necessity coefficient of U* is high but its fuzzy density is low. The subjects in this group seem to consider that the use of nuclear energy leads to '*negative impacts of large scale technology*' and '*potential for threat*'. They seem to consider that '*fringe benefits*' is a relevant aspect but not a very important one. This is why they answer that nuclear energy is unfavorable. Figure 5 shows the structure of the attitude of the FRG CON GROUP. Dotted lines mean that the fuzzy density or the necessity coefficient are low.

Table 23. Structure of evaluation of the object: FRG CON GROUP

Aspects	Fuzzy densities	Overlapping coefficients	Necessity coefficients	Performance index
T*	0.25	−0.23	0.83	0.15
U*	0.09	−0.14	0.87	
V*	0.17	−0.64	0.46	
W*	0.02	−0.57	0.44	

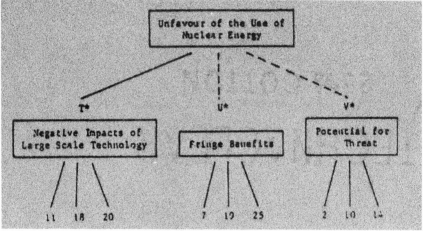

Fig. 5. Model of public attitude: FRG CON GROUP.

4.5. Comparative remarks

(i) PRO GROUPS

The subjects in the JAPAN PRO GROUP, the PHILIPPINE PRO GROUP and the FRG PRO GROUP seem to consider in common that 'economic progress' ('technical & economic progress' in the PHILIPPINE PRO GROUP) is an important aspect for the use of nuclear energy.

The subjects in the JAPAN PRO GROUP seem to consider that the use of nuclear energy leads to benefits such as 'fringe benefits' and 'economic progress'. On the contrary, the subjects in the PHILIPPINE PRO GROUP seem to consider that on one hand nuclear energy leads to 'technical & economic progress' and on the other hand it leads to risk such as 'potential for threat'. So the degree of favorability in the Philippine group is not so high as that in the Japanese group. These results may reflect the fact that there is no nuclear power plant in operation but only one under construction in the Philippines, whereas there are 24 nuclear power plants operating in Japan [8] (status in 1982).

The responders in the FRG PRO GROUP seem to consider that the use of nuclear energy gives them benefits such as *'economic progress'*. However, they are also rather anxious about nuclear energy even if they are not so anxious as the responders in the PHILIPPINE PRO GROUP. So the degree of favorability in the FRG PRO GROUP is between those of Japan and the Philippines. There are about 15 nuclear power plants operating in the FRG [8] (status in 1982). From the above it may be interpreted that the Japanese students give strong support to the use of nuclear energy; on the other hand, the Philippine students weak support and the degree of the FRG students' support is in the middle of them.

(ii) CON GROUPS

We can see that the structures of both groups, PHILIPPINE CON GROUP and FGR CON GROUP, are similar to each other. The subjects in both groups seem to have risk considerations like *'negative impacts of large scale technology'* as the reason against the use of nuclear energy. It can be seen that two attributes belong to this aspect in both structures: 11, "exposes people to hazards which they cannot influence by any actions of their own" and 18 "leads to accidents which affect large numbers of people at the same time". Attribute 19, "is a long term solution to energy needs", belongs to the aspect *'fringe benefits'* in both the structures. The Philippine and the German students seem to consider that this aspect is a relevant factor for the use of nuclear energy but not an important one. After all, it may be considered that they are not necessarily wholly against the use of nuclear energy.

5. Conclusions

This paper has suggested fuzzy measure analysis to model a human evaluation process and find its structure. The public attitude towards the use of nuclear energy has been analyzed by using data obtained from a questionnaire. It has been shown that fuzzy measure analysis enables us to improve the primary structure found by factor analysis and to get the simplified structures of pro/con groups in three countries, Japan, the Philippines, and the FRG. And further, it has also been shown that this analysis enables us to obtain interpretations which cannot be obtained through factor analysis only. It becomes easy to explain the structures of the attitudes by defining overlapping and necessity coefficients. A fuzzy integral model is very useful when additivity in the evaluation and/or independence of the attributes are not assured. The results of the presented analysis show the good performance of the fuzzy integral models.

Fuzzy measure analysis is useful for a structural analysis, i.e., selection of the relevant attributes and the interpretation of the structure.

Appendix I. Statements in questionnaire

The use of nuclear energy
1. improves our standard of living
2. restricts personal freedom through rigorous security measures
3. promotes my nation's industrial development
4. is harmful to future generations
5. leads to technological progress
6. requires management of dangerous wastes
7. helps to conserve natural resources
8. provides a source of threats from terrorists
9. uses up valuable land
10. leads to dependency on small groups of specialists
11. exposes people to hazards which they cannot influence by any actions of their own
12. assures the economic independence of my country
13. has an impact on people's health
14. postpones the development of alternative energy sources
15. involves a technology which is usable is a tool in international politics
16. has a long-term impact on climate
17. provides a cheap energy source
18. leads to accidents which affect large numbers of people at the same time
19. is a long-term solution to energy needs
20. leads to environmental pollution,
21. restricts options for future societal development
22. increases my nation's prestige
23. leads to consumption-oriented society
24. concentrates power in big industrial enterprises
25. leads to increased employment
26. stimulates scientific and technical research
27. reduces the need to conserve energy
28. leads to diffusion of knowledge for construction of weapons
29. involves hazardous agents which cannot be detected by man's senses
30. leads to a more even distribution of income among, nations

Appendix II

In this Appendix we shall show the results of factor analysis and fuzzy measure analysis using another effective evaluation. A new effective evaluation of an attribute e'_{ij} is defined by

$$e'_{ij} = [e_{ij}(b_{ij} + 0.5) + 10.5]/21, \quad 0 \leq e'_{ij} \leq 1. \tag{A.1}$$

That is, the size of the shift is 0.5 instead of 0.75. Eqs. (13) and (A.1) have the same properties as discussed in Section 4.2. Let the normalization of w_{ij} and F_j be the same as before.

The difference between the results of factor analysis using Eq. (13) and those using Eq. (A.1) is seen at two points. Tables A.1 and A.2 show this difference. In both tables, (a) shows the results using Eq. (13) and (b) those using Eq. (A.1).

(1) It seems to be appropriate that the aspect F in the PHILIPPINE PRO GROUP in Table A.1(b) is labeled as *'technical progress'* rather than *'technical & economic progress'* since attributes 17 and 23 which belong to F in Table A.1(a) do not appear in F in Table A.1(b). These attributes are concerned with *'economics'*. But attributes 5, 7, and 26 concerned with *'technical progress'* appear in common in this aspect.

In the FRG CON GROUP in Table A.2(a) factor 1 is the aspect *'negative impacts of large scale technology'* and factor 3 is *'potential for threat'*. On the other hand, the factors 1 and 3 are reversed in Table A.2(b). Nevertheless, attributes, 8, 11, 18, 20 and 28 belong in common to the aspect *'negative impacts of large scale technology'* and attributes 2, 10, 13, 14 and 27 also appear in common in the aspect *'potential for threat'*.

That is, the results of factor analysis using Eq. (13) and Eq. (A.1) are almost the same.

The final results of fuzzy measure analysis using Eq. (A.1) are now shown in Table A.3. Figures A.1 through A.5 show the comparison between (a) the results using Eq. (13) and (b) those using Eq. (A.1).

Table A.1. *Comparison between results of factor analysis using effective evaluations (13) and (A.1): PHILIPPNE PRO GROUP*

(a) Using Eq. (13)			(b) Using Eq. (A.1)		
Aspects	Attributes	Common factor loadings	Aspects	Attributes	Common factor loadings
Negative	11	0.779	Negative	11	0.758
Impacts of	12	0.700	Impacts of	12	0.672
Large Scale	18	0.815	Large Scale	18	0.816
Technology	20	0.633	Technology	20	0.629
	24	0.738		24	0.737
E	29	0.779	E	29	0.768
Technical	5	0.628	Technical	3	0.407
and Economic	7	0.624	Progress	5	0.672
Progress	17	0.430		7	0.608
	23	0.406	F	26	0.573
F	26	0.509			
Impact on	19	0.761	Impact on	19	0.776
Society.	21	0.832	Society	21	0.826
G	25	0.454	G	25	0.417
Potential	2	0.604	Potential	2	0.605
for Threat	8	0.425	for Threat	9	0.458
	9	0.477		13	0.458
H	13	0.460	H	15	0.540
	15	0.532			

Table A.2. Comparison between results of factor analysis using effective evaluations (13) and (A.1): FRG CON GROUP

(a) Using Eq. (13)			(b) Using Eq. (A.1)		
Aspects	Attributes	Common factor loadings	Aspects	Attributes	Common factor loadings
Negative	4	0.754	Potential	2	0.640
Impacts of	8	0.530	for Threat	10	0.728
Large Scale	11	0.650		13	0.629
Technology	18	0.737	T	14	0.693
	20	0.690		15	0.598
T	28	0.599		27	0.584
Fringe	7	0.500	Fringe	7	0.482
Benefits	12	0.649	Benefits	12	0.623
	19	0.731		19	0.740
U	25	0.469	U	25	0.441
	26	0.479			
Potential	2	0.642	Negative	8	0.495
for Threat	10	0.719	Impacts of	11	0.640
	13	0.625	Large Scale	18	0.729
V	14	0.676	Technology	20	0.701
	27	0.610		24	0.505
			V	28	0.565
Progress in	3	0.644	Progress in	3	0.669
National	5	0.578	National	5	0.637
Development	17	0.447	Development	17	0.516
W	22	0.519	W	22	0.470

Table A.3. Results of fuzzy measure analysis

Group	Aspects	Attributes	g	n	ξ	J
(a) JAPAN PRO	B	1, 7, 17, 30	0.85	−0.57	0.91	0.14
	D*	3, 5, 12	0.85	−0.57	0.92	
(b) PHILIPPINE PRO	F	3, 5, 7, 26	0.72	−0.47	0.87	0.16
	G	19, 21, 25	0.84	−0.47	0.93	
(c) PHILIPPINE CON	K*	11, 18, 29	0.17	−0.16	0.87	0.14
	L	3, 5, 22, 26	0.05	−0.42	0.60	
	M*	17, 19, 30	0.06	+0.43	−	
(d) FRG PRO	P*	13, 18, 20	0.79	−0.67	0.86	0.12
	Q*	1, 3, 5, 12, 17	0.83	−0.75	0.88	
	R	4, 8, 11, 16	0.67	−0.91	0.70	
(e) FRG CON	T*	2, 10, 14, 15	0.18	−0.66	0.46	0.15
	U	7, 12, 19, 25	0.03	−0.42	0.59	
	V*	11, 20, 28	0.25	−0.66	0.50	
	W*	3, 5, 17	0.01	−0.42	0.59	

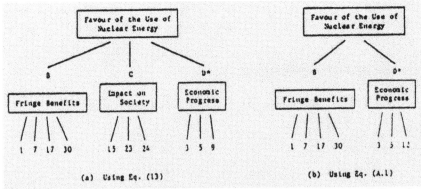

Fig. A.1. Comparison between results using effective evaluations (13) and (A.1):JAPAN PRO GROUP.

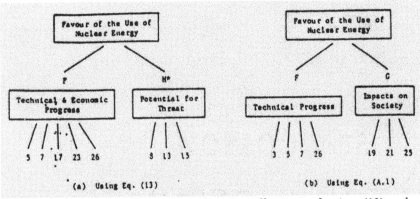

Fig. A.2. Comparison between results using effective evaluations (13) and (A.1): PHILIPPINE PRO GROUP.

(i) JAPAN PRO GROUP

The aspect C, '*impact on society*', in Figure A.1(a) is not seen in Figure A.1(b). But it is found that the two aspects '*fringe benefits*' and '*economic progress*' appear in both structures. The results in Figure A.1 show that the responders in this group are in favor of the use of nuclear energy for '*fringe benefits*' and '*economic progress*'. That is, it can be considered that the structure in Figure A.1(a) is not so different from that in Figure A.1(b).

(ii) PHILIPPINE PRO GROUP

The aspect '*impacts on society*' instead of the aspect '*potential for threat*' appears in the structure in Figure A.2(b). However, both structures have the same tendency that on one hand the responders in this group seem to consider benefits such as '*technical (& economic) progress*', and on the other hand they seem to consider

358 *Fuzzy Modeling and Control: Selected Works of M. Sugeno*

risk such as 'potential for threat' or 'impacts on society'. This tendency reflects the fact that only 8 respondents out of 55 answer 'very favorable' on the use of nuclear energy.

(iii) PHILIPPINE CON GROUP

It is found that both structures in Figure A.3 are almost the same.

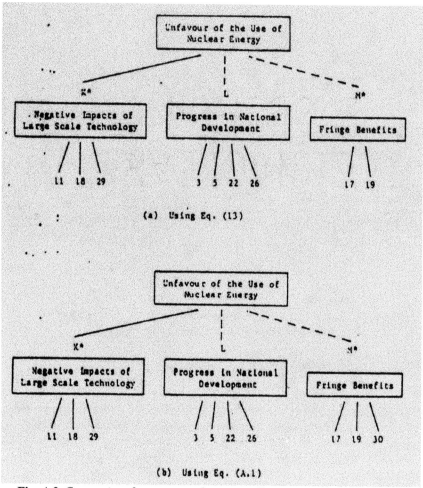

Fig. A.3. Comparison between results using effective evaluations (13) *and* (A.1): PHILIPPINE CON GROUP.

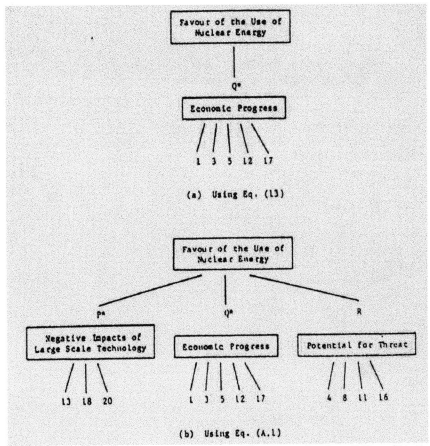

Fig A.4. *Comparison between results using effective evaluations* (13) *and* (A.1): FRG PRO GROUP.

(iv) FRG PRO GROUP

The aspects '*negative impacts of large scale technology*' and '*potential for threat*' appear in the structure in Figure A.4(b) though these aspects are not seen in Figure A.4(a). The grades of subjective importance of risk considerations such as '*negative impacts of large scale technology*' (the aspect P*) and '*potential for threat*' (the aspect R) in Table A.3 are higher than those of the risk in Table 21. Tables 21 and A.3 differ at these points. However, the grade of '*economic progress*' is highest as is shown in Tables 21 and A.3. Both structures in Figure A.4 show that the responders in this group seem to be in favor of the use of nuclear energy for '*economic progress*'.

(v) FRG CON GROUP

The aspect *'progress in national development'* appears in the structure in Figure A.5(b). But the subjective importance of this aspect is very small as is shown in Table A.3. We do not drop this aspect since its necessity coefficient is the biggest among those of the aspects in this group. It is found that except these points both results in Figure A.5 have almost the same structures.

From the above considerations it can be considered that another size of the shift in the effective evaluation, which satisfies (1) through (6) discussed in Section 4.2, affects the results only a little.

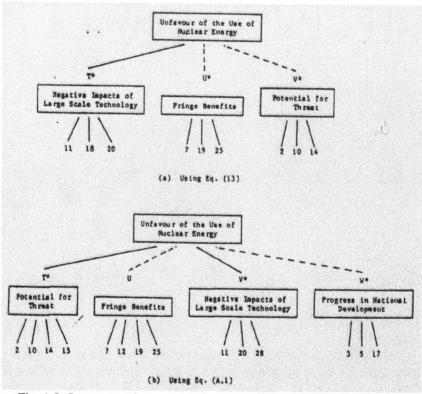

Fig. A.5. Comparison between results using effective evaluations (13) *and* (A.1): FRG CON GROUP.

References

[1]. T. L. Fine, *Theories of Probability* (Academic Press, New York, 1973).
[2]. M. Fishbein, An investigation of the relationships between beliefs about an object and the attitude toward that object, *Human Relations* 16 (1963) 233-240.
[3]. M. Fishbein and I. Ajzen, *Belief, Attitude, Intention and Behaviour, An Introduction to Theory and Research* (Addison-Wesley, Reading, MA, 1975).
[4]. K. Ishii and M Sugeno, A model of human evaluation process using fuzzy measures, *Internat. J. Man-*

Machine Stud. 22 (1985) 19-38.
[5]. F. Niehaus and E. Swaton, *Risk assessment and public acceptance on nuclear power*, Progetto Vese, Valutazione Effectti Ambientali e Socioeconomici dei Sistemi Energetici (ENEA, Rome,Dec. 1981).
[6] M. Sugeno, *Theory of fuzzy integrals and its applications*, Thesis, Tokyo Institute of Technology(1974).
[7]. M. Sugeno, *Fuzzy measures and fuzzy integrals: A survey*, in: M. M.Gupta, G.N. Saridis and B.R. Gaines, Eds., Fuzzy Automata and Decision Processes (North-Holland, Amsterdam, 1977) 89-102.
[8]. E. Swaton and 0. Renn, *Attitude towards nuclear power: A comparison between three countries*, WP.84-11, IIASA (Feb. 1984).
[9]. S.T. Wierzchon, An algorithm for identification of fuzzy measures, *Fuzzy Sets and Systems* 9(1983) 9-73.

Pseudo-Additive Measures and Integrals[*]

M. Sugeno and T. Murofushi
Department of Systems Science,
Tokyo Institute of Technology,
4259 Nagatsula, Midori-ku,
Yokohama 227, Japan

Submitted by L. Zadeh
Received May 17, 1985

Abstract

We suggest pseudo-additive measures based on a pseudo-addition and discuss integrals with respect to pseudo-additive measures. A pseudo-additive measure is a special type of fuzzy measures. To define an integral, a multiplication corresponding to a pseudo-addition is introduced. The resulting integral is an extension of the Lebesgue integral. In this context, Radon-Nikodym-like theorems are shown.

1. Introduction

The concept of fuzzy sets suggested by Zadeh [8] is a mathematical expression of sets without precise boundaries and has been applied to various problems in engineering. A fuzzy set A on X is characterized by assigning the grade of belonging to A to each element x in X.

Sugeno [6] suggested the concept of fuzzy measures: another mathematical expression of fuzziness in contrast to fuzzy sets. A fuzzy measure on X is characterized by assigning the grade of certainty of "$x \in A$" to each subset A of X, where x is an unknown element of X. Fuzzy measures are set functions with monotonicity which do not necessarily have additivity. These are more flexible than probability measures and applicable in ambiguous circumstances. Sugeno [6] further suggested the fuzzy integral which is an integral with respect to fuzzy measures. The fuzzy integral has very nice properties since it is defined only by order relation. These two concepts have also been applied for various problems (Sugeno [6], Seif and Aguilar-Martin [5], and Ishii and Sugeno [2]) so far.

We now state the definitions of a fuzzy measure and a fuzzy integral. Let $(\Omega, \mathcal{A})$ be a measurable space. A set function $\mu \; \mathcal{A}: \to [0, 1]$ is said to be a fuzzy measure iff the following holds:

(F1) $\mu(\phi) = 0, \mu(\Omega)$
(F2) A, B and $\mathcal{A} \subset B \Rightarrow \mu(A) \leq \mu(B)$,
(F3) $\{A_n\} \subset \mathcal{A}$ and $A_n \uparrow A \Rightarrow \mu(A_n) \uparrow \mu(A)$,
(F4) $\{A_n\} \subset \mathcal{A}$ and $A_n \downarrow A \Rightarrow \mu(A_n) \downarrow \mu(A)$.

The fuzzy integral of a measurable function $h: \to \Omega[0,1]$ is defined by

$$\int h(\omega) \circ \mu(\cdot) = \sup_{\alpha \in [0,1]} \min[\alpha, \mu(\{\omega \mid h(\omega) > \alpha\})].$$

By the definition, the fuzzy measure is an extension of the probability measure. The fuzzy integral is, however, not an extension of the Lebesgue integral. So the question how to define an extended Lebesgue integral with respect to the fuzzy measures arises. It seems very difficult to answer this question since the fuzzy measures have monotonicity only. But it is possible to consider an extension of the Lebesgue integral together with a class of the fuzzy measures satisfying some special conditions.

In this paper we consider the condition that, for some binary operation $\hat{+}$,

$$\mu(A \cup B) = \mu(A) \hat{+} \mu(B) \text{ whenever } A \cap B = \phi.$$

This condition is called $\hat{+}$ decomposability by Weber [7]. He defines an extended Lebesgue integral with respect to the $\hat{+}$ decomposable fuzzy measures in case that $\hat{+}$ is a continuous Archimedean t-conorm. His integral is defined by

$$\int f \hat{+} \mu = g^* \left[\int f d\bar{\mu} \right]$$

where g^* is a pseudo-inverse of an additive generator g of $\hat{+}$ and is an ordinary measure satisfying $\mu = g^* \circ \bar{\mu}$. If $\hat{+}$ has no additive generator, that is, $\hat{+}$ is non-Archimedean, then this integral cannot be defined. In this paper we shall deal with a more general case. We consider a non-Archimedean $\hat{+}$ and a multiplication corresponding to $\hat{+}$. Thus, the resulting integrals include not only the Lebesgue integrals but also Sugeno's fuzzy integrals with respect to the F-additive fuzzy measures.

We consider a set function μ on a σ-algebra $\mathcal{A}$ of sets satisfying the following conditions:

(S1) $\mu(\phi) = 0$,
(S2) $A, B \in \mathcal{A}$ and $A \subset B \Rightarrow \mu(A) \leq \mu(B)$,
(S3) $A, B \in \mathcal{A}$ and $A \cap B = \phi \Rightarrow \mu(A \cup B) = \mu(A) \hat{+} \mu(B)$,
(S4) $\{A_n\} \subset \mathcal{A}$ and $A_n \uparrow A \Rightarrow \mu(A_n) \uparrow \mu(A)$,

The condition $\mu(\Omega) = 1$ is not essential and we remove (F4), i.e., the continuity from above since (F4) does not fit (S3). The set function is not always a fuzzy measure in the sense of Sugeno.

We first define a pseudo-addition to be a binary operation $\hat{+}$ characterized by the above four conditions and show that it is represented by a family of one-place functions. Then we define a pseudo-additive measure to be a set function satisfying the above conditions if $\hat{+}$ in (S3) is a pseudo-addition. We show that, for some pseudo-additive measure, the universal set can be partitioned by the values of a pseudo-additive measure. To define an integral of a function with respect to a pseudo-additive measure, we should set the following requirement. The integral of a function

$$f(\omega) = \begin{cases} a, & \omega \in A \\ 0, & \omega \notin A \end{cases}$$

with respect to a pseudo-additive measure μ depends on only a and $\mu(A)$, that is, there is a two-place function I such that

$$\hat{\int} f \, d\mu = I(a, \mu(A)).$$

Here we denote $I(a, x)$ by $a \hat{\cdot} x$ and define the extended Lebesgue integral with respect to a pseudo-additive measure by replacing the ordinary addition + and the ordinary multiplication . with $\hat{+}$ and $\hat{\cdot}$, respectively, in the definition of the Lebesgue integral. The resulting integral has the properties similar to the Lebesgue integral. For example, a monotone convergence theorem and Radon-Nikodym-like theorems hold.

2. Pseudo-Additions

For the consistency with the conditions (S1)-(S4), it is necessary that a binary operation $\hat{+}$ satisfies the following:

(P1) $x \hat{+} 0 = 0 \hat{+} x = x$,

(P2) $(x \hat{+} y) \hat{+} z = (x \hat{+} y) \hat{+} z$,

(P3) $x \leq x'$ and $y \leq y' \Rightarrow x \hat{+} y \leq x' \hat{+} y'$,

(P'4) $x_n \uparrow x$ and $y_n \uparrow y \Rightarrow x_n \hat{+} y_n \uparrow x \hat{+} y$.

For the sake of simplicity we put a stronger condition

(P4) $x_n \to x$ and $y_n \to y \Rightarrow x_n \hat{+} y_n \to x \hat{+} y$.

In place of (P'4) (P4) is very natural for an operation $\hat{+}$ on $[0, +\infty]$. We call a binary operation $\hat{+}$ on $[0, +\infty]$ satisfying (P1)-(P4) a pseudo-addition: the ordinary addition satisfies those conditions.

A pseudo-addition can be represented by a family of one-place functions.

Definition 2. 1. Let $\{(\alpha_k, \beta_k) : k \in K\}$ be a family of disjoint open intervals in $[0, +\infty]$ indexed by a countable set K. For each $k \in K$, associate a continuous and strictly increasing function

$$g_k : [\alpha_k, \beta_k] \to [0, +\infty].$$

We say that a binary operation $\hat{+}$ has a representation

iff
$$\{\langle(\alpha_k, \beta_k), g_k\rangle : k \in K\}$$

$$x \hat{+} y = \begin{cases} g_k^*(g_k(x) + g_k(y)), & (x,y) \in [\alpha_k, \beta_k]^2 \\ \max(x, y), & \text{otherwise,} \end{cases}$$

where g_k^* is the pseudo-inverse of g_k, defined by
$$g_k^*(x) = g_k^{-1}(\min(x, g_k(\beta_k))).$$

For example, the ordinary addition + has the representation $\{\langle(0, +\infty), u\rangle\}$ where $u(x) = x$, $\forall x \in [0, +\infty]$, and the binary operation max has the representation ϕ, that is, it has no (α_k, β_k).

The next theorem follows from [3] and [4].

Theorem 2.1. *A binary operation is a pseudo-addition iff it has a representation* $\{\langle(\alpha_k, \beta_k), g_k\rangle : k \in K\}$.

As a corollary of this, we obtain that a pseudo-addition is commutative. Throughout the rest of the paper, $\hat{+}$ is used as a pseudo-addition and $\{\langle(\alpha_k, \beta_k), g_k\rangle : k \in K\}$ is a representation of $\hat{+}$. The set of all idempotent elements with respect to $\hat{+}$ is denoted by I, that is, $I = \{x : x \hat{+} x = x\}$. Obviously I is a closed set and
$$I = [0, +\infty] - \bigcup_{k \in K} (\alpha_k, \beta_k).$$

We write
$$\overset{n}{\underset{i}{\hat{\sum}}} x_i = x_1 \hat{+} \cdots \hat{+} x_n,$$

and
$$\overset{\infty}{\underset{i=1}{\hat{\sum}}} x_i = \lim_{n \to \infty} \overset{n}{\underset{i=1}{\hat{\sum}}} x_i,$$

Definition 2.2. A half open interval $(\alpha_k, \beta_k]$ is called nilpotent iff, for each $x \in (\alpha_k, \beta_k]$, there is a positive integer n such that $nx = \beta_k$, formally,
$$\overset{n}{\underset{i=1}{\hat{\sum}}} x = \beta_k.$$

It is easy to show that $(\alpha_k, \beta_k]$ is nilpotent iff $g_k(\beta_k) < +\infty$.

Obviously $\{0\} \cup (\alpha_k, \beta_k]$ is a submonoid of $([0, +\infty], \hat{+})$ for every $k \in K$. For every $k \in K$, we introduce a function
$$\bar{g}_k : \{0\} \cup (\alpha_k, \beta_k] \to [0, +\infty]$$
defined by

$$\bar{g}_k(x) = \begin{cases} g_k^*(x), & x > 0 \\ 0, & x = 0. \end{cases}$$

then the pseudo-addition on $\{0\} \cup (\alpha_k, \beta_k]$ is expressed by
$$x \mathbin{\hat{+}} y = \bar{g}_k^*(\bar{g}_k(x) + \bar{g}_k(y)),$$
where $\bar{g}_k^*$ is defined by
$$\bar{g}_k^*(x) = \begin{cases} g_k^*(x), & x > 0 \\ 0, & x = 0. \end{cases}$$

These functions $\bar{g}_k$ and $\bar{g}_k^*$ are used later.

3. Pseudo-Additive Measures

Let $(\Omega, \mathscr{A})$ be a measurable space.

Definition 3. 1. A set function $\mu: \mathscr{A} \to [0, +\infty]$ is said to be a pseudoadditive measure (with respect to $\hat{+}$) iff μ satisfies the following conditions:
(1) $\mu(\phi) = 0$,
(2) $A, B \in \mathscr{A}$ and $A \cap B = \phi \Rightarrow \mu(A \cup B) = \mu(A) \mathbin{\hat{+}} \mu(B)$,
(3) $\{A_n\} \subset \mathscr{A}$ and $A_n \uparrow A \Rightarrow \mu(A_n) \uparrow \mu(A)$,

We write a pseudo-additive measure with respect to $\hat{+}$ as a $\hat{+}$ measure for short and call the triplet $(\Omega, \mathscr{A}, \mu)$ a $\hat{+}$ measure space. Obviously the ordinary measure is the pseudo additive measure with respect to the ordinary addition $+$. In the sequel we shall write, $\sum_n, \bigcup_n, \bigcap_n$, etc. in place of, $\sum_{n=1}^{\infty}, \bigcap_{n=1}^{\infty}, \bigcap_{n=1}^{\infty}$, etc.

By the definition, a $\hat{+}$ measure μ is motone: $A, B \in \mathscr{A}$ and $A \subset B \Rightarrow \mu(A) \leq \mu(B)$. It is also σ-pseudo-additive: $\{A_n\}$ is a disjoint sequence of sets in $\mathscr{A} \Rightarrow \mu(\bigcup_n A_n) = \hat{\sum}_n \mu(A_n)$ It is easy to show that a $\hat{+}$ measure is not always continuous from above. However, the next theorem holds.

Theorem 3.1. *If μ is a $\hat{+}$ measure on $\mathscr{A}$ and if $\{A_n\}$, is a decreasing sequence of sets in $\mathscr{A}$ such that $\lim_{n \to \infty} \mu(A_n)$ is not idempotent or equal to zero, then*
$$\lim_{n \to \infty} \mu(A_n) = \mu\left(\bigcap_n A_n\right).$$

Proof. If $\lim_{n \to \infty} \mu(A_n) = 0$, then $\lim_{n \to \infty} \mu(A_n) = \mu(\bigcap_n A_n) = 0$ follows from the fact that $0 \leq \mu(\bigcap_n A_n) \leq \mu(A_n)$ for every n.

Assume that $a = \lim_{n \to \infty} \mu(A_n)$ is not idempotent. Then there is $k \in K$ such that $a \in (\alpha_k, \beta_k)$, and there exists a positive integer N such that, $\mu(A_N) \in (\alpha_k, \beta_k)$. $\{A_N - A_n\}$ is an increasing sequence and

$$\mu(A_n) = \mu(A_N) \hat{+} \mu(A_N - A_n) \quad \text{for } N \le n.$$

Thus, as $n \to \infty$,

$$\mu(A_N) = a \hat{+} \mu\left(A_N - \bigcap_n A_n\right).$$

On the other hand, obviously

$$\mu(A_N) = \mu\left(\bigcap_n A_n\right) \hat{+} \mu\left(A_N - \bigcap_n A_n\right).$$

Therefore, if $\mu(A_N - \bigcap_n A_n) < \alpha_k$, then

$$a = \mu(A_N) = \mu\left(\bigcap_n A_n\right).$$

Furthermore, if $\mu(A_N - \bigcap_n A_n) \ge \alpha_k$, then

$$a = g_k^*\left(g_k \mu(A_N)) - g_k\left(\mu\left(A_N - \bigcap_n A_n\right)\right)\right) = \mu\left(\bigcap_n A_n\right).$$

4. Decomposition Theorem

We define a concept substitute for σ-finiteness.

Definition 4.1. Let μ be a $\hat{+}$ measure on $(\Omega, \mathscr{A})$. μ is said to be σ-decomposable iff it holds that if $\mathscr{P}$ is a class of mutually disjoint non-null sets in $\mathscr{A}$, then $\mathscr{P}$ is countable.

Proposition 4.1. *σ-finiteness of an ordinary measure μ implies its σ-decomposability.*

Proof. Let $\mathscr{P}$ be a class of mutually disjoint non-null sets. Since μ is of σ-finite, there exists a sequence $\{B_n\}$ such that $\Omega = \bigcup_n B_n$, and
$$\mu(B_n) < +\infty \quad \text{for } n = 1, 2, \cdots$$
We write
$$\mathscr{P}_{n,m} = \{A \in \mathscr{P} : \mu(B_n)/2^m < \mu(A \cap B_n) \le \mu(B_n)/2^{m-1}\} \quad \text{for } n, m = 1, 2, \cdots$$

Since $\mathscr{P}_{n,m}$ has at most $2^m - 1$ elements, $\mathscr{P} = \bigcup_{n,m} \mathscr{P}_{n,m}$ is countable.

The converse of this proposition does not hold because, if $\mathscr{A} = \{\Omega, \phi\}$ and $\mu(\Omega) = +\infty$, then μ is not σ-finite but σ-decomposable. We will show the precise relation between σ-finiteness of an ordinary measure and its σ-decomposability.

For every $k \in K$ we define $\mathscr{W}_k(\mu)$ (or $\mathscr{W}_k$) to be a class of all the sets $A \in \mathscr{A}$ with the following properties:

(WK1) $B \in \mathscr{A}$ and $B \subset A \Rightarrow \mu(B) \in \{0\} \cup (\alpha_k, \beta_k]$,

(WK2) there exists a sequence $\{B_n\} \subset \mathscr{A}$, such that $A = \bigcup_n B_n$ and $\mu(B_n) < B_k$ for $n = 1, 2,....$

Similarly $\mathscr{W}_I(\mu)$ (or $\mathscr{W}_I$) is defined to be a class of all the sets $A \in \mathscr{A}$ satisfying

(WI) $B \in \mathscr{A}$ and $B \subset A \Rightarrow \mu(B) \in I$

For $A, B \in \mathscr{A}$, we denote $\mu(B - A) = 0$ by $A \subset B[\mu]$, and $\mu((B-A) \cup (A-B)) = 0$ by $A = B [\mu]$.

Let $\mathscr{C} \subset \mathscr{A}$ and $M \in \mathscr{C}$. We say that M is μ-maximal in $\mathscr{C}$ iff $C \subset M[\mu]$ for every $C \in \mathscr{C}$.

Lemma 4.2. *If μ is a σ-decomposable $\hat{+}$ measure on $\mathscr{A}$ and if $\mathscr{C}$ is a subclass of $\mathscr{A}$ satisfying the following conditions*:

(L1) $\mathscr{C} \in \phi$,
(L2) $C \in \mathscr{C}$ and $D \subset C[\mu] \Rightarrow D \in \mathscr{C}$,

(L3) $\{C_n\} \subset \mathscr{C} \Rightarrow C[\mu] \Rightarrow \bigcup_n C_n \in \mathscr{C}$
then $\mathscr{C}$ has a μ-maximal element.

Proof. Let D be a set of all classes consisting of mutually disjoint non-null sets in $\mathscr{C}$. If $D = \phi$, then any element $M \in \mathscr{C}$ is μ-maximal in $\mathscr{C}$. Let us assume $D \neq \phi$. D is inductively ordered with respect to class inclusion. By Zorn's lemma there exists a maximal element $\mathscr{D}_0$ in D. Since μ is σ-decomposable, $\mathscr{D}_0$ is countable, so if $M = \bigcup \mathscr{D}_0$, then $M \in \mathscr{C}$.

We show that M is μ-maximal in $\mathscr{C}$. Let us assume that there exists $C \in \mathscr{C}$ such that $\mu(C-M) > 0$. Since $C-M \in \mathscr{C}$, $\mathscr{D}_0 \cup \{C-M\}$ is greater than $\mathscr{D}_0$ in D. This contradicts the fact that $\mathscr{D}_0$ is maximal in D.

Theorem 4.3 (Decomposition theorem). *If μ is σ-decomposable $\hat{+}$ measure on $(\Omega, \mathscr{A})$, then the followings hold*:

(1) $\mathscr{W}_I$ and $\mathscr{W}_k$ $(k \in K)$ have μ-maximal elements $\mathscr{W}_I$ and $\mathscr{W}_k$ respectively.

(2) $W_k \cap W_{k'} = \phi[\mu]$ for $k \neq k'$, and W_I and $W_k = \phi[\mu]$ for every $k \in K$

(3) $\Omega = \bigcup_{k \in K} W_k \cup W_I [\mu]$.

Proof. (1) It is easy to show that $\mathscr{W}_I$ and $\mathscr{W}_k$ satisfy the conditions (L1)-(L3) in Lemma 4.2. Hence $\mathscr{W}_I$ and $\mathscr{W}_k$, have μ-maximal elements, W_I and W_k respectively.

Note that
$$A \in \mathscr{W}_I \Leftrightarrow A \subset W_I[\mu],$$
and
$$A \in \mathscr{W}_k \Leftrightarrow A \subset W_k[\mu], \quad \forall k \in K.$$

(2) Let $A = W_k \cap W_l$. Since $A \in \mathscr{W}_k$, there exists $\{B_n\} \subset \mathscr{A}$, such that $\cup_n B_n = A$ and $\mu(B_n) \in \{0\} \cup (\alpha_k, \beta_k)$ for $n = 1, 2, \ldots$. On the other hand, the fact that $B_n \subset A \in \mathscr{W}_l$, implies $B_n \in \mathscr{W}_l$, that is, $\mu(B_n) \in I$. Therefore, $\mu(B_n) = 0$ for $n = 1, 2, \ldots$. So we have $\mu(A) = 0$. Similarly $W_k \cap W_{k'} = \phi[\mu]$ for $k \neq k'$.

(3) Let $X = \Omega - [\cup_{k \in K} W_k \cup W_I.]$ If $X \in \mathscr{W}_I$, then $X \subset W_I[\mu]$ and so, $\mu(X) = 0$. We now assume that $X \notin \mathscr{W}_I$, then there exists a subset A of X such that $\mu(A) \in (\alpha_k, \beta_k)$ for some $k \in K$. We write $\mathscr{C} = \{B \in \mathscr{A}: B \subset A \text{ and } \mu(B) \leq \alpha_k\}$. It is easy to check that $\mathscr{C}$ satisfies (L1)-(L3) in Lemma 4.2. Let M be a μ-maximal element in $\mathscr{C}$ and let $B = A - M$. Obviously $\mu(B) = \mu(A) > 0$ and $B \in \mathscr{W}_k$. Since $B \subset X$, this contradicts the definition of X.

By this theorem, the relation between σ-finiteness of an ordinary measure and its σ-decomposability is made clear.

Proposition 4.4. *If μ is an ordinary measure on $(\Omega, \mathscr{W})$, then μ is σ-finite iff μ is σ-decomposable and $W_I = \phi[\mu]$.*

Proof. Suppose that μ is σ-finite. By Proposition 4.1, μ is σ-decomposable. Since $I = \{0, +\infty\}$ in this case, obviously $W_I = \phi[\mu]$.

Conversely suppose that μ is σ-decomposable and $W_I = \phi[\mu]$. In the representation of the ordinal addition, we have $K = \{0\}$ and $(\alpha_0, \beta_0) = (0, +\infty)$. By the theorem, we have $\Omega = W_0 \cup W_I = W_0[\mu]$ that is $\Omega \in \mathscr{W}_0$. It follows from the property (WK2) of $\mathscr{W}_0$, that μ is σ-finite.

Lastly we show a correspondence of a $\hat{+}$ measure with an ordinary measure.

Theorem 4.5. *Let $(\Omega, \mathscr{A}, \mu)$ be a $\hat{+}$ measure space. If Ω has the property (WK1) for some $k \in K$, there exists an ordinary measure $\overline{\mu}$ such that. $\mu = \overline{g}_k^* \circ \mu$. Moreover if Ω has the properties (WK1) and (WK2), i.e., $\Omega \in \mathscr{W}_k$ then $\overline{\mu}$ is σ-finite and unique.*

Proof. Let
$$\mathscr{B}_0 = \{A \in \mathscr{A} : \mu(A) < \beta_k\},$$
$$\mathscr{B} = \left\{\bigcup_n A_n \in \mathscr{B}_0, \text{ for } n = 1, 2, \cdots\right\},$$
and
$$\mathscr{C} = \mathscr{A} - \mathscr{B}.$$

If $\{A_n\}$, $\{B_m\}$ are mutually disjoint sequences of sets in $\mathscr{B}_0$ such that $\cup_n A_n = \cup_m B_m$, then

$$\sum_n \bar{g}_k^* \circ \mu(A_n) = \sum_n \bar{g}_k^* \circ \mu\left(\bigcup_m (A_n \cap B_m)\right)$$

$$= \sum_n \bar{g}_k \left[\hat{\sum_m} \mu((A_n \cap B_m))\right]$$

$$= \sum_n \bar{g}_k \circ \bar{g}_k^* \left[\sum_m \bar{g}_k \circ \mu(A_n \cap B_m)\right]$$

$$= \sum_{n,m} \bar{g}_k \circ \mu(A_n \cap B_m)$$

$$= \sum_{n,m} \bar{g}_k \circ \mu(B_m).$$

Therefore we can well define $\bar{\mu}$ by

$$\bar{\mu}(A) = \begin{cases} \sum_n \bar{g}_k \circ \mu(A_n), & \text{for } A = \bigcup_n A_n \\ & \text{where } \{A_n\} \subset \mathcal{B}_0 \text{ is mutually disjoint} \\ +\infty, & \text{for } A \in \mathcal{C} \end{cases}$$

We show that $\bar{\mu}$ is an ordinary measure on $\mathcal{A}$. Obviously $\bar{\mu}(\phi) = 0$. Let $\{A_n\}$ be a disjoint sequence of the sets in $\mathcal{A}$. If $\bigcup_n A_n \in \mathcal{C}$, then there exists a positive integer m such that $A_m \in \mathcal{C}$. Thus $\bar{\mu}(\bigcup_n A_n) = +\infty = \sum_n \bar{\mu}(A_n)$. Let us assume that $\bigcup_n A_n \in \mathcal{B}$, then there exists a disjoint sequence $\{B_m\} \subset \mathcal{B}_0$ such that $\bigcup_n A_n = \bigcup_m B_m$. Since $A_n \cap B_m \in \mathcal{B}$ for $n, m = 1, 2, \ldots$,

$$\sum_n \bar{\mu}\left(\bigcup_n A_n\right) = \bar{\mu}\left(\bigcup_m (B_m)\right)$$

$$= \sum_m \bar{g}_k \circ \mu(B_m)$$

$$= \sum_n \bar{g}_k \circ \mu\left(\bigcup_n (A_n \cap B_m)\right)$$

$$= \sum_{n,m} \bar{g}_k \circ \mu(A_n \cap B_m)$$

$$= \sum_n \bar{\mu}(A_n).$$

It follows from the definition of $\bar{\mu}$ that $\mu = \bar{g}_k^* \circ \bar{\mu}$. If $\Omega \in \mathcal{W}_k$, then $\mathcal{B} = \mathcal{A}$, so it follows that $\bar{\mu}$ is σ-finite and unique.

5. Multiplications

It is natural to assume that an integral of a function
$$f(\omega) = \begin{cases} a, & \omega \in A \\ 0, & \omega \notin A \end{cases}$$
with respect to a $\hat{+}$ measure μ depends on only a and $\mu(A)$. We introduce a binary operation $\hat{\cdot}$ called a multiplication and express the integral of f by $a \hat{\cdot} \mu(A)$. Furthermore we require that the indefinite integral with respect to a $\hat{+}$ measure is also a $\hat{+}$ measure. So we set up the following conditions for:

(M1) $a \hat{\cdot} (x \hat{+} y) = (a \hat{\cdot} x) + (a \hat{\cdot} y)$,
(M2) $a \le b \Rightarrow a \hat{\cdot} x \le b \hat{\cdot} x$,
(M3) $a \hat{\cdot} x = 0 \Leftrightarrow a = 0$ or $x = 0$,
(M4) there exists a left identity element, that is, $\exists\, e \in [0, +\infty]$, $\forall x \in [0, +\infty]$, $e \hat{\cdot} x = x$,
(M5) $0 < a < +\infty$, $a_n \to a$. and $x_n \to x \Rightarrow a_n \hat{\cdot} x_n \to a \hat{\cdot} x$, and $(+\infty) \hat{\cdot} x = \lim_{a \to +\infty} a \hat{\cdot} x$.

We call a left operation $\hat{\cdot}$ on $([0, +\infty], \hat{+})$ satisfying (M1)-(M5) a multiplication corresponding to $\hat{+}$. For example, the ordinary multiplication, denoted ., is one of the multiplications corresponding to the ordinary addition +, and both min and . are multiplications corresponding to max.

The next theorem shows the structure of a multiplication on $[\alpha_k, \beta_k]$. The proof is shown in the Appendix.

Theorem 5.1. *If $\hat{\cdot}$ is a multiplication corresponding to $\hat{+}$, then there exists a family of nondecreasing continuous functions $\{h_k: k \in K\}$ such that*
$$a \hat{\cdot} x = g_k^*(h_k(a) g k(x)) \quad \text{for } a > 0 \text{ and } x \in [\alpha_k, \beta_k]$$
$$h_k(e) = 1, \quad 0 < h_k(a) < +\infty \quad \text{for } 0 < a < +\infty,$$
and if $(\alpha_k, \beta_k]$ is nilpotent, then $h_k(a) = 1$ for $a \le e$.

We cannot characterize the structure of a multiplication on I, the set of all idempotent elements. But by the theorem if $\overline{\bigcup_{k \in K} [\alpha_k, \beta_k]} = [0, +\infty]$, then, for every idempotent element x
$$a \hat{\cdot} x = \begin{cases} 0, & a = 0 \\ x, & a > 0. \end{cases}$$
In this case the converse of the theorem holds.

Proposition 5.2. *Let e be a number in $(0, +\infty]$ and let $\{h_k : k \in K\}$ be a family of nondecreasing continuous functions on $(0, +\infty]$ such that $h_k(e) = 1$, $0 < h_k(a) < +\infty$, for $0 < a < +\infty$, and, if $(\alpha_k, \beta_k]$ is nilpotent, $h_k(a) = 1$ for $a \le e$. If*

$$a \hat{\cdot} x = \begin{cases} 0, & a = 0 \\ x, & a > 0 \text{ and } x \in I \\ g_k^*(h_k(a)g_k(x)), & a > 0 \text{ and } x \in [\alpha_k, \beta_k], \end{cases}$$

then $\hat{\cdot}$ is a multiplication corresponding to $\hat{+}$.

Proof. It is easy to check that $\hat{\cdot}$ satisfied the conditions (M1)-(M5).

If $\hat{+}$ has no nilpotent interval, there is a multiplication with good properties.

Proposition 5.3. *If $\hat{+}$ has no nilpotent interval, and if*

$$a \hat{\cdot} x = \begin{cases} 0, & a = 0 \\ x, & a > 0 \text{ and } x \in I \\ g_k^*(a \cdot g_k(x)), & a > 0 \text{ and } x \in [\alpha_k, \beta_k], \end{cases}$$

then $\hat{\cdot}$ is a multiplication corresponding to $\hat{+}$ with the following properties:

$$(a+b) \hat{\cdot} x = (a \hat{\cdot} x) \hat{+} (b \hat{\cdot} x)$$
$$(ab) \hat{\cdot} x = a \hat{\cdot} (b \hat{\cdot} x).$$

Proof. Trivial.

Throughout the rest of the paper $\hat{\cdot}$, is used as a multiplication corresponding to $\hat{+}$, and $\{h_k : k \in K\}$ is a family of functions satisfying the conditions in Theorem 5.1. We further define $h_k(0) = 0$, $\forall k \in K$ for convenience.

The next proposition is used in Sections 6 and 7.

Proposition 5.4. (1) *If $a_j \in [0, +\infty]$ and $x_j \in \{0\} \cup (\alpha_k, \beta_k]$ for $j = 1, 2, \ldots, n$, then*

$$\hat{\sum}_{j=1}^{n} a_j \hat{\cdot} x_j \in \{0\} \cup (\alpha_k, \beta_k]$$

and

$$\hat{\sum}_{j=1}^{n} a_j \hat{\cdot} x_j = \overline{g}_k^*\left(\sum_{j=1}^{n} h_k(a_j) \overline{g}_k(x_j)\right).$$

(2) *If $a_j \in [0, +\infty]$ and $x_j \in I$ for $j = 1, 2, \ldots, n$, then*

$$\hat{\sum}_{j=1}^{n} a_j \hat{\cdot} x_j \in I$$

Proof. (1) The first assertion is trivial. We have

$$\overset{n}{\underset{j=1}{\wedge}\Sigma} a_j \hat{\ast} x_j = \bar{g}_k^* \left(\sum_{j=1}^{n} \bar{g}_k \circ \bar{g}_k^*(h_k(a_j)\bar{g}_k(x_j)) \right).$$

Hence if $h_k(a_j)\bar{g}_k(x_j) < \bar{g}_k(\beta_k)$ for every j, then we obtain

$$\overset{n}{\underset{j=1}{\wedge}\Sigma} a_j \hat{\ast} x_j = \bar{g}_j^* \left(\sum_{j=1}^{n} h_k(a_j)\bar{g}_k(x_j) \right).$$

Let us assume that $h_k(a_{j0})\bar{g}_k(x_{j0}) \geq \bar{g}_k(\beta_k)$ for some j_0. Then

$$\sum_{j=1}^{n} \bar{g}_k \circ \bar{g}_k^*(h_k(a_j)\bar{g}_k(x_j)) \geq \bar{g}_k(\beta_k).$$

and

$$\sum_{j=1}^{n} h_k(a_j)\bar{g}_k(x_j) \geq \bar{g}_k(\beta_k).$$

Therefore,

$$\overset{n}{\underset{j=1}{\wedge}\Sigma} a_j \hat{\ast} x_j = \beta_k = \bar{g}_k^* \left(\sum_{j=1}^{n} h_k(a_j)\bar{g}_k(x_j) \right).$$

(2) Trivial.

6. Integration

Now we define the integrals with respect to the $\hat{+}$ measures. We develop a theory in a similar manner to the ordinary integral theory.

Let $(\Omega, \mathcal{A}, \mu)$ be a $\hat{+}$ measure space.

Definition 6. 1. Let

$$f(\omega) = \begin{cases} a_j, & \omega \in A_j \ \ j = 1, 2, \cdots, n \\ 0, & \text{otherwise,} \end{cases}$$

where $A_j \in \mathcal{A}$, $0 \leq a_j < +\infty$ for $j = 1, 2, \ldots, n$, and $A_i \cap A_j = \phi$ for $i \neq j$. We define the integral of f over $B \in \mathcal{A}$ to be

$$\overset{\wedge}{\int_B} f \, d\mu = \overset{n}{\underset{j=1}{\wedge}\Sigma} a_j \hat{\ast} \mu(A_j \cap B).$$

Let f be a nonnegative measurable function on Ω and let $\{f_n\}$ be a sequence of nonnegative measurable simple functions such that $f_n(\omega) \uparrow f(\omega)$ for every $\omega \in B$. We define the integral of f over B to be

$$\overset{\wedge}{\int_B} f \, d\mu = \lim_{n \to \infty} \overset{\wedge}{\int_B} f_n \, d\mu.$$

It is easy to show that this integral is well defined. Note that the definition depends on the choice of a multiplication.

This integral has the same properties as Lebesgue's one. We define the characteristic function of $A \in \mathscr{A}$ by χ_A:

$$\chi_A(\omega) = \begin{cases} e, & \omega \in A \\ 0, & \omega \notin A. \end{cases}$$

Proposition 6.1. Let $A, B \in \mathscr{A}$ and let f and g be nonnegative measurable functions on Ω.

(1) $f \leq g$ a.e. $\Rightarrow \hat{\int_\Omega} f \, d\mu \leq \hat{\int_\Omega} f \, d\mu$.

(2) $A \cap B = \phi \quad \hat{\int_{A \cup B}} f \, d\mu = \hat{\int_A} f \, d\mu \hat{+} \hat{\int_B} f \, d\mu$

(3) $\hat{\int_\Omega} \chi_A \, d\mu = \mu(A)$.

(4) $\hat{\int_\Omega} f \, d\mu = 0 = f \Rightarrow 0$ a.e.

(5) $\hat{\int_A} f \, d\mu = \hat{\int_{\Omega \chi A}} \hat{\cdot} f \, d\mu$

(6) *The monotone convergence theorem; if* $\{f_n\}$ *is a sequence of nonnegative measurable functions on* Ω *such that*

$$f_n(\omega) \uparrow f(\omega) \text{ a.e}$$

then

$$\hat{\int_\Omega} f \, d\mu = \lim_{n \to \infty} \hat{\int_\Omega} f_n \, d\mu.$$

(7) If $v(A) = \hat{\int_A} f \, d\mu$ for every $A \in \mathscr{A}$, then v is a $\hat{+}$ measure on $(\Omega, \mathscr{A})$.

We omit the proof. By Proposition 5.3, the next proposition holds.

Proposition 6.2. Assume that $\hat{+}$ has no nilpotent interval, and further assume

$$a \hat{\cdot} x = \begin{cases} 0, & a = 0 \\ x, & a > 0 \text{ and } x \in I \\ g_k^*(ag_k(x)), & a > 0 \text{ and } x \in [\alpha_k, \beta_k] \end{cases}$$

Let f and g be nonnegative measurable functions, and let $0 \leq a, b \leq \infty$, then

(1) $\hat{\int}(af+bg)\,d\mu = a \hat{\cdot} \hat{\int} f\,d\mu \,\hat{+}\, b \hat{\cdot} \hat{\int} g\,d\mu,$

(2) if $v(a) = \hat{\int}_A f\,d\mu$ for every $A \in \mathscr{A}$ then

$$\hat{\int} g\,dv = \hat{\int} g \hat{\cdot} f\,d\mu,$$

Last, we show a relation between an integral with respect to some $\hat{+}$ measure and a Lebesgue integral.

Theorem 6.3. *If Ω has the property (WK1) for some $k \in K$, and if $\overline{\mu}$ is an ordinary measure such that, $\mu = \overline{g}_k^* \circ \overline{\mu}$, then, for every nonnegative measurable function f on Ω,*

$$\hat{\int} f\,d\mu = \overline{g}_k^*\left[\int h_k \circ f\,d\overline{\mu}\right]$$

Proof. By the left continuity of $\overline{g}_k^*$ and the monotone convergence theorem, it is sufficient to prove the theorem for a simple function f.

If $f(\omega) = 0$ for almost all $\omega \in \Omega$, it is trivial. So let

$$f(\omega) = \begin{cases} a_j, & \omega \in A_j \text{ for } j = 1, 2, \cdots, n \\ 0, & \text{otherwise,} \end{cases}$$

where $0 < a_j < +\infty$ for $j = 1, 2, \ldots, n$, and $A_i \cap A_j = \phi$ for $i \neq j$. Then we have

$$\hat{\int} f\,d\mu = \hat{\sum}_{j=1}^{n} a_j \hat{\cdot} \mu(A_j)$$

$$= \overline{g}_k^*\left(\sum_{j=1}^{n} h_k(a_j)\overline{g}_k \circ \mu(A_j)\right)$$

and

$$\overline{g}_k^*\left(\int h_k \circ f\,d\overline{\mu}\right) = \overline{g}_k^*\left(\sum_{j=1}^{n} h_k(a_j)\overline{\mu}(A_j)\right).$$

If $\overline{g}_k \circ \mu(A_j) = \overline{\mu}(A_j)$ for every j, the theorem holds. So let us assume that

$$\overline{g}_k \circ \mu(A_{j0}) \neq \overline{\mu}(A_{j0}) \quad \text{for some } j_0$$

In this case $(\alpha_k, \beta_k]$ is nilpotent. By the proof of Theorem 4.5, we have $\overline{g}_k \circ \mu \leq \overline{\mu}$ and $\mu(A_{j0}) = \beta_k$. Thus,

$$\sum_{j=1}^{n} h_k(a_j)\overline{\mu}(A_j) \geq \sum_{j=1}^{n} h_k(a_j)\overline{g}_k \circ \mu(A_j)$$
$$\geq h_k(a_{j0})\overline{g}_k \circ \mu(A_{j0})$$
$$\geq g_k \beta_k.$$

Therefore,

$$\hat{\int} f \, d\mu = \beta_k = \overline{g}_k^* \left(\int h_k \circ f \, d\overline{\mu} \right)$$

The proof is complete.

7. Radon-Nikodym-like Theorem I

We now prove a Radon-Nikodym-like theorem for the integral defined with a certain multiplication.

By Lemma 4.2, the next lemma holds.

Lemma 7.1. *If μ and ν are $\hat{+}$ measures on $(\Omega, \mathcal{A})$, and if μ is σ-decomposable, then the family of ν-null sets, element N.*

We call N a μ-maximal ν-null set.

Throughout this section, we assume that a multiplication satisfies the following conditions:

(A1) $a \hat{\cdot} x = x$ for $a > 0$, $x \in I$,

(A2) $\lim_{a \to +0} a \hat{\cdot} \alpha = \alpha_k$ for $x \in (\alpha_k, \beta_k)$ if $(\alpha_k, \beta_k]$ is not nilpotent

and

$(+\infty) \hat{\cdot} x = \beta_k$ for $x \in (\alpha_k, \beta_k), k \in K$

Theorem 7.2 (Radon-Nikodym-like theorem I). *If μ and ν are $\hat{+}$ measures on $(\Omega, \mathcal{A})$, and if μ is σ-decomposable, then there exists a function f such that, for every $A \in \mathcal{A}$*

$$\hat{\int}_A f d\mu = \nu(A),$$

iff the following conditions are satisfied:

(1) $\mu(A) = 0 \Rightarrow \nu(A) = 0$,
(2) $A \in \mathcal{W}_k(\mu) \Rightarrow \nu(A) \in \{0\} \cup (\alpha_k, \beta_k], \forall k \in K$,
(3) $(\alpha_k, \beta_k]$ *is nilpotent and* $A \in \mathcal{W}_k(\mu) \Rightarrow \mu(A - N) \leq \nu((A - N)$
(4) $A \in \mathcal{W}_I(\mu) \Rightarrow \mu(A - N) = \nu(A - N)$,

where N is a μ-maximal ν-null set.

In the rest of this section we prove this theorem. First, we consider $\mathcal{W}_k(\mu)$.

Lemma 7.3. Let $(\alpha_k, \beta_k]$ be not nilpotent. If μ and v are $\hat{+}$ measures on $(\Omega, \mathcal{A})$, if μ is σ-decomposable, and if $\Omega \in \mathcal{W}_k(\mu)$, then there exists a function f such that, for every $A \in \mathcal{A}$

$$\hat{\int_A} f \, d\mu = v(A)$$

iff the following conditions are satisfied:
   (1) $\mu(A) = 0 \Rightarrow v(A) = 0$,
   (2) $v(A) \in \{0\} \cup (\alpha_k, \beta_k]$, for every $A \in \mathcal{A}$,

Proof. The necessity is trivial, so we prove the sufficiency. By Theorem 4.5 there exist ordinary measures $\bar{\mu}$ and $\bar{v}$ such that $\mu = \bar{g}_k^* \circ \bar{\mu}$, and $v = \bar{g}_k^* \circ \bar{v}$, and $\bar{\mu}$ is σ-finite. Since $\bar{v}$ is absolutely continuous with respect to $\bar{\mu}$ there exists a function $\bar{f}$ such that, for every $A \in \mathcal{A}$ $\int_A \bar{f} d\bar{\mu} = \bar{v}(A)$. By the assumption (A2) the range of h_k is $[0, +\infty]$, hence there is a measurable function f such that $\bar{f} = h_k \circ f$. Therefore for every $A \in \mathcal{A}$

$$\hat{\int_A} f \, d\mu = \bar{g}_k^* \left[\int_A \bar{f} \, d\bar{\mu} \right]$$
$$= \bar{g}_k^*(\bar{v}(A))$$
$$= v(A).$$

Lemma 7.4. Let $(\alpha_k, \beta_k]$ be nilpotent. If μ and v are $\hat{+}$ measures on $(\Omega, \mathcal{A})$, if μ is σ-decomposable, and if $\Omega \in \mathcal{W}_k(\mu)$, then there exists a function f such that, for every $A \in \mathcal{A}$

$$\hat{\int_A} f \, d\mu = v(A)$$

iff the following conditions are satisfied:
   (1) $\mu(A) = 0 \Rightarrow v(A) = 0$,
   (2) $v(A) \in \{0\} \cup (\alpha_k, \beta_k]$, for every $A \in \mathcal{A}$,
   (3) if N is a μ-maximal v-null set, $v(A - N) \geq \mu(A - N)$ for every $A \in \mathcal{A}$.

Proof. Let $\bar{\mu}$ and $\bar{v}$ be ordinary measures such that $\mu = \bar{g}_k^* \circ \bar{\mu}$, and $v = \bar{g}_k^* \circ \bar{v}$. First, suppose that $\hat{\int_A} f \, d\mu = v(A)$ for every $A \in \mathcal{A}$. (1) and (2) are trivial. Since $h_k(a) > 1$ for $a > 0$, and since $f(\omega) > 0$ for almost all $\omega \in \Omega - N$.

$$v(A-N) = \hat{\int}_{A\_N} f\, d\mu$$
$$= \bar{g}_k^*\left[\int_{A-N} h_k \circ f\, d\bar{\mu}\right]$$
$$\geq \bar{g}_k^*\left[\int_{A-N} d\bar{\mu}\right]$$
$$= \bar{g}_k^*(\bar{\mu}(A-N))$$
$$= \mu(A-N)$$

Conversely suppose (1)-(3). Similarly to the previous lemma we obtain a function $\bar{f}$ such that, for every $A \in \mathscr{A}$ $\int_A \bar{f} d\bar{\mu} = \bar{v}(A)$. Since $\bar{\mu}(A-N) \leq \bar{v}(A-N)$ for every $A \in \mathscr{A}$ for almost all $\omega \in \Omega - N$. And obviously $\bar{f}(\omega) = 0$ for almost all $\omega \in N$. Therefore there is a measurable function f such that for every $A \in \mathscr{A}$ there holds

$$\hat{\int}_A f d\mu = v(A)$$

Next we consider $\mathscr{W}_l(\mu)$.

Lemma 7.5. *If $A \in \mathscr{W}_l(\mu)$ and if $f(\omega) > 0$ for almost all $\omega \in A$, then*

$$\hat{\int}_A f d\mu = \mu(A)$$

Proof. By the assumption (A1), for every $a \in (0, +\infty]$,

$$\hat{\int}_A a\, d\mu = a \hat{\cdot} \mu(A) = \mu(A)$$

Hence for every positive integer n,

$$\hat{\int}_A f\, d\mu = \hat{\int}_{A\cap\{f\geq 1/n\}} f\, d\mu \,\hat{+}\, \hat{\int}_{A\cap\{f\geq 1/n\}} f\, d\mu$$
$$\geq \hat{\int}_{A\cap\{f\geq 1/n\}} 1/n\, d\mu$$
$$= \mu(A\cap\{f\geq 1/n\}),$$

so we have

$$\hat{\int}_A f\, d\mu \geq \lim_{n\to\infty}\mu(A\cap\{f\geq 1/n\}) = \mu(A).$$

On the other hand,

$$\int_A^\wedge f \, d\mu \leq \int_A^\wedge (+\infty) \, d\mu = \mu(A).$$

The proof is complete.

Lemma 7.6. *If μ and v are measures on $(\Omega, \mathscr{A})$, and if $\Omega \in \mathscr{W}_f(\mu)$, then there exists a function f such that, for every $A \in \mathscr{A}$,*

$$\int_A^\wedge f \, d\mu = v(A)$$

iff $v(A - N) = \mu(A - N)$ for every $A \in \mathscr{A}$, where N is a μ-maximal v-null set.

Proof. If $\int_A^\wedge f \, d\mu = v(A)$ for every $A \in \mathscr{A}$, then since $f(\omega) > 0$ for almost all $\omega \in \Omega - N$,

$$v(A - N) = \int_{A-N}^\wedge f \, d\mu = v(A)$$

for every $A \in \mathscr{A}$.

On the other hand, if $v(A - N) = \mu(A - N)$ for every $A \in \mathscr{A}$ and if f is a measurable function such that $f(\omega) = 0$ for $\omega \in N$, and $f(\omega) > 0$ for $\omega \in \Omega - N$, then

$$\int_A^\wedge f \, d\mu = v(A)$$

for every $A \in \mathscr{A}$.

Now Theorem 7.2 follows from these lemmas and Theorem 4.3.

8. Radon-Nikodym-Like Theorem II

If $\overline{\bigcup_{k \in K} [\alpha_k, \beta_k]} = [0, +\infty]$, then the assumption (A1) in Section 7 is satisfied. However, for example, if $\hat{+}$ is max, then (A1) means that, for every $x \in [0, +\infty]$,

$$a \hat{\cdot} x = \begin{cases} 0, & a = 0 \\ x, & a > 0 \end{cases}$$

and this multiplication is not natural. So in this section we prove a Radon-Nikodym-like theorem for max-measures. The max-measure is called F-additive in [6].

Let μ be a σ-decomposable max-measure on $(\Omega, \mathscr{A})$ and let v be a max-measure on $(\Omega, \mathscr{A})$ such that $v(A) = 0$ whenever $\mu(A)=0$. Let $a \in [0,+\infty]$. Lemma 4.2 implies that a class of all the sets $A \in \mathscr{A}$ satisfying $v(B) \geq a \hat{\cdot} \mu(B)$ for every $B \in \mathscr{A}$ and $B \subset A$ has a μ-maximal set, and we denote it by $[v/\mu](a)$.

Lemma 8.1. *If $0 \leq a \leq b \leq +\infty$, and if $A \subset [v/\mu](a) - [v/\mu](b)$ and $A \in \mathscr{A}$, then $a \hat{\cdot} \mu(A) \leq v(A) \leq b \hat{\cdot} \mu(A)$.*

Proof. Let $A \subset [v/\mu](a) - [v/\mu](b)$ It is sufficient to prove that $v(A) \leq b \hat{\cdot} \mu(A)$. Let $\mathscr{C}$ be a class of all those sets $C \in \mathscr{A}$ that $C \subset A$ and $v(C) \leq b \hat{\cdot} \mu(C)$. Similarly to the proof of Lemma 4.2, we obtain a set $M \in \mathscr{C}$ such that, if $C \in \mathscr{C}$ and $C \cap M = \emptyset$, then $\mu(C) = 0$. Therefore we have that $A - M \subset [v/\mu](b)[\mu]$. By the definition of A, $\mu(A - M) = 0$. Hence $v(A) \leq b \hat{\cdot} \mu(A)$.

We say that a number x is multiplicatively finite, m-finite for short, if $\lim_{a \to +0} a \hat{\cdot} x = 0$. A measurable set A is called m-finite (with respect to μ) if $\mu(A)$ is m-finite, and a max-measure μ on $(\Omega, \mathscr{A})$ is called σ-m-finite if Ω is a countable union of m-finite sets.

Theorem 8.2 (Radon-Nikodym-like theorem II). *If μ is a σ-m-finite σ-decomposable max-measure on $(\Omega, \mathscr{A})$ and if v is a max-measure on $(\Omega, \mathscr{A})$, then there exists a function f such that*

$$\overset{\wedge}{\int_A} f \, d\mu = v(A) \text{ for every } A \in \mathscr{A}$$

iff for every m-finite set A with respect to μ
$$v(A) \leq (+\infty) \hat{\cdot} \mu(A).$$

Moreover if every positive m-finite number is right reducible (i.e., if x is positive m-finite and if $a \hat{\cdot} x = b \hat{\cdot} x$ then $a = b$), the function f is unique in the sense of a.e.

Proof. If $A \in \mathscr{A}$ and if $\overset{\wedge}{\int_A} f \, d\mu = v(A)$, then $v(A) \leq \overset{\wedge}{\int_A}(+\infty) \, d\mu = (+\infty) \hat{\cdot} \mu(A)$.

Conversely suppose that $v(A) \leq (+\infty) \hat{\cdot} \mu(A)$ for every m-finite set A. We may assume that Ω is m-finite. We write
$$H_n^j = [v/\mu](j/2^n) - [v/\mu]((j+1)/2^n) \text{ for } n, j = 0, 1, 2, \cdots,$$
and
$$H_\infty = [v/\mu](+\infty) \left(= \Omega - \bigcup_j H_n^j[\mu] \right).$$

We define for $n = 0, 1, 2, \ldots$,
$$f_n(\omega) = \begin{cases} j/2^n, & \omega \in H_n^j \text{ for } j = 0, 1, 2, \cdots \\ +\infty, & \omega \in H_\infty \end{cases}$$
and
$$f(\omega) = \lim_{n \to \infty} f_n(\omega)$$
We prove that, for every $A \in \mathscr{A}$.
Let $A \in \mathscr{A}$

$$\hat{\int_A} f\, d\mu = \lim_{n\to\infty} \hat{\int_A} f_n\, d\mu b$$

$$= \lim_{n\to\infty}\left[\hat{\int_{A\cap H_\infty}} f_n\, d\mu \;\hat{+}\; \sum_j \hat{\int_{A\cap H_n^j}} f_n\, d\mu\right]$$

$$= \lim_{n\to\infty}\left[(+\infty)\stackrel{\cdot}{\cdot}\mu(A\cap H_\infty) \;\hat{+}\; \sum_j (j/2^n)\stackrel{\cdot}{\cdot}\mu(A\cap H_n^j)\right]$$

$$= \lim_{n\to\infty}\left[v(A\cap H_\infty) \;\hat{+}\; \sum_j v(A\cap H_n^j)\right]$$

$$= v(A).$$

We show that $\hat{\int_A} f\, d\mu \geq v(A)$.

$$v(a) = v(A\cap H_\infty) \;\hat{+}\; \sum_j v(A\cap H_0^j)$$

$$= \max\{v(A\cap H_\infty), \sup_j v(A\cap H_0^j)\}.$$

Hence, if $v(A) = v(A\cap H_\infty)$, then

$$\hat{\int_A} f\, d\mu \geq \int_{A\cap H_\infty} f\, d\mu$$

$$= (+\infty)\stackrel{\cdot}{\cdot} \mu(A\cap H_\infty)$$

$$= v(A\cap H_\infty)$$

$$= v(A).$$

So let us assume that. $v(A) > v(A\cap H_\infty)$ and $v(A) > c$ There is an integer j_0, such that $v(A\cap H_0^{j_0}) > c$. Since

$$A\cap H_n^j = (A\cap H_{n+1}^{2j}) \cup (A\cap H_{n+1}^{2j+1})[\mu]$$

and

$$v(A\cap H_n^j) = \max\{v(A\cap H_{n+1}^{2j}), v(A\cap H_{n+1}^{2j+1})\}$$

we obtain a sequence $\{j_n\}$, satisfying

$$A\cap H_m^{j_m} \subset A\cap H_n^{j_n}[\mu] \quad \text{for } m \geq n$$

and

$$v(A\cap H_n^{j_n}) = v(A\cap H_0^{j_0}) \quad \text{for } n = 1, 2, \cdots$$

Let $x_n = \mu(A \cap H^{jn}_n)$, $a_n = j_n/2^n$ and $b_n = (j_n + 1)/2^n$ for $n = 1, 2, \ldots$. By Lemma 8.1, $a_n \hat{\cdot} x_n \leq v(A \cap H^{jn}_n) \leq b_n \hat{\cdot} x_n$. The sequence $\{x_n\}$ of nonnegative numbers converges since it is nonincreasing. Obviously $\{a_n\}$ and $\{b_n\}$ converge to the same number. And x_n is m-finite for $n = 1, 2, \ldots$. It follows from these facts that $\lim_{n \to \infty} a_n \hat{\cdot} x_n = v(A \cap H^{j0}_0)$ and, hence, there is an integer m such that $(j_m/2^m) \hat{\cdot} \mu(A \cap H^{jm}_m) > c$. Therefore,

$$\hat{\int_A} f \, d\mu \geq A \hat{\int_A} f_m \, d\mu$$

$$\geq (j_m/2^m) \hat{\cdot} \mu(A \cap H^{jm}_m)$$

$$> c$$

and we have that $\hat{\int_A} f \, d\mu \geq v(A)$.

We now prove the second assertion. Suppose that $\hat{\int_A} f \, d\mu = \hat{\int_A} g \, d\mu$ for every $A \in \mathcal{A}$. Let P be a set of all pair (r, s) of nonnegative rational numbers such that $r > s$, and let

$$A_{r,s} = \{\omega : f(\omega) > r > s > g(\omega)\}$$

for every $(r, s) \in P$. Since

$$\hat{\int_{A_{r,s}}} f \, d\mu \geq \hat{\int_{A_{r,s}}} r \, d\mu = r \hat{\cdot} \mu(A_{r,s}),$$

and

$$\hat{\int_{A_{r,s}}} g \, d\mu \geq \hat{\int_{A_{r,s}}} s \, d\mu = s \hat{\cdot} \mu(A_{r,s}),$$

it follows from the assumption that $\mu(A_{r,s}) = 0$. Therefore, $\mu(\{\omega : (f(\omega) > g(\omega)\}) = 0$, and similarly $\mu(\{\omega : (g(\omega) > f(\omega)\}) = 0$, hence $f = g$ a.e.

If $\hat{\cdot}$ is min, the integral with respect to a max-measure is the fuzzy integral in the sense of Sugeno [6]:

$$\hat{\int_A} f \, d\mu = \int_A f \circ \mu = \sup_{\alpha \in [0, +\infty]} \min\{\alpha, \mu(\{\omega : f(\omega) > \alpha\})\}.$$

In this case every number in $[0, +\infty]$ is m-finite. We have:

Corollary 8.3. *Let a multiplication $\hat{\cdot}$ be min. If μ it is a σ-decomposable max-measure on $(\Omega, \mathcal{A})$ and if v is a max-measure on $(\Omega, \mathcal{A})$, then there exists a function f such that*

$$\hat{\int_A} f \, d\mu = v(A) \quad \text{for every } A \in \mathcal{A}$$

iff $v(A) \leq \mu(A)$ for every $A \in \mathcal{A}$.

If a multiplication is the ordinary one, then m-infiniteness is equivalent to finiteness.

Corollary 8.4. *Let a multiplication $\hat{*}$ be a ordinary one. If μ is σ-decomposable σ-finite max-measure on $(\Omega, \mathscr{A})$ and if v is a max-measure on $(\Omega, \mathscr{A})$, there exists a function f such that*

$$\overset{\wedge}{\int_A} f \, d\mu = v(A) \quad \text{for every } A \in \mathscr{A}$$

iff $v(A) = 0$ whenever $\mu(A) = 0$. Moreover the function f is a.e. unique.

References

[1] J. ACZEL, "Lecture on Functional Equations and Their Applications," Academic Press, New York, 1966.
[2] K. ISHII and M. SUGENO, A model of human evaluation process using fuzzy measure, *Internat. J. Man-Machine Stud.* **22** (1985), 19-38.
[3] C. H. LING, Representation of associative functions, *Publ Math. Debrecen* **12** (1965), 189-212.
[4] P. S. MOSTERT AND A. L. SHIELDS, On the structure of semigroups on a compact manifold with boundary, *Ann. of Math.* **65** (1957), 117-143.
[5] A. SEIF AND J. AGUILAR-MARTIN, Multi-group classification using fuzzy correlation, *Fuzzy Sets and Systems* **3** (1980), 109-122.
[6] M. SUGENO, "Theory of Fuzzy Integrals and Its Applications," Doctoral thesis, Tokyo Inst. of Tech., Tokyo, 1974.
[7] S. WEBER, $\perp$-Decomposable measures and integrals for Archimedean t-conorms $\perp$, *J. Math. Anal. Appl* **101** (1984), 114-138.
[8] L. A. ZADEH, Fuzzy sets, *Inform. and Control* **8** (1965), 338-353

Appendix

Here we prove Theorem 5.1. First, we prove a sequence of lemmas. Assume that a composition $(a, x) \mapsto a \hat{*} x$ has the properties (M1)-(M5).

Lemma A.1. *If x is idempotent, then so is $a \hat{*} x$.*

Proof. It follows from (M1) that
$$(a \hat{*} x) \hat{+} (a \hat{*} x) = a \hat{*} (x \hat{+} x) = a \hat{*} x.$$

Lemma A.2. $a \hat{*} \alpha_k = \alpha_k$ *for every $a \geq e$ where e is a left identity.*

Proof. If $e = +\infty$, it is trivial. So assume that $e < +\infty$, and assume that there exists a number $a \in (e, +\infty)$ such that $a \hat{*} \alpha_k \neq \alpha_k$. By (M2) we have that $a \hat{*} \alpha_k > \alpha_k = e \hat{*} \alpha_k$. Then it follows from (M5) that there exists a number a_0, such that $e < a_0 < a$ and $\alpha_k < a_0 \hat{*} \alpha_k < \beta_k$. This contradicts Lemma A.1. Therefore, $a \hat{*} \alpha_k = \alpha_k$ for $e \leq a < +\infty$, and $(+\infty) \hat{*} \alpha_k = \lim_{a \to +\infty} a \hat{*} \alpha_k = \alpha_k$

Lemma A.3. $a \hat{\ } \beta_k = \beta_k$ for $0 < a \le e$.

Proof. Similar to Lemma A.2.

Lemma A.4. (1) $\hat{\sum}_{j=1}^n x = g_k^*(ng_k(x))$, $\forall x \in (\alpha_k, \beta_k]$.

(2) $\hat{\sum}_j x = \beta_k$, $\forall x \in (\alpha_k, \beta_k)$.

(3) $a \hat{\ } (\hat{\sum}_j x) = \hat{\sum}_j a \hat{\ } x$, $\forall a, x \in [0, +\infty]$.

Proof. Trivial.

Lemma A.5. $a \hat{\ } \beta_k = \beta_k$ for $a \ge e$.

Proof. Let $e \le a < +\infty$. Since $a \hat{\ } \beta_k \ge e \hat{\ } \beta_k = \beta_k$ and $a \hat{\ } \alpha_k = \alpha_k$ there exists a number $x \in (\alpha_k, \beta_k]$ such that $a \hat{\ } x = \beta_k$. Hence

$$a \hat{\ } \beta_k = a \hat{\ } \left(\hat{\sum}_j x \right) = \left(\hat{\sum}_j a \hat{\ } x \right) = \left(\hat{\sum}_j \beta_k \right) = \beta_k$$

and

$$(+\infty) \hat{\ } \beta_k = \lim_{a \to +\infty} a \hat{\ } \beta_k = \beta_k$$

Lemma A.6. $a \hat{\ } \alpha_k = \alpha_k$ for $0 < a \le e$.

Proof. Let us assume that there exists a number a such that $0 < a \le e$ and $a \hat{\ } \alpha_k \ne \alpha_k$. The monotonicity of implies that $a \hat{\ } \alpha_k < \alpha_k$. Since $a \hat{\ } \beta_k = \beta_k$ there exists a number $x \in (\alpha_k, \beta_k]$ such that $a \hat{\ } x = \alpha_k$. Hence,

$$a \hat{\ } \beta_k = a \hat{\ } \left(\hat{\sum}_j x \right) = \left(\hat{\sum}_j a \hat{\ } x \right) = \left(\hat{\sum}_j \alpha_k \right) = \alpha_k$$

This contradicts the fact that $a \hat{\ } \beta_k = \beta_k$.

Proof of Theorem 5.1. By previous lemmas we obtain the fact that if $a \in (0, +\infty)$ and $x \in (\alpha_k, \beta_k]$ then $a \hat{\ } x \in [\alpha_k, \beta_k]$.

We define a function $f_k : (0, +\infty) \to (\alpha_k, \beta_k]$ by:

$$f_k(a) = \min\{x : a \hat{\ } x = \beta_k\}.$$

Let $a \in (0, +\infty)$, $x, y \in [\alpha_k, \beta_k]$ and $g_k(x) + g_k(y) < g_k(f_k(a))$. We obtain that
$$g_k(x) + g_k(y) < g_k(f_k(a))$$
$$\Rightarrow x \hat{+} y < f_k(a)$$
$$\Rightarrow a \hat{\cdot} (x \hat{+} y) < \beta_k$$
$$\Rightarrow (a \hat{\cdot} x) \hat{+} (a \hat{\cdot} y) < \beta_k$$
$$\Rightarrow g_k(a \hat{\cdot} x) + g_k(a \hat{\cdot} y) < g_k(\beta_k)$$
Therefore it follows from (M1) that
$$g_k(a \hat{\cdot} g_k^*(g_k(x) + g_k(y))) = g_k(a \hat{\cdot} x) + g_k(a \hat{\cdot} y).$$
We introduce into this equation the notations
$$u = g_k(x), \quad v = g_k(y), \quad E_a(s) = g_k(a \hat{\cdot} g_k^*(s)).$$
We have
$$E_a(u + v) = E_a(u) + E_a(v).$$
Since the continuity of g_k and (M5) imply the continuity of E_a, there exists a function h_k such that
$$E_a(u) = h_k(a) u \quad \text{for } 0 \leq u < g_k(f_k(a)).$$
so we have that
$$a \hat{\cdot} x = g_k^*(h_k(a) g_k(x)) \quad \text{for } \alpha_k \leq x < f_k(a).$$
By (M5) and the continuity and monotonicity of g_k, the above equation holds for $\alpha_k \leq x \leq \beta_k$ (M2) and (M5) imply that h_k is continuous and nondecreasing. So we define $h_k(+\infty) = \lim_{a \to +\infty} h_k(a)$ and then we obtain that
$$a \hat{\cdot} x = g_k^*(h_k(a) g_k(x)).$$
for $a \in (0, +\infty)$, and $x \in [\alpha_k, \beta_k]$.

We next show that $0 < h_k(a) < +\infty$ for $0 < a < +\infty$. Let $0 < a < +\infty$. If $h_k(a) = 0$, then
$$a \hat{\cdot} \beta_k = g_k^*(h_k(a) g_k(\beta_k)) = \alpha_k$$
and this contradicts that $a \hat{\cdot} \beta_k = \beta_k$. If $h_k(a) = +\infty$, then, for every $x \in [\alpha_k, \beta_k]$,
$$a \hat{\cdot} x = g_k^*(h_k(a) g_k(x)) = g_k^*(+\infty) = \beta_k$$
hence $a \hat{\cdot} \alpha_k = \lim_{x \to \alpha_k + 0} a \hat{\cdot} x = \beta_k$, and this contradicts that $a \hat{\cdot} \alpha_k = \alpha_k$.

By (M4), we have that $h_k(e) = 1$.

Last, we prove that $h_k(a) = 1$ for $0 < a \leq e$ if is nilpotent. Let $(\alpha_k, \beta_k]$ be nilpotent. It is sufficient to prove that, if $a \in (0, e]$ and $x \in [\alpha_k, \beta_k]$, then $a \hat{\cdot} x = x$. Assume that there are numbers $a \in (0, e]$ and $x \in [\alpha_k, \beta_k]$ such that $a \hat{\cdot} x \neq x$. Since, $a \hat{\cdot} x \leq e \hat{\cdot} x = x$ we have, $a \hat{\cdot} x < x$ that is, $g_k(a \hat{\cdot} x) < g_k(x)$.

If $y = g_k^*(g_k(\beta_k) - g_k(x))$ then

$$g_k(a \hat{\cdot} x) + g_k(a \hat{\cdot} y) \le g_k(a \hat{\cdot} x) + g_k(y)$$
$$= g_k(a \hat{\cdot} x) + g_k(\beta_k) - g_k(x)$$
$$< g_k(\beta_k)$$

therefore,
$$\beta_k > (a \hat{\cdot} x) \hat{+} (a \hat{\cdot} y) = a \hat{\cdot} (x \hat{+} y) = a \hat{\cdot} \beta_k = \beta_k$$

This is a contradiction.
The proof is now complete.

A Model of Human Evaluation Process Using Fuzzy Measure[*]

K. Ishii and **M. Sugeno**
*Department of Systems Science,
Tokyo Institute of Technology,
4259 Nagatsuta, Midori-ku, Yokohama, 227 Japan*

Received 27 October 1983

Abstract

This paper presents a mathematical model for the human subjective evaluation process using fuzzy integrals based on the general fuzzy measure. First, the concept of overlap coefficient is introduced to describe the properties of a given fuzzy measure and to characterize the human evaluation process. Further, a new algorithm is developed to identify the fuzzy measure with full degrees of freedom. Lastly, the model and the identification scheme are applied to two practical examples: prediction of wood strength by an experienced inspector and trouble evasive actions taken by a computer game player.

1. Introduction

Human decision-making process plays a significant role in our society. Many industrial processes rely on trained professionals for operation and control, while human ability to make decisions remains the key essence in fields such as education and management. The objective of this paper is to devise a mathematical model that simulates how experienced individuals integrate information, evaluate it, and decide on their actions.

Conventionally, the linear combination form was used as the mathematical model to approximate the human evaluation process. This so-called linear model

---

[*] © 1985 *Academic Press. Retyped with written permission from Int. J. Man-Machine Studies,* **22** (*1985*) *19-38.*

is obviously inadequate, since human subjective evaluation does not always hold linearity. Sugeno (1974) presented the theory of fuzzy measures and fuzzy integrals as a means to express fuzzy systems and further proposed to use his theory in modeling human subjective evaluation process. While the conventional Lebesgue measures assume additivity, fuzzy measures assume only monotonicity and thus are very general. Hence, human subjective scales can be better approximated using fuzzy measures than using the additive ones. Sugeno also constructed the λ-measure as a special case of fuzzy measures. Here, the measure is constrained by a parameter λ, which describes the degree of additivity the measure holds. Thus, the degree of freedom of a λ-measure is much smaller than that of an unconstrained one. While fuzzy measures are scales of human subjective evaluation, fuzzy integrals can be interpreted as operations that quantify human evaluation based on these measures.

Once the form of the mathematical model is chosen, our next problem is that of identifying the parameters of the model from the human input/output data. Many studies have been done on identification of fuzzy measures in recent years. Sugeno (1974) formalized it as an optimization problem of minimizing the error between the human evaluation and the fuzzy integral model output. Sekita and Tabata (1977a) considered identification as a minimization problem of the difference between human subjective scales and fuzzy measures. Wierzchon (1983) introduced a mapping from λ-measures to probability measures and used the least squares method.

While there have been many studies on the theoretical aspects of fuzzy measures, research on its application has been limited to only a few. Sugeno (1974) applied his fuzzy integral model to examples such as subjective evaluation of female faces and grading similarity of patterns. Seif and Aguilar-Martin (1980) applied the concept of conditional fuzzy measures to pattern recognition. Sekital and Tabata (1977b) adopted fuzzy integrals for evaluation of human health index.

Most of the research done until today was based on the λ-measure because of its mathematical soundness, modest degrees of freedom, and ease of identification. However, with the λ-measure, the interaction between each information element is characterized by a single parameter λ; hence, the full power of fuzzy measure is not utilized. Further, most identification schemes proposed are based on optimization techniques, and because the fuzzy integrals are not differentiable with respect to fuzzy measures, this approach turns out to be extremely cumbersome and slow.

Considering the above difficulties, this paper proposes the two following ideas. First, the use of the general fuzzy measure is proposed to express human subjective scales instead of the λ-measure. Further, as scales that describe the properties of a given fuzzy measure, the overlap coefficient m_{ij} and the necessity coefficient ξ_j are defined and used to characterize the human subjective evaluation process. Second, a new algorithm based on the model reference adaptive approach is developed to identify the fuzzy measure with full degrees of freedom. Through computer simulation, the developed scheme, namely, Fuzzy measure Learning Identification Algorithm (FLIA), is shown to be much faster than, and superior to, the conventional identification schemes.

In addition, the proposed model and FLIA are applied to two practical examples: prediction of lumber strength by an experienced inspector and trouble evasive actions taken by a computer game player. Through these examples, the fuzzy integral model based on the general fuzzy measure is shown to accurately extract the characteristics of the human subjective evaluation process. Further, the validity of the coefficients, m and ξ, is confirmed.

2. Fuzzy measures and fuzzy integrals

This chapter explains the basic properties of fuzzy measures and fuzzy integrals. Also, as scales that describe the properties of a given fuzzy measure, the overlap coefficient m_{ij} and the necessity coefficient ξ_j are defined.

Definition 1. Consider a set of n information elements $K=\{s_1, s_2, \ldots, s_n\}$. If a set function $g: 2^K \to [0, 1]$ has the following properties listed below, then g is called a fuzzy measure.

(1) $g(\phi)=0$ and $g(K) = 1$ (1)

(2) $A, B \in 2^K$ and $A \subset B \to g(A) \le g(B)$.   (2)

Listed below are three special cases of fuzzy measures; each measure is subject to its respective constraint on g.

(i) Probability measure:
$$A, B \in 2^K \text{ and } A \cap B = \phi \to g(A \cup B) = g(A) + g(B). \quad (3a)$$

(ii) F-additive measure:
$$A, B \in 2^K \to g(A \cup B) = g(A) \vee g(B) \quad (3b)$$

(iii) λ-measure:
$$A, B \in 2^K \, A \cap B = \phi \to g(A \cup B) = g(A) + g(B) + \lambda g(A)g(B) \quad (3c)$$
where $\lambda \in [-1, \infty)$.

Now, the following inequality will hold for any two elements s_i and s_j, $s_i \ne s_j$.

$$1 \ge g(\{s_i, s_j\}) \ge g(\{s_i\}) \vee g(\{s_j\}) \quad (4)$$

In other words, $g(\{s_i, s_j\})$ can take any value as long as it satisfies Eq. (4). In order to describe the properties of a given fuzzy measure, we introduce the concept of coupling and overlap between elements.

Definition 2. Consider a fuzzy measure space $(K, 2^K, g)$. The coupling coefficient μ_{AB} is defined as follows.

$$\mu_{AB} = \frac{g(A \cup B) - [g(A) + g(B)]}{g(A) \wedge g(B)} \quad (5)$$

where $A, B \in 2^K$ and $A \cap B = \phi$ and $A, B \ne \phi$.

The coupling μ_{AB} indicates the degree of additivity that the fuzzy measure holds between subsets A and B.

The following cases may be considered.

$\mu_{AB} = -1$: The fuzzy measure is F-additive between A and B.

$\mu_{AB} < 0$: The fuzzy measure exhibits tendency of offset between A and B.
$\mu_{AB} = 0$: The fuzzy measure is additive between A and B.
$\mu_{AB} > 0$: The fuzzy measure exhibits amplification between A and B.

Definition 3. The overlap coefficient $m_{ij} \in [-1, 1)$ is expressed as follows

$$m_{ij} = \begin{cases} \mu_{ij} & i \neq j \text{ and } \mu_{ij} \leq 0 \\ \dfrac{\mu_{ij}}{\mu_{ij}+1} & i \neq j \text{ and } \mu_{ij} \geq 0 \\ 0 & i = j \end{cases} \quad (6)$$

where $\mu_{ij} = \mu_{\{s_i\}\{s_j\}}$ and $s_i, s_j \in K$.

The overlap coefficient m_{ij}, with its range of $[-1, 1)$, is a symmetrical scale of the degree of additivity that g holds between elements s_i and s_j, e.g., $m_{ij} = 0$ indicates that g is additive between s_i and s_j.

Definition 4. Overlap coefficient matrix M.
$$M = (m_{ij}). \quad (7)$$

Definition 5. The degree of overlap η_j of each element s_j is defined as follows:

$$\eta_j = \left(\sum_{i=1}^{n} m_{ij}^3\right) \bigg/ (n-1) \qquad j = 1 \sim n. \quad (8)$$

The coefficient η_j shows the average overlap between the s_j element and other elements. A negative η_j shows that s_j is a redundant information element, while a positive η_j indicates s_j's independence.

Definition 6. The necessity coefficient $\xi_j \in [-1, 0]$ of the element s_j is defined as follows.

$$\xi_j = \eta'_j [1 - g(\{s_j\})] \quad (9)$$

where

$$\eta'_j = \begin{cases} \eta_j & \eta_j \leq -0.4 \\ 0 & \eta_j > -0.4 \end{cases}.$$

If $\xi_j = 0$, s_j is an absolutely necessary element. If ξ_j is close to -1, s_j is a redundant information.

Definition 7. Consider a fuzzy measure space $(K, 2^K, g)$. The fuzzy integral of a function $h: K \to [0, 1]$ with respect to g is defined as follows.

$$\fint_k h(s) \cdot g(\cdot) = \bigvee_{j=1}^{n} [h(s_j) \wedge g(K_j)] \quad (10)$$

where
$$h(s_j) \le h(s_j + 1) \quad \text{for } 1 \le j \le n-1 \tag{11}$$
$$K_j = \{s_1, s_2, \cdots, s_j\} \quad (j = 1, 2, \cdots, n) \tag{12}$$

Here the function $h(s_j)$ is monotonically decreasing with respect to j. If this is not so, one should simply rearrange the index to achieve Eq. (11).

As can be seen in Sugeno's paper, Lebesgue integrals and fuzzy integrals have similar mathematical properties. However, while Lebesgue integrals can be visualized as functionals that compute the area under a function, fuzzy integrals should be regarded as functionals that characterize a function in a totally different way.

3. The model of human evaluation process

Our goal here is to devise a mathematical model that approximates the human evaluation process. The traditional model for this purpose is the linear model:

$$\tilde{z} = \sum_{i=1}^{n} a_i h(s_i) \tag{13}$$

where $K = \{s_1, s_2, \ldots, s_n\}$: set of information elements; $h : K \to [0, 1]$: function that quantify information; a_i: constant parameter; $\tilde{z}$: model output.

This model turns out to be quite inadequate because human evaluation does not always hold linearity. Further, the coefficients, namely, a_i's are difficult to interpret since they are merely partial derivatives of $\tilde{z}$ with respect to $h(s_i)$.

Barkan (1982) used his fuzzy integration based on λ-measure to model the process and showed the feasibility of his model. However, with this approach, the interaction between each information input is characterized by a single parameter λ, and the full power of fuzzy measure is not utilized.

In this paper, we propose to use the fuzzy integration based on the general fuzzy measure. This integration is defined as:

$$\tilde{z} = \smallint_K h(s_i) \cdot g(\cdot). \tag{14}$$

With this model, the increased degrees of freedom allow us to approximate the human subjective evaluation process more accurately than before. The fuzzy measure $g(\cdot)$, the overlap coefficient matrix M, and the necessity coefficient ξ_i characterize the process well; they can generally be interpreted in the following manner:

(1) $g(\{s_i\})$ $s_i \in K$ indicates the importance of s_i as an information input. The term "importance" implies the total evaluation of the sensitivities of the s_i's, and accuracy in reaching the output z.

(2) The overlap coefficient matrix M shows the interaction between each information element. A small m_{ij} implies that the level of importance does not increase even when s_i and s_j are put together, i.e., they possess large overlap. Conversely, if m_{ij} is close to +1, the importance is amplified when s_i and s_j are put together.

(3) The necessity coefficient ξ_i reflects the importance of an information element s_i and its overlap with other elements. $\xi_i = -1$ indicates a low importance and large overlap, hence, implies that s_i is not necessary. On the other hand, $\xi_i = 0$ indicates the opposite and, thus, implies that s_i cannot be dropped.

4. Fuzzy measure learning identification algorithm (FLIA)

This section deals with the identification of the fuzzy measure from human input/output data; that is, to find the fuzzy measure that best approximates the human evaluation process when it is modeled by Eq. (14).

The conventional identification schemes based on optimization techniques turn out to be extremely cumbersome and slow, especially with the increased degrees of freedom associated with our model.

Here, we take a totally different approach based on the model reference adaptive identification. The concept of the developed scheme, namely, the Fuzzy measure Learning Identification Algorithm (FLIA), is shown in Figure 1; we set an initial fuzzy measure arbitrarily and adjust them adaptively as a set of data is processed.

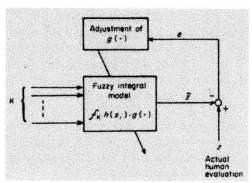

Fig.1. Concept of FLIA.

4.1. *The Description of the Algorithm*

Consider identifying the fuzzy measure $g(\cdot)$ from m sets of data, $\{h_j(s_1),\cdots,h_j(s_n),z_j\}$ $j = 1 \sim m$. The actual identification is carried out as follows.

(1) Initialize the fuzzy measure $g(\cdot)$ on $(K, 2^K)$; In this process, $2^n - 2$ values must be set. Naturally, $g(\cdot)$ must satisfy Eq. (1) and Eq. (2). Note that $g(\cdot)$ possess $\sum_{i=1^n}^{n-1} C_i = 2^n - 2$ degrees of freedom.

(2) For a set of data, $\{h_j(s_1),\cdots,h_j(s_n),z_j\}$, carry out the following adjustments. Note that, for each set of data, the index has to be altered so that

$$h(s_i) \geq h(s_i + 1) \quad 1 \leq i \leq n-1 \tag{15}$$

Also, we use the following notation:

$$\tilde{z} = \int_K h(s_i) \cdot g(\cdot) \tag{16}$$

$$e = z - \tilde{z} \tag{17}$$

$$\bar{i} = \min\{i \mid \tilde{z} = h(s_i) \wedge g(\{s_1, s_2, \cdots, s_i\})\} \tag{18}$$

$$\mathscr{F}_i = (\{s_1\}, \{s_1, s_2\}, \cdots, \{s_1, s_2, \cdots, s_i\}) \tag{19}$$

⟨1⟩ When |e| < TOL: No adjustment. Here, TOL: a positive real value, to be explained.

⟨2⟩ When $e \geq$ TOL (see Figure 2)

$$\forall E \in \mathscr{F} \bar{i} \text{ s.t. } g(E) < z \quad g(E) \leftarrow g(E) + [z - g(E)]\alpha \tag{20}$$

subject to constraint,

$$E, E' \in 2^K \text{ and } E \subset E' \rightarrow g(E) \leq g(E') \tag{21}$$

here, α: adaptive gain, to be explained. Note: No adjustment for E s.t. $z \leq g(E)$.

⟨3⟩ When $e <$ TOL:

$$\forall E \in \mathscr{F} \bar{i} \text{ s.t. } g(E) > z \quad g(E) \leftarrow g(E) + [z - g(E)]\alpha \tag{22}$$

subject to constraint, Eq. (20). Note: No adjustment for E s.t. $z \geq g(E)$.

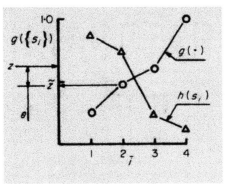

Fig. 2. Error e vs. $\bar{i}$.

The constant TOL designates the dead band of FLIA and prevents the unnecessary chattering of $g(\cdot)$ due to the noise that contaminates the raw data. The value of TOL depends on the problem concerned but is usually set according to the following equation:

$$\text{TOL} = 0.2 \times \sigma_w \tag{23}$$

where σ_w^2: variance of the predicted noise.

The adaptive gain has a significant effect on the convergence properties of FLIA. Here, α is dynamically adjusted as follows:

$$\alpha = \alpha_1 \cdot \alpha_2 \cdot \alpha_3 \tag{24}$$

$$\alpha_1(i) = \frac{i(n-1)}{(n/2)^2} : \quad \text{distribution gain} \quad (25)$$

$$\alpha_2(i) = (\beta^p)^{i-1} : \quad \text{subset gain} \quad (26)$$

$$\alpha_3 = \gamma \wedge \sigma_e^2/\xi : \quad \text{adaptive gain} \quad (27)$$

where σ_e^2: RMS (Root Mean Square) of the error e; p: iteration index, to be explained; β, γ, ξ :constants, to be explained.

The distribution gain α_1 allows the measure g(E), $|E| \cong n/2$ to be adjusted greatly because of the wide degree of freedom at $|E| \cong n/2$ (see Figure 3). The subset gain α_2 is for adjustments of g(E), $E \in \mathscr{F}_{i-1}$ which do not directly influence $\tilde{z}$. The adaptive gain α_3 reduces the amount of adjustment on g(E) when FLIA begins to converge. Empirically, the constants are set at $\gamma = 2$, $\xi = 3\sigma_w^2$.

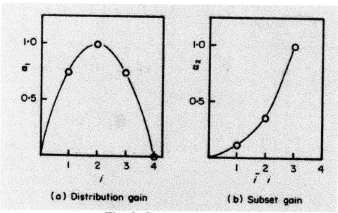

Fig. 3. Gain parameters.

When there are m sets of data, we process them through FLIA repetitively (see Figure 4). The data sets are reordered for each iteration. The integer p is the index of this repetitive identification. Eq. (26) implies that, in the beginning, FLIA adjusts even the g(E) that do not directly influence $\tilde{z}$; but after a few cycles of identification, only the g(E), $|E| = \bar{i}$ is adjusted.

4.2. The Computer Simulation of FLIA

In order to verify the performance of FLIA, some data were generated artificially using a known fuzzy measure with $n = 4$; we had 81 sets of data, $\{h_j(s_1), h_j(s_2), h_j(s_3), h_j(s_4), z_j\}$ $j = 1 \sim 81$, where,

$$z_j = \int_K h_j(s_i) \cdot g_{\text{true}}. \quad (28)$$

Here, g_{true} is the known measure. FLIA was tested to see if it will converge to the true measure starting from two different initial measures. Table 1 shows the true measure g_{true} and two initial measures g_{INI}.

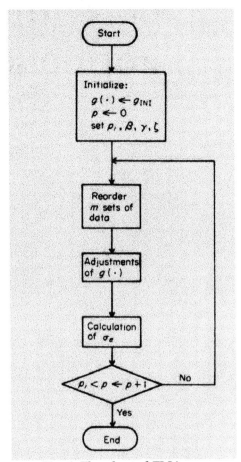

Fig. 4. *Flowchart of FLIA.*

Figure 5 shows the learning identification process during the first cycle, while Table 2 shows the fuzzy measures after 10 iterations of FLIA. For each initial measure, the root mean value of the model output error,

$$\sigma_e = \sqrt{\sum_{j=1}^{81}(z_j - \tilde{z}_j)^2 / 81} \qquad (29)$$

was of the order of 10^{-10} after 10 iterations of FLIA.

There exist some minor identification errors for $g(E)$, $|E| = 3$, but this is due to the deficiency on the number of data. With more data, these values can be improved. Note that FLIA shows excellent convergence for $g(E)$, $|E| = 1, 2$, that are important in calculating the overlap coefficient.

Table 1. *Fuzzy measures used for computer simulation*

(a) $g_1 = g_{true}$

$G(1) = 0\cdot 1000$	$G(1, 2) = 0\cdot 3000$	$G(1, 2, 3) = 0\cdot 5000$
$G(2) = 0\cdot 2105$	$G(1, 3) = 0\cdot 3235$	$G(1, 2, 4) = 0\cdot 8667$
$G(3) = 0\cdot 2353$	$G(1, 4) = 0\cdot 7333$	$G(1, 3, 4) = 0\cdot 8824$
$G(4) = 0\cdot 6667$	$G(2, 3) = 0\cdot 4211$	$G(2, 3, 4) = 0\cdot 9474$
	$G(2, 4) = 0\cdot 8070$	
	$G(3, 4) = 0\cdot 8235$	

(b) g_2: a probability measure

$G(1) = 0\cdot 2500$	$G(1, 2) = 0\cdot 5000$	$G(1, 2, 3) = 0\cdot 7500$
$G(2) = 0\cdot 2500$	$G(1, 3) = 0\cdot 5000$	$G(1, 2, 4) = 0\cdot 7500$
$G(3) = 0\cdot 2500$	$G(1, 4) = 0\cdot 5000$	$G(1, 3, 4) = 0\cdot 7500$
$G(4) = 0\cdot 2500$	$G(2, 3) = 0\cdot 5000$	$G(2, 3, 4) = 0\cdot 7500$
	$G(2, 4) = 0\cdot 5000$	
	$G(3, 4) = 0\cdot 5000$	

(c) g_3

$G(1) = 0\cdot 1500$	$G(1, 2) = 0\cdot 4500$	$G(1, 2, 3) = 0\cdot 7500$
$G(2) = 0\cdot 2000$	$G(1, 3) = 0\cdot 5000$	$G(1, 2, 4) = 0\cdot 8000$
$G(3) = 0\cdot 3000$	$G(1, 4) = 0\cdot 5500$	$G(1, 3, 4) = 0\cdot 8500$
$G(4) = 0\cdot 4000$	$G(2, 3) = 0\cdot 6000$	$G(2, 3, 4) = 0\cdot 9000$
	$G(2, 4) = 0\cdot 6500$	
	$G(3, 4) = 0\cdot 7000$	

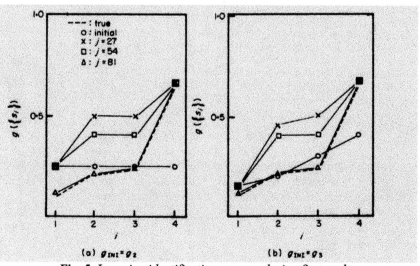

Fig. 5. *Learning identification process during first cycle.*

Table 2. The identified fuzzy measures after 10 iterations of FLIA

(a) $g_1 = g_{true}$

$G(1) = 0.1000$	$G(1,2) = 0.3000$	$G(1,2,3) = 0.5000$
$G(2) = 0.2105$	$G(1,3) = 0.3235$	$G(1,2,4) = 0.8667$
$G(3) = 0.2353$	$G(1,4) = 0.7333$	$G(1,3,4) = 0.8824$
$G(4) = 0.6667$	$G(2,3) = 0.4211$	$G(2,3,4) = 0.9474$
	$G(2,4) = 0.8070$	
	$G(3,4) = 0.8235$	

(b) g_2: a probability measure

$G(1) = 0.2500$	$G(1,2) = 0.5000$	$G(1,2,3) = 0.7500$
$G(2) = 0.2500$	$G(1,3) = 0.5000$	$G(1,2,4) = 0.7500$
$G(3) = 0.2500$	$G(1,4) = 0.5000$	$G(1,3,4) = 0.7500$
$G(4) = 0.2500$	$G(2,3) = 0.5000$	$G(2,3,4) = 0.7500$
	$G(2,4) = 0.5000$	
	$G(3,4) = 0.5000$	

(c) g_3

$G(1) = 0.1500$	$G(1,2) = 0.4500$	$G(1,2,3) = 0.7500$
$G(2) = 0.2000$	$G(1,3) = 0.5000$	$G(1,2,4) = 0.8000$
$G(3) = 0.3000$	$G(1,4) = 0.5500$	$G(1,3,4) = 0.8500$
$G(4) = 0.4000$	$G(2,3) = 0.6000$	$G(2,3,4) = 0.9000$
	$G(2,4) = 0.6500$	
	$G(3,4) = 0.7000$	

From these results, we can draw the following conclusions on the performance of FLIA.

(1) FLIA has fast convergence. For the example above, to reduce σ_e down to 10^{-4} FLIA consumed roughly 1/10 of computation time as compared to an identification scheme based on the complex method, an optimization technique. For the complex method, the degrees of freedom was held down to 4 by considering g_{true} as a λ-measure.

(2) Since with FLIA, the fuzzy measure is directly adjusted by the model output error e, the characteristics of the raw data are well reflected on the identified measure, and it can easily be interpreted.

(3) When the number of information elements is deficient, or when the human input/output data is obviously not suited to fuzzy integral models, the identified measure may chatter.

(4) If the number of data is deficient compared to the degrees of freedom of $g(\cdot)$, the measure may not converge to the true value, depending on the initial measure.

Problems associated with (3) and (4) also apply to the conventional methods. As a whole, one can clearly see the superiority of FLIA. When applying FLIA on real human data, one should always use the same initial measure in order to establish a standard reference on which one can compare the identified fuzzy measures. In this paper, we adopt g_2 (Table 1), the probability measure, as our standard initial fuzzy measure.

5. Application to prediction of wood strength

5.1. *The Problem*

The prediction of wood strength is a difficult problem because of the material's complexity and irregularity. Barkan (1982) carried out some experimental work in pursuit of a good prediction for the material's ultimate bending strength. One of the authors participated in this project, in which the following measurements were taken:
  (1) radius of curvature under a set uniform bending moment;
  (2) density;
  (3) grain angle;
  (4) moisture content.

The first measurement was a newly developed one; a specimen was subjected to a four-point loading as shown in Figure 6, and the local radii of curvature were measured at nine points along the span of uniform bending stress. Thus, the radius of curvature profile was available for use to predict the ultimate strength. Experiments were carried out for sixty 1" x 2" x 30" pine beams: thirty clear specimens and thirty knotted ones. After the measurements were taken, each specimen was actually brought to a fracture, and the ultimate bending stress was recorded.

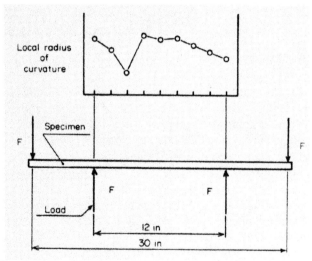

Fig. 6. Experimental setup to measure wood bending strength.

Experienced inspectors predict the ultimate strength and grade the specimens quite accurately from the given measurements. Here, we model the evaluation process of the inspectors using fuzzy integration as in Figure 7; the radius of curvature profile is evaluated from its maximum, minimum, mean and maximum gradient, and this evaluation of the profile is considered together with density, grain angle and moisture content to make the total evaluation, the prediction of ultimate bending stress.

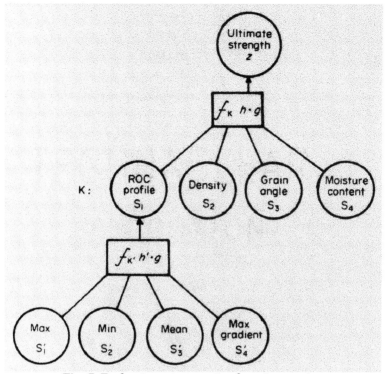

Fig. 7. *Evaluation process using fuzzy integration.*

Let us focus on the latter half of the evaluation process. Let the set of information elements be

$$K = \{s_1, s_2, s_3, s_4\}$$

where s_1: radius of curvature profile; s_2: density; s_3: grain angle; s_4: moisture content. The fuzzy integral model can be expressed as follows:

$$\tilde{z} = \int_K h(s_i) \cdot g(\cdot) \tag{30}$$

where $\tilde{z}$: prediction of z.
The function $h : K \to [0, 1]$ is assigned as follows:

$$h(s_1) = z' \tag{31}$$

$$h(s_i) = \frac{x_i}{\bar{x}_i - \underline{x}_i} \quad i = 2 \sim 4 \tag{32}$$

where z': evaluation of the radius of curvature profile [-]; x_2: density [lb/in^3]; x_3: cosine of the grain angle [-]; x_4: 100-moisture content [%]; $\bar{x}_i$ and $\underline{x}_i$ are the maximum and the minimum values among the sample specimens, respectively. Also,

$$z = US/15{,}000 \text{ psi} \tag{33}$$

where US: ultimate bending strength [psi].

5.2. Results and Discussions

The fuzzy measures $g(\cdot)$ were identified separately for the clear and knotted specimens. Tables 3 and 4 show the identified measures and the overlap coefficient matrix M. Figure 8 shows the $g(\{s_i\})$ and the necessity coefficients ξ for both cases.

Table 3. The identified fuzzy measure-clear wood

$G(1) = 0.4000$	$G(1,2) = 0.7000$	$G(1,2,3) = 0.9000$
$G(2) = 0.5000$	$G(1,3) = 0.7673$	$G(1,2,4) = 0.0500$
$G(3) = 0.5796$	$G(1,4) = 0.5140$	$G(1,3,4) = 0.7809$
$G(4) = 0.4993$	$G(2,3) = 0.6972$	$G(2,3,4) = 0.6972$
	$G(2,4) = 0.5000$	
	$G(3,4) = 0.5796$	

$$M = \begin{pmatrix} 0.0000 & -0.5000 & -0.5309 & -0.9632 \\ -0.5000 & 0.0000 & -0.7648 & -1.0000 \\ -0.5309 & -0.7648 & 0.0000 & -1.0000 \\ -0.9632 & -1.0000 & -1.0000 & 0.0000 \end{pmatrix}$$

Table 4. The identified fuzzy measure-knotted wood

$G(1) = 0.6000$	$G(1,2) = 0.6462$	$G(1,2,3) = 0.7084$
$G(2) = 0.1837$	$G(1,3) = 0.7084$	$G(1,2,4) = 0.8500$
$G(3) = 0.3409$	$G(1,4) = 0.6500$	$G(1,3,4) = 0.8353$
$G(4) = 0.1837$	$G(2,3) = 0.3409$	$G(2,3,4) = 0.3409$
	$G(2,4) = 0.1837$	
	$G(3,4) = 0.3409$	

$$M = \begin{pmatrix} 0.0000 & -0.7485 & -0.6820 & -7278 \\ -0.7485 & 0.0000 & -1.0000 & -1.0000 \\ -0.6820 & -1.0000 & 0.0000 & -1.0000 \\ -0.7278 & -1.0000 & -1.0000 & 0.0000 \end{pmatrix}$$

Note: $g(1, 2) = g(\{s_1, s_2\})$ etc.

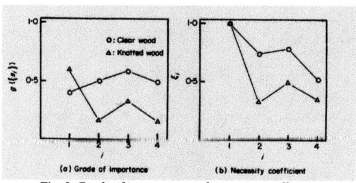

Fig. 8. Grade of importance and necessity coefficient.

The process shown in Figure 7 was also approximated by the linear model

$$\tilde{z} = \sum_{i=1}^{4} g(\{s_i\}) \times h(s_i) \tag{34}$$

and the λ-model (fuzzy integral model based on the λ-measure). The identification of parameters was done by linear regression for the linear model and by the complex method for the λ-model. The comparison is summarized in Table 5.

Table 5. Comparison of the models

Wood	Model	$g(\{s_1\})$	$g(\{s_2\})$	$g(\{s_3\})$	$g(\{s_4\})$	λ	σ_e
Clear	Linear	0·0876	0·3642	0·2714	0·2294	—	0·0796
	λ	0·2566	0·2230	0·3014	0·2790	−0·1467	0·0493
	Fuzzy	0·4000	0·5000	0·5796	0·4992	—	0·0477
Knotted	Linear	0·7779	0·0598	0·1050	0·0338	—	0·0911
	λ	0·5691	0·0245	0·0456	0·0593	3·3626	0·0964
	Fuzzy	0·6000	0·1837	0·3409	0·1837	—	0·0893

Table 6. Degradation of model performance when an information element is dropped

Wood	Dropped element	σ_e	$\sigma_e - \sigma_{e_{opt}}$	ξ_i
Clear	None	$0·04777 = \sigma_{e_{opt}}$		
	s_1	0·07572	0·02795	0·0
	s_2	0·05031	0·00254	−0·2620
	s_3	0·05194	0·00417	−0·2238
	s_4	0·04834	0·00057	−0·4830
Knotted	None	$0·08934 = \sigma_{e_{opt}}$		
	s_1	0·17557	0·08622	0·0
	s_2	0·08937	0·00003	−0·6583
	s_3	0·10102	0·01168	−0·5091
	s_4	0·09163	0·00229	−0·6491

Further, in order to verify the validity of ξ, the process was modeled without one of the input elements, and the degradation of σ_e was observed. Table 6 summarizes the results.

From these results the following points were observed:
(1) The obtained parameters, $g(\{s_i\})$ and ξ agree well with our intuitive accounts on the measured elements. The radius of curvature profile s_1 was thought to be less important for the clear wood, which has few grain irregularities, and indeed, $g(\{s_1\})$ for the clear wood is quite low at 0.4. Rather, the grain angle seems to be the most important element.

The necessity coefficient of s_1, however, turns out to be the highest, since s_1 reflects the quality of wood that other elements do not extract. For the knotted wood, on the other hand, s_1, which is sensitive to grain irregularities and knots, is the most important and necessary element.

(2) From Table 5, one can clearly see the superiority of the fuzzy integral model over the other models in approximating the process. σ_e is improved by up to 40% against the linear model, and 7% against the λ-model. Note that the coefficients of the linear model are difficult to physically interpret. For example, the coefficient of s_i for the clear wood is extremely low despite s_1's high necessity.

(3) The degradation of the model performance when one of the input elements is dropped coincides with the necessity coefficient ξ. From Table 6, one can see that a low ξ implies a small degradation, $\sigma_e - \sigma_{e_{opt}}$ and vice versa. This result verifies the validity of ξ as a scale that indicates the necessity of the input elements.

6. Application to analysis of plant operator actions

6.1. *The Problem*

Human operators of various engineering plants play a significant role in coping with emergencies. In this example, we tried to simulate such a situation on a computer game (see Figure 9). The player has to control the flying object (FO) along the set altitude from the right of the screen to the left. The horizontal component of the velocity v_x is constant, and the vertical component v_y can be varied according to the player's command c_0. If FO crashes into a mountain, the player is disqualified. In a cloud, FO becomes invisible, and some disturbance contaminates the velocity command. If FO goes into an air pocket, the altitude is suddenly changed by a random amount.

Here, we assume that $|c_0|$ reflects the player's evaluation of the FO's degree of emergency, z. We can model this subjective evaluation process as in Figure 10.

The set of information elements are as follows:
$$K = \{s_1, s_2, s_3, s_4\}$$
where s_1: danger signal from the nearest mountain; s_2: danger signal from the nearest cloud; s_3: danger signal from the nearest air pocket; s_4: danger signal due to the offset from the set altitude; $z = |c_0|$: perceived degree of emergency $\in [0, 1]$.

The fuzzy integral model can be expressed as follows:
$$\tilde{z} = \int_K h(s_1) \cdot g(\cdot) \tag{35}$$

The function $h: K \to [0, 1]$ is assigned as follows:
$$h(s_1) = 1 \wedge (\alpha_1/d_m) \tag{36}$$
$$h(s_2) = 1 \wedge (\alpha_2/d_c) \tag{37}$$
$$h(s_3) = 1 \wedge (\alpha_3/d_a) \tag{38}$$
$$h(s_4) = 1 \wedge (|e_0|/\alpha_4) \tag{39}$$

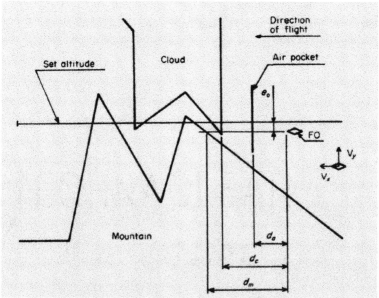

Fig. 9. *Characteristics of a computer game.*

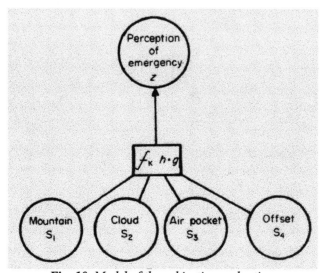

Fig. 10. *Model of the subjective evaluation.*

where d_m: horizontal distance to the nearest mountain; d_c: horizontal distance to the nearest cloud; d_a: horizontal distance to the nearest air pocket; e_0: offset from the set altitude; $\alpha_1 \sim \alpha_4$: constants that are determined by preliminary experiments.

We asked two people to play the game. Player A was well acquainted with the game and decided to evade only the mountains, ignoring other obstacles. Player B was a beginner and evaded every obstacle he faced. Their trajectories are shown in Figures 11 and 12. The data were collected at 63 points, equally spaced, along the trajectory.

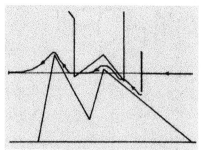

Fig. 11. Trajectory of Player A.   *Fig. 12. Trajectory of Player B.*

6.2. The Results and Discussions

Figure 13 summarizes the identified results, $g(\{s_i\})$ and ξ. The comparison with the linear model is shown in Table 7.

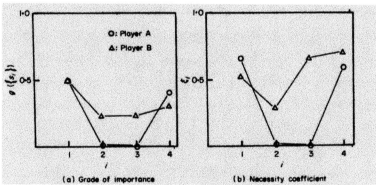

Fig. 13. Grade of importance and necessity coefficient.

Table 7. Comparison of the models

Player	Model	$g(\{s_1\})$	$g(\{s_2\})$	$g(\{s_3\})$	$g(\{s_4\})$	σ_e
A	Linear	0·6609	0·2078	−0·1437	0·3674	0·1951
	Fuzzy	0·5000	0·0133	0·0001	0·3983	0·1835
B	Linear	0·6220	0·2602	0·6791	0·1983	0·1846
	Fuzzy	0·5000	0·2278	0·2278	0·3113	0·1614

In addition, attempts were made to reproduce the trajectory of both players by a computer, using the identified fuzzy integral model. Here, $|c_0| = z$ were calculated according to Eq. (35), and sign (c_0) was decided as follows. First a function sig : $K \to \{-1, 1\}$ is assigned so that it designates the direction that FO should take to evade each obstacle. For Figure 9,

$$\text{sig}(s_1) = 1$$
$$\text{sig}(s_2) = -1$$
$$\text{sig}(s_3) = -1$$
$$\text{sig}(s_4) = \begin{cases} -1 & e_0 > 0 \\ 1 & e_0 \le 0 \end{cases}$$

Then, sign (c_0) is assigned as follows:
$$\text{sign}(c_0) = \text{sig}(s_i) \tag{40}$$
where
$$i = \{j \mid h(s_j) \wedge g(\{s_j\})\} = \bigvee_{k=1}^{n}[h(s_k) \wedge g(\{s_k\})] \tag{41}$$

Hence,
$$c_0 = \text{sign}(c_0) \times \int_K h(s_i) \cdot g(\cdot) \tag{42}$$

The calculation were made at 630 points, equally spaced, along the CRT screen. The reproduced trajectories for a different setting to Figure 9 are shown on Figures 14 and 15.

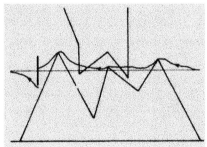

Fig. 14. Trajectory for Player A.

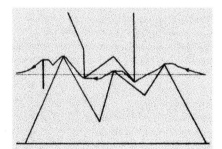

Fig. 15. Trajectory for Player B.

From these results, the following points were noted:
(1) The identified parameters, $g(\{s_i\})$ and ξ reflect each player's strategy quite well. For Player A, who totally ignored the cloud s_2 and the airpocket s_3, both $g(\{s_i\})$ and ξ for these elements turned out to be very small. On the other hand, Player B, who evaded everything, assigned roughly equal $g(\{s_i\})$ and ξ for all elements.
(2) The model performance was improved over that of the linear model by $6 \sim 14\%$ in σ_e. Again, the coefficients of the linear model seem to have little significance; in particular, the coefficient of s_3 for Player A taking a negative value is hard to interpret.
(3) The reproduced trajectories by the fuzzy integral model clearly display each player's strategy even in a different setting. This fact implies that the fuzzy integral model has accurately extracted the characteristics of the players.

7. Conclusions

In this paper, a mathematical model for the human evaluation process was devised, based on fuzzy integration; the general fuzzy measure was adopted for the model instead of the conventional λ-measure thus enabling a more refined evaluation of each information element. For this purpose, the overlap coefficient m and the necessity coefficient ξ were defined.

Further, in order to deal with the increased degrees of freedom of the fuzzy integral model, Fuzzy measure Learning Identification Algorithm was developed, and the validity of the scheme was shown.

Finally, the proposed model and FLIA were applied to two practical examples, and their feasibility was confirmed. The identified model and coefficients were found to characterize the human evaluation process quite accurately.

References

[1]. Barkan, P. (1982). Weyerhaeuser gift in aid of research on wood strength. *Status Report*. Design Div., Stanford University.

[2]. Dubois, D. and Prade, H. (1980). *Fuzzy Sets and Systems: Theory and Applications*, pp.125-146. New York: Academic Press.

[3]. Self, A. and Aguilar-Martin, J. (1980). Multi-group classification using fuzzy correlation. *Fuzzy Sets and Systems*, 3, 109-123.

[4]. Sekita, Y. and Tabata, Y. (1977a). *A Consideration on Identifying Fuzzy Measures*. Athens: XXIII Intn. Mtg. of the Institute of Management Sciences.

[5]. Sekita, Y. and Tabata, Y. (1977b). A health status index model using a fuzzy approach. *Discussion Paper*. Faculty of Economics, Osaka University.

[6]. Sugeno, M. (1974). Theory of fuzzy integrals and its applications. *Dr. Thesis*. Tokyo Institute of Technology.

[7]. Sugeno, M. and Terano, T. (1977). A model of learning based on fuzzy information. *Kybernetes*, 6, 157-166.

[8]. Wierzchon, S. T. (1983). An algorithm for identification of fuzzy measure. *Fuzzy Sets and Systems*, 9, 69-73.

Appendix

Professor Sugeno has made a large number of contributions to the field of Fuzzy Sets and Fuzzy Systems Theory. There are many papers that he has published in English language journals and an equal number or more that appear in Japanese. The papers selected for this edited volume are only a small number of papers which demonstrate the breadth and depth of his work. We thought it would be proper to list all his contributions in an Appendix for the benefit of students and researchers who might want to pursue this further.

List of Publications of M. Sugeno

Theory of Fuzzy Integrals and Its Applications, Ph.D. Thesis, Tokyo Institute of Technology, 1974

Journal Papers

[1]. A Model of Learning based on Fuzzy Information, *Kybernetes*, 6, 157/166, 1977 (with T. Terano).
[2]. Fuzzy Systems with Underlying Deterministic Systems, *Summary of Papers on General Fuzzy Problems*, 3, 25/52, 1977.
[3]. Generalized Truth Value in Truth Qualification, *Summary of Papers on General Fuzzy Problems*, 4, 5/9, 1978 (with H. Imaoka).
[4]. Fuzzy Control based on a New Algorithm-Experimental Study, *Summary of Papers on General Fuzzy Problems*, 5, 39/45, 1979 (with S. Masuda and T. Terano).
[5]. On Fuzzy Sets Manipulating Language, *Summary of Papers on General Fuzzy Problems*, 6, 1/6, 1980 (with M. Ito).
[6]. Diagnostic Aspects of Fuzzy Reasoning in Radiological Interpretation, *Summary of Papers on General Fuzzy Problems*, 6, 25/31, 1980 (with T. Nishioka and E. Okada).
[7]. Application of Fuzzy Reasoning to Dental Diagnosis, *Summary of Papers on General Fuzzy Problems*, 7, 6/10, 1981 (with T. Onisawa et al.).
[8]. Applying Dynamic Programming to Fuzzy Problems, *Summary of Papers on General Fuzzy Problems*, 7, 16/20, 1981 (with D. Tang et al.).
[9]. The Factors Field and Background Structure for Fuzzy Subsets, *J. of Fuzzy Mathematics*, 2, 2, 45/54, June 1982 (with P. Z. Wang).

[10]. L-Fuzzy Category, *Fuzzy Sets and Systems*, **11**, 1, 43/64, 1983 (with M. Sasaki).
[11]. Multi-Dimensional Fuzzy Reasoning, *Fuzzy Sets and Systems*, **9**, 3, 313/325, 1983 (with T. Takagi).
[12]. Fuzzy Relational Equations and the Inverse Problems, *Fuzzy Sets and Systems*, **15**, 1, 79/90, 1985 (with C. P. Pappis).
[13]. Fuzzy Identification of Systems and Its Applications to Modelling and Control, *IEEE Trans. System, Man, and Cybernetics*, **15**,1,116/132,1985 (with T. Takagi).
[14]. State-Transition Categories, *Fuzzy Sets and Systems*, **15**, 2, 269/283, 1985 (with M. Sasaki).
[15]. A Model of Human Evaluation Process Using Fuzzy Measure, *Int. J. of Man-Machine Studies*, **22**, 19/38, 1985 (with K. Ishii).
[16]. Fuzzy Control of Model Car, *Fuzzy Sets and Systems*, **16**, 2, 103/113, 1985 (with M.Nishida).
[17]. Fuzzy Modelling and Control of Multilayer Incinerator, *Fuzzy Sets and Systems*, **18**, 329/346, 1986 (with G. T. Kang).
[18]. Fuzzy Measure Analysis of Public Attitude Towards the Use of Nuclear Energy, *Fuzzy Sets and Systems*, **20**, 259/289, 1986 (with T. Onisawa).
[19]. Pseudo-Additive Measures and Integrals, *J. Math. Appl.*, **122**, 197/222, 1987 (with T. Murofushi).
[20]. Structure Identification of Fuzzy Model, *Fuzzy Sets and Systems*, 28, 1, 15/33, 1988 (with G. T. Kang).
[21]. An Interpretation of Fuzzy Measure and Choquet's Integral as an Integral with Respect to Fuzzy Measure, *Fuzzy Sets and Systems*, **29**, 2, 201/227, 1989 (with T. Murofushi).
[22]. Fuzzy Algorithmic Control of a Model Car by Oral Instructions, *Fuzzy Sets and Systems*, **32**, 32, 2, 207/219, 1989 (with T. Murofushi).
[23]. Application of Fuzzy Control Logic for Dead-Time Processes in a Glass Melting Furnace, *Fuzzy Sets and Systems*, **38**, 3, 251/265, 1990, (with S. Aoki, and S. Kawaji).
[24]. Fuzzy t-Conorm Integral with Respect to Fuzzy Measures, *Fuzzy Sets and Systems*, **42**, 57/71, 1991 (with T. Murofushi).
[25]. A Theory of Fuzzy Measures: Representations, the Choquet Integral, and Null Sets, *J. Math. Anal. Appl.*, **159**, 532/549, 1991 (with T. Murofushi).
[26]. Successive Identification of a Fuzzy Model and Its Applications to Prediction of a Complex System, *Fuzzy Sets and Systems*, **42**, 315/334, 1991 (with K. Tanaka).
[27]. A Study on Subjective Evaluations of Printed Color Images, *J. Approximate Reasoning*, **5**, 3, 213/222, 1991 (with K. Tanaka).
[28]. Stability Analysis and Design of Fuzzy Control Systems, *Fuzzy Sets and Systems*, **45**, 2, 135/156, 1992 (with K. Tanaka).
[29]. Fuzzy Measure of Fuzzy Events Defined by Fuzzy Integrals, *Fuzzy Sets and Systems*, **50**, 3, 293/313, 1992 (with M. Grabisch and T. Murofushi).
[30]. A Fuzzy Logic based Approach to Qualitative Modeling, *IEEE Trans. Fuzzy Systems*, **1**, 1, 7/31, 1993 (with T. Yasukawa).

[31]. Continuous-From-Above Possibility Measures and F-Additive Fuzzy Measures on Separable Metric Spaces: Characterization and Regularity, *Fuzzy Sets and Systems*, **54**, 351/354, 1993 (with T. Murofushi).
[32]. Some Quantities Represented by the Choquet Integral, *Fuzzy Sets and Systems*, **56**, 229/235, 1993 (with T. Murofushi).
[33]. Learning based on Linguistic Instructions Using Fuzzy Theory, *J. of Japan Society for Fuzzy Theory and Systems*, **4**, 6, 677/700, 1992 (with G. K. Park).
[34]. A Review and Comparison of Five Reasoning Methods, *Fuzzy Sets and Systems*, **57**, 257/294, 1993 (with H. Nakanishi and I. B. Turksen).
[35]. An Approach to Linguistic Instruction based Learning, *Int. J. of Uncertainty, Fuzziness and Knowledge-Based Systems*, **1**, 1, 19/56, 1993 (with G. K. Park).
[36]. Non-monotonic Fuzzy Measures and the Choquet Integral, *Fuzzy Sets and Systems*, **64**, 1, 73/86, 1994 (with T. Murofushi and M. Machida).
[37]. A Hierarchical Decomposition of Choquet Integral Model, *Int. J. of Uncertainty, Fuzziness and Knowledge-Based Systems*, **3**, 1, 1/15, 1995 (with K. Fujimoto).
[38]. A Clusterwise Regression-type Model for Subjective Evaluation, *J. of Japan Society for Fuzzy Theory and Systems*, **7**, 2, 291/310, 1995 (with S. H. Kwon).

(in Japanese).

[39]. Fuzzy Measure and Fuzzy Integral, *Trans. of the Society for Instrument and Control Engineers*, **8**, 2, 92/102, 1972.
[40]. Constructing Fuzzy Measure and Grading Similarity of Patterns by Fuzzy Integral, *Trans. of the Society for Instrument and Control Engineers*, **9**,3, 111/118, 1973.
[41]. Inverse Operation of Fuzzy Integrals and Fuzzy Measures, *Trans. of the Society for Instrument and Control Engineers*, **11**, 1, 32/37, 1975.
[42]. Fuzzy Decision Making Problems, Trans. of the Society for Instrument and Control Engineers, **11**, 6, 85/90, 1975.
[43]. Fuzzy Reasoning for Image-Based Diagnosis, *Radioactive Ray Image Research*, **10**, 3, 165/169, 1980 (with T. Nishioka et al.).
[44]. Application of Fuzzy Reasoning to Dental Diagnosis, *Trans. of the Society for Instrument and Control Engineers*, **18**, 7, 46/52, 1982 (with T. Onisawa et al.).
[45]. An Approach to Failure Analysis Using Fuzzy Theory, *Trans. of the Society for Instrument and Control Engineers*, **20**, 6, 28/35, 1984 (with T. Onisawa).
[46]. Self-Organising Fuzzy Controller, *Trans. of the Society for Instrument and Control Engineers*, **28**, 8, 50/56, 1984 (with T. Yamazaki).
[47]. Application of Fuzzy Reasoning to Water Purification Process, *Systems and Control*, **28**, 10, 45/52, 1984 (with O. Yagishita and O. Itoh).
[48]. An Approach to Failure Analysis Using Fuzzy Unreliability, *Trans. of the*

Society for Instrument and Control Engineers, **21**, 8, 65/71, 1985 (with T. Onisawa).
[49]. Fuzzy Modelling, *Trans. of the Society for Instrument and Control Engineers*, **23**, 6, 106/108, 1987 (with G. T. Kang).
[50]. Fuzzy Identification of Ill-Defined Systems, *Trans. of the Institute of Electrical Engineers of Japan*, **109-C**, 5, 367/374, 1989 (with A. Ohsato and T. Sekiguchi).
[51]. An Expert System for Investment Using Fuzzy Reasoning, *Information Processing*, **30**, 8, 963/969, 1989 (with B. Kaneko).
[52]. Hierarchical Decomposition of Choquet Integral Systems, *J. of Japan Society for Fuzzy Theory and Systems*, **4**, 4, 161/164, 1992 (with T. Murofushi).
[53]. Learning based on Linguistic Instructions Using Fuzzy Theory, J. of Japan Society for Fuzzy Theory and Systems, **4**, 6, 158/175, 1992 (with G. K. Park).
[54]. Stability Analysis of Fuzzy Systems and Construction Procedure for Lyapunov Functions, *Trans. of the Japan Society of Mechanical Engineers*, **C58**, 550, 80/86, 1993 (with K. Tanaka).
[55]. Contextualized Metaphors - A Relevance Based Model of Metaphor Interpretation, *J. of Natural Language Processing*, **101**, 13, 97/104, 1994 (with A. Utsumi).
[56]. An Approach to Social Simulation based on Linguistic Information - An application to the forecast of foreign exchange rate changes, *J. of Japan Society for Fuzzy Theory and Systems*, **6**, 4, 701/719, 1994 (with I. Kobayashi).
[57]. Qualitative System Description Based on Knowledge and Numerical Data and Its Application to the Design of a Fuzzy Controller, *J. of Japan Society for Fuzzy Theory and Systems*, **6**, 4, 720/735, 1994 (with T. Yasukawa).
[58]. Intelligent Fuzzy Computing - Marriage of fuzzy theory and functional linguistics, *J. of Computational Statistics*, **7**, 2, 189/198, 1994.
[59]. A Study on Data Fusion Based on Functional Linguistic Theory, *J. of Japan Society for Fuzzy Theory and Systems*, **8**, 1, 187/203, 1996 (with Y. Fukuyama).
[60]. A Computational Model of Interpreting Contextualized Metaphors Using Relevance Theory, *J. of Information Processing Society*, **37**, 6, 1017/1029, 1996 (with A. Utsumi).

Invited Conference Lectures

[1]. A New Approach based on Fuzzy Sets Concept to Fault Tree Analysis and Diagnosis of Failure at Nuclear Power Plants, IAEA *Seminar on Diagnosis of and Response to Abnormal Occurrences at Nuclear Power Plants*, Dresden, June 1984 (with T. Onisawa and Y.Nishiwaki).
[2]. Advances in Fuzzy Measures : Theory and Applications, *First International Conference on Fuzzy Information Processing*, Hawaii, July 1984.

[3]. For What the Fuzzy Theory Advances - Toward Subjectification of Science from Scientification of Subjectivity, *4th Fuzzy System Symposium*, Tokyo, May 1988 (in Japanese).

[4]. Fuzzy Control Principle, Practice and Perspective, *First IEEE International Conference on Fuzzy Systems*, San Diego, March 1992.

[5]. Categories of Uncertainty and Its Modalities, *Joint Japanese-European Symposium on Fuzzy Systems*, Berlin, October 1992.

[6]. Fuzzy Control of an Unmanned Helicopter, *Second IEEE International Conference on Fuzzy Systems*, San Francisco, March 1993.

[7]. Toward Intelligent Computing, *Fifth International Fuzzy Systems Association World Congress*, Seoul, 1993.

[8]. Qualitative Modelling based on Numerical Data and Knowledge Data, and Its Application to Control, *3rd IEEE International Conference on Computational Intelligence*, Orlando, June 1994.

[9]. Intelligent Computing, *Fifth International Conference on Information Processing and Management of Uncertainty in Knowledge-Based Systems*, Paris, July 1994.

[10]. Intelligent Fuzzy Computing, *Second Conference of the Pacific Association for Computational Linguistics*, Brisbane, April 1995.

[11]. Navigation of an Unmanned Helicopter, *Sixth IFSA World Congress*, Sao Paulo, July 1995.

[12]. Stability Issue of Fuzzy Control Systems, *1996 Biennial Conference of the North American Fuzzy Information Processing Society*, Berkeley, June 1996.

Conference Papers

[1]. An Approach to the Identification of Human Characteristics by Applying the Fuzzy Integral, *3rd IFAC Symposium. on Identification and System Parameter Estimation*, The Hague, 1973 (with T. Terano).

[2]. Subjective Evaluation of Fuzzy Objects, *IFAC Symposium.*, Budapest, 1974 (with Y. Tsukamoto and T. Terano).

[3]. Fuzzy Systems with Underlying Deterministic Systems, *The 4th Meeting of the European Working Group for Fuzzy Sets*, Stockholm, Nov.1976.

[4]. Fuzzification of LAleph - 1 and Its Application to Control, *International Conference on Cybernetics and Society*, Tokyo, 1978 (with Y. Tsukamoto and T. Takagi).

[5]. A Model of Dialog based on Fuzzy Set Concept, *6th International Joint Conference on Artificial Intelligence*, Tokyo, 1979 (with Imaoka).

[6]. Planning in Management by Fuzzy Dynamic Programming, *IFAC Symposium.*, Marseille, 1983 (with T. Terano and Y. Tsukamoto).

[7]. Derivation of Fuzzy Control Rules from Human Operator's Control Actions, ibid., 1983 (with T. Takagi).

[8]. Accidents and Human Factors, *Symposium. on the Risks and Benefits of Energy Systems*, Julich (Germany)., April 1984 (with Y. Nishiwaki et al.).

[9]. Fuzzy Parking Control of Model Car, *23rd IEEE Conference on Decision and Control*, Las Vegas, December 1984 (with K. Murakami).

[10]. Fuzzy Algorithmic Control of a Model Car by Oral Instructions, *Second IFSA Congress*, Tokyo, 1987 (with T, Murofushi et al.).
[11]. Choquet's Integral as an Integral Form for a General Class of Fuzzy Measures, ibid., 1987 (with T. Murofushi).
[12]. Expert System for Investment, International Workshop on Fuzzy System Applications, Iizuka, 225/227, 1988 (with B. Kaneko).
[13]. Application of Fuzzy Control Logic for Dead-Time Processes in a Glass Melting Furnace, *Proc. of 3rd IFSA Congress*, Seattle, 1989 (with S. Aoki and S. Kawachi).
[14]. Null Sets with Respect to Fuzzy Measures, ibid., 1989 (with T. Murofushi).
[15]. Stability Analysis of Fuzzy Systems Using Lyapunov's Direct Method, *Proc. of NAFIPS'90*, Toronto, 1990 (with K. Tanaka).
[16]. A Dynamic Memory Model for Scene Understanding, *Proc. of 6th Fuzzy System Symposium*, 211/214, 1990 (with W. Zhang).
[17]. Fuzzy Integrals and Dual Measures, *Sino-Japan Joint Meeting on Fuzzy Sets and Systems*, Beijing, 1990 (with M. Grabisch).
[18]. Knowledge based Scene Understanding, *Proc. of 7th Fuzzy System Symposium*, 419/422, 1991 (with W. Zhang).
[19]. An Alternative Matching for Fuzzy Sets, ibid., 229/233, 1991 (with A. Ralescu).
[20]. Linguistic Modeling based on Numerical Data, Proc. of 3rd IFSA *World Congress*, 264/267, Brussels, 1991 (with T. Yasukawa).
[21]. Fast Stability Checking Algorithm for Fuzzy Dynamical Systems, *Joint Hungarian-Japanese Symposium on Fuzzy Systems & Applications*, Budapest, 1991 (with K. Tanaka).
[22]. Helicopter Flight Control based on Fuzzy Logic, Proc. of 1st *International Fuzzy Engineering Symposium*, 1120/1121, Yokohama, 1991 (with T, Murofushi et al.).
[23]. Conceptual Fuzzy Sets, ibid., 261/272, 1991 (with T. Takagi and T. Yamaguchi).
[24]. Non- Additivity of Fuzzy Measures Representing Preferential Dependence, Proc. of 2nd International Conference on Fuzzy Logic and Neural Network, Iizuka, 1992 (with T. Murofushi).
[25]. Fuzzy Modeling, *Joint Japanese-European Symposium on Fuzzy Systems*, Berlin, October 1992 (with T. Yasukawa).
[26]. Fuzzy Hierarchical Control of an Unmanned Helicopter, *Proc. of 5th IFSA Congress*, 179/182, Seoul, July 1993 (with M. Guriffin and A. Bastian).
[27]. Social System Simulation based on Systemic Functional Linguistic Theory, ibid., 679/682, 1993, (with I. Kobayashi).
[28]. An Approach to Linguistic Instruction based Learning and Its Application to Helicopter Flight Control, ibid., 1082/1085, 1993 (with G.K. Park).
[29]. Issues for Blending Fuzzy Controllers in Hierarchical Systems, *Proc. of First Asian Fuzzy Systems Symposium*, Singapore, Nov.1993 (with M. Griffin and G. Walker).

[30]. A Relevance based Approach to Comprehending Contextualized Metaphors, *15th Int. Conference on Computational Linguistics*, Kyoto, 1994 (with A. Utsumi).
[31]. Qualitative Modeling based on Numerical Data and Knowledge Data, and its Application to Control, IEEE *World Congress on Computational Intelligence*, Orlando, June 1994 (with T.Yasukawa).
[32]. Fuzzy Programs based on Context Situation, FUZZ-IEEE/IFES '95, Yokohama, March 1995 (with J. Nishino).
[33]. A New Approach to Time Series Modeling with Fuzzy Measures and the Choquet Integral, ibid (with S. H. Kwon).
[34]. A Study on Data Fusion based on Systemic Functional Linguistic Theory, ibid (with Y. Fukuyama).
[35]. Hierarchical Decomposition Theorems for Choquet Integral Models, ibid. (with K. Fujimoto and T. Murofushi).
[36]. A Similarity Measure of Fuzzy Attributed Graphs and its Application to Object Recognition, FUZZ-IEEE '96, New Orleans, September 1996 (with W-J. Liu).
[37]. An Analysis of a Fuzzy Algorithmic Text from the Viewpoint of Daily Language, ibid (with I. Kobayashi).

(in Japanese).

[38]. On Fuzzy Non-Deterministic Problem (I)., (11)., *Proc. of 10th SICE Conference*, 285/288, 1971.
[39]. Similarity Degree of Patterns based on Fuzzy Integral, *Proc. of 11th SICE Conference*, 123/124, 1972.
[40]. Macroscopic Search of a Maze, ibid., 429/430, 1972 (with T. Saitoh and T. Terano).
[41]. Fuzzy Decision Making Process in Imperfect Information Game, ibid., 431/432, 1972 (with S. Koizumi and T. Terano).
[42]. Classification of Patterns by Fuzzy Integral, *Proc. of 15th Joint Conference on Automatic Control*, 335/336, 1972.
[43]. Subjective Evaluation of Faces, *Proc. of 16th Joint Conference on Automatic Control*, 319/320, 1973 (with H. Katoh, T. Terano).
[44]. Decision Making Process in a Game, *Proc. of 13th SICE Conference*, 73/74, 1974 (with S. Koizumi and T. Terano).
[45]. On Fuzzy Decision Making Problems, ibid., 83/84, 1974 (with T. Sano and T. Terano).
[46]. Conditional Fuzzy Measures and Their Applications, Proc. of 17th Joint Conference on Automatic Control, 41/42, 1974 (with T. Terano).
[47]. Macroscopic Search of Maximal of an Objective Function with Multiple Peaks, *Proc. of 18th Joint Conference on Automatic Control*, 185/186, 1975 (with T. Terano and M. Nakayama).
[48]. Robustness of Fuzzy Control, *Proc. of 19th SICE Conference*, 601/602, 1980 (with T. Takagi).
[49]. Programming Language for Fuzzy Data Processing, *Proc. of First Knowledge Engineering Symposium*, 11/16, 1983 (with O. T. Kang).

[50]. Design of Fuzzy Controller and Its Applications, ibid., 39/44, 1983 (with T. Takagi).
[51]. Fuzzy Control of a Model Car, ibid., 143/148, 1983 *(with* M. Nishida).
[52]. Fuzzy Dynamic Programming and Its Application to Planning, *Proc. of 23rd SICE Conference*, 1983 (with T. Terano).
[53]. Attitude Analysis for Use of Nuclear Energy by Fuzzy Measures, *Proc. of 10th SICE Systems Symposium*, 207/212, 1984 (with T. Onisawa and Y. Nishiwaki).
[54]. Fuzzy Modeling and Fuzzy Control of a Multilayer Incinerator, *Proc. of 10th SICE Systems Symposium*, 363/368, 1984 (with O. T. Kang).
[55]. Human Reliability at a View of Fuzziness, *Proc. of 11th SICE Systems Symposium*, 35/40, 1985 (with T. Onisawa).
[56]. Macroscopic Commands Driven Control of a Model Car, ibid., 313/316, 1985 (with K. Katayama).
[57]. On Modalities of Uncertainties based on Classification of Words, *Proc. of 2nd Fuzzy Systems Symposium*, 148/153, 1986.
[58]. Choquet Integral as an Integral Form for a General Class of Fuzzy Measure, *Proc. of 3rd Fuzzy Systems Symposium*, 31/36, 1987 (with T. Murofushi).
[59]. Verbal Control of a Model Car, ibid., 49/54, 1987 (with K. Katayama and T. Mon).
[60]. Identification of Ill-defined Systems by Means of Convexly Combined Fuzzy Relational Equations, ibid., 89/94, 1987 (with A. Ohsato).
[61]. Fuzzy Modeling and Analysis of Systems, *Proc. of 13th SICE Systems Symposium*, 81/84, 1987.
[62]. Expert System for Investment, *Proc. of 4th Fuzzy Systems Symposium*, 187/191, 1988 (with B. Kaneko).
[63]. A Study on Subjective Evaluations of Color Printing Images, ibid., 229/234, 1988 (with K. Kanaka).
[64]. Fuzzy t-Conorm Integral - Generalization of the Fuzzy Integral and the Choquet Integral, ibid., 345/350, 1988 (with T. Murofushi).
[65]. A Consideration on On-line Fuzzy Modeling, Proc. of 14th SICE Systems Symposium, 71/74, 1988 (with T. Tanaka).
[66]. On Analysis and Design of Fuzzy Control Systems, *Proc. of 5th Fuzzy Systems Symposium*, 127/132, 1989 (with T. Tanaka).
[67]. Prediction of the Motion of an Object Based on Fuzzy Inference, ibid., 95/100, 1989 (with T. Tatematsu and T. Murofushi).
[68]. A New Method of Choosing the Number of Clusters for Fuzzy C-Means Method, ibid., 247/252, 1989 (with Y. Fukuyama).
[69]. A Study on Hovering of Model Helicopter using Fuzzy Control, *Proc. of 6th Fuzzy Systems Symposium*, 559/560, 1990 (with J. Nishino and T. Murofushi).
[70]. An Approach to Simulation of Linguistic Models based on Fuzzy Theory, ibid., 169/172, 1990 (with M. Asami).
[71]. Fuzzy Integral with Respect to Dual Measures and Its Applications to

Multi-Attribute Pattern Recognition, ibid., 205/209, 1990 (with M. Grabish).
[72]. Multi-Attribute Utility Functions Represented by the Choquet Integral, ibid., 147/150, 1990 (with T. Murofushi).
[73]. Fuzzy Control of a Simulated Helicopter, Proc. of 7th Fuzzy Systems Symposium, 35/38, 1991 (with T. Hyakutake and T. Murofushi).
[74]. Linguistic Modeling based on Numerical Data, ibid., 83/86, 1991 (with T. Yasukawa).
[75]. Modification of a Fuzzy Measure According to Enlarging and Abridging a Frame of Discernment - An Interpretation of Subnormal Fuzzy Measures, ibid., 541/544, 1991 (with T. Murofushi).
[76]. A Study on Understanding of Natural Language based on Knowledge Base Using Fuzzy Theory, ibid., 541/544, 1991 (with I. Kobayashi).
[77]. Fuzzy Learning Control based on Linguistic Instructions, ibid., 619/922, 1991 (with O. K. Park).
[78]. A Hierarchical Structured Fuzzy Controller for Unmanned Helicopter, *Proc. of 8th Fuzzy Systems Symposium*, 53/56, 1992 (with J. Nishino and H. Miwa).
[79]. A Management Method of Fuzziness Using Background Knowledge in Fuzzy Algorithm, ibid., 349/352, 1992 (with M. Kitajima).
[80]. A Model based Design of Qualitative Control Rules, ibid., 533/536, 1992 (with T. Yasukawa).
[81]. Learning based on Linguistic Instructions Using Fuzzy Theory, ibid., 561/564, 1992 (with O. K. Park).
[82]. An Approach to Interpretation of Linguistic Model based on Functional Linguistic Theory, ibid., *589/592*, 1992 (with I. Kobayashi).
[83]. Hierarchization of Evaluation Structure Using the Choquet Integral, *Proc. of 20th Conference of The Behaviormetric Society of Japan*, 1992 (with T. Murofushi and K. Fujimoto).
[84]. Fuzzy Flight Control of an Unmanned Helicopter, *Proc. of 9th Fuzzy Systems Symposium*, 37/40, 1993 (with Y. Saitoh and I. Hirano).
[85]. An Implementation of Execution System for Fuzzy Algorithms, ibid., 509/512, 1993 (with J. Nishino).
[86]. Linguistic System Description based on Qualitative Knowledge Data and Numerical Data, ibid., 517/520, 1993 (with T. Yasukawa).
[87]. A Hierarchical Evaluation Model Using Choquet Integral (2)., ibid., 673/676, 1993 (with K. Fujimoto and T. Murofushi).
[88]. A Study on Image Understanding of Story based on Functional Linguistic Theory, ibid., 733/736, 1993 (with Y. Fukuyama).
[89]. A Clusterwise Regression-type Model of Subjective Evaluation, *Proc. of 10th Fuzzy Systems Symposium*, 313/316, 1994 (with S. H. Kwon).
[90]. FAPS-2, the Interpreter based on Context for Fuzzy Programs, ibid., 559/562, 1994 (with J. Nishino).
[91]. Intelligent Flight Control of Unmanned Helicopter, ibid., 745/748, 1994 (with H. Akagi and S. Tani).
[92]. A Study on Image Understanding based on Functional Linguistic Theory,

ibid., 783/786, 1994 (with Y. Fukuyama).
[93]. A Hierarchical Evaluation Model Using Choquet Integral (3)., ibid., 805/808, 1994 (with K. Fujimoto and T. Murofushi).
[94]. Approach to Definition of Fuzzy Straight Line, *Proc. of 11th Fuzzy System Symposium*, 45/48, 1995 (with W. J. Liu).
[95]. A Study on Construction of Situation Database, ibid., 133/136 (with T. Fujimoto).
[96]. OPS and Image Guided Landing of an Unmanned Helicopter, ibid., 389/392 (with I. Hirano and S. Kotsu).
[97]. Fuzzy Measures and Choquet Integral on Locally Compact Hausdorff Space, ibid., 431/434 (with Y. Narukawa and T. Murofushi).
[98]. Visual Representation of Fuzzy Measure, ibid., 445/448 (with M. Miyaoka and T. Murofushi).
[99]. A Verbal GO Game Model based on the Context of Situation, ibid., 599/600 (with J. Nishino).
[100]. Two Approaches to Construction of Situation Database, *Proc. of 12th Fuzzy System Symposium*, 59/62, 1996 (with T. Fujimoto).
[101]. An Approach to Computing using Linguistic Information, 63/66, ibid (with I. Kobayashi).
[102]. Applications of Fuzzy Sets to Case-Based Reasoning, 463/466, ibid (with W-J. Liu).
[103]. Tracking Control of an Unmanned Helicopter, 585/588, ibid (with S. Nakamura).
[104]. Intelligent Navigation of an Unmanned Helicopter, 589/592, ibid (with S. Tani and M. Nakamura).

Papers in Edited Volumes

[1]. *Conditional Fuzzy Measures and Their Applications*, in: Fuzzy Sets and Their Applications to Cognitive and Decision Processes, 151/170, 1975 (with T. Terano).
[2]. *Macroscopic Optimization by Using Conditional Fuzzy Measures*, in: Fuzzy Automata and Decision Processes, 197/208, 1977 (with T. Terano).
[3]. *Analytical Representation of Fuzzy Systems*, ibid., 177/189, 1977 (with T. Terano).
[4]. *Macroscopic Optimization by Using Conditional Fuzzy Measures*, ibid., 197/208, 1977 (with T. Terano).
[5]. *Recognition of Linguistically Instructed Path to Destination*, in: Approximate Reasoning in Decision Analysis 341/350, 1982 (with H. Imaoka and T. Terano).
[6]. *A New Approach to Design of Fuzzy Controller*, in: Advances in Fuzzy Sets, Possibility Theory, and Applications, 325/334, 1983 (with T. Takagi).
[7]. *An Experimental Study on Fuzzy Parking Control Using a Model Car*, in: Industrial Applications of Fuzzy Control, 125/138, 1985 (with K. Murakami).

[8]. *A Microprocessor based Fuzzy Controller for Industrial Purposes*, ibid., 231/239, 1985 (with T. Yamazaki).
[9]. *Application of Fuzzy Reasoning to the Water Purification Process*, ibid., 19/39, 1985 (with 0. Yagishita and 0. Itoh).
[10]. *Toward Intelligent Computing, in: Future Directions of Fuzzy Theory and Systems*, World Scientific, 1995

Survey/Lecture/Introductory Papers

[1]. Fuzzy Measures and Fuzzy Integrals - A Survey, in: Fuzzy Automata and Decision Processes, 89/102, 1977
[2]. An Introductory Survey on Fuzzy Control, *Information Sciences*, **36**, 59/83, 1985

(in Japanese).

[3]. Development of Fuzzy Sets Theory, *Systems and Control*, **19**, 5, 3/8, 1975.
[4]. Application of Fuzzy Set and Logic to Control, *J. of the Society of Instrument and Control Engineers*, **18**, 2, 8/18, 1979.
[5]. Imaginable Uncertainty, *Mathematical Science*, **191**, 10/14, 1979.
[6]. Fuzzy Reasoning, *Operations Research*, **26**, 12, 692/697, 1981.
[7]. Applications of Fuzzy Theory, *The Journal of The Institute of Electronics and Communication Engineers of Japan*, **65**, 12, 1264/1266, 1982.
[8]. Fuzzy Control, *J. of the Society for Instrument and Control Engineers*, **22**, 1, 84/86, 1983.
[9]. Fuzzy Theory, Probability and Statistics, *J. of the Society for Instrument and Control Engineers*, **22**, 1, 160/165, 1983 (with A. Ohsumi and Y. Morita).
[10]. Fuzzy Theory (I)., ibid., **22**, 1, 171/174, 1983.
[11]. Fuzzy Theory (II)., ibid., **22**, 4, 42/46, 1983.
[12]. Fuzzy Theory (III)., ibid., **22**, 5, 38/42, 1983.
[13]. Fuzzy Theory (IV)., ibid., **22**, 6, 50/55, 1983.
[14]. Philosophy of Fuzziness, *Systems and Control*, **28**, 7, 3/6, 1984.
[15]. Fuzzy Control, *Systems and Control*, **28**, 7, 18/22, 1984 (with T. Yamazaki).
[16]. Real Applications of Fuzzy Control, *Nikkei Mechanical*, **6**, 18, 54/65, 1984.
[17]. Fuzzy Systems, *J. of the Society for Instrument and Control Engineers*, **24**, 7, 75/76, 1985.
[18]. Fuzzy Logic and Mathematics, *J. of Science Kagaku Asahi*, **536**, 24/26, 1985.
[19]. Fuzzy Control, *The Journal of The Institute of Electronics and Communication Engineers of Japan*, **69**, 5, 476/478, 1986.
[20]. Recent Applications of Fuzzy Theory, *Mathematical Science*, **284**, 5/9, 1987.

[21]. Fuzziness a la carte, *Advertizing*, **32**, 6, 34/41, 1987.
[22]. What fuzzy theory aims at, *Technology and Economics*, **248**, 2/12, 1987.
[23]. Fuzzy Theory, *Mathematical Seminar*, **27**, 1, 50/55, 1988.
[21]. Philosophy on Fuzzy Words, *Bulletin of Science University of Tokyo*, **5**, 3, 8/11, 1988.
[22]. Image of Future Fuzzy Computers, *Computer Today*, **25**, 4/10, 1988.
[23]. Make a computer approach to human : Fuzzy Theory, *Spectrum*, **1**, 8, 8/17, 1988.
[24]. What is fuzzy theory?, *Ushio*, **8**, 50/61, 1988.
[25]. Why is fuzzy theory coming up ?, *Trigger*, **9**, 85/88, 1988.
[26]. Fuzzy Theory and Its Applications, *Chemistry and Biology*, **26**, 9, 562/567, 1988.
[27]. Theory of Fuzziness : Two Kinds of Fuzziness, *BUTSURI*, **43**, 11, 834/840, 1988.
[28]. Fuzzy Control of a Model Car, *J. of the Robotics Society of Japan*, **6**, 6, 68/73, 1988 (with Murofushi).
[29]. Systems Engineering and Fuzziness, *Jap. J. of Behaviormetrics*, **16**, 1, 1988.
[30]. Fuzzy Theory, *Mathematical Seminar*, **27**, 1, 21/24, 1989.
[31]. Fuzzy Theory and Its Application to Control, *OYO BUTURI*, **58**, 1, 40/45, 1989.
[32]. Fuzzy Theory and Its Applications, *KINOH-ZAIRYO*, **9**, 2, 49/55, 1989.
[33]. Fuzzy Control and Its Industrial Applications, *J. of Japan Res. Assn. Text. End-Uses*, **30**, 4, 23/27, 1989.
[34]. New Development of Fuzzy Control, *J. of the Society for Instrument and Control Engineers*, **28**, 11,943/945, 1989.
[35]. Industrial Applications of Fuzzy Theory, *J. of Japan Society for Fuzzy Theory and Systems* (with K. Hirota).
[36]. Fuzzy Theory : Applications and Perspectives, *Technology and Economics*, **275**, 39/50, 1990.
[37]. Fuzzy Control and Its Applications, *Iron and Steel*, **76**, 3, 329/336, 1990.
[38]. Introduction to Fuzzy Measure Theory (I)., *J. of Japan Society for Fuzzy Theory and Systems*, **1**, 5, 174/181, 1990 (with Murofushi).
[39]. Introduction to Fuzzy Measure Theory (II)., ibid., 2, 3, 370/381, 1990 (with Murofushi).
[40]. Fuzzy Technology, *Trans. of the Institute of Electrical Engineers of Japan*, **111**, 1, 3/7, 1991.
[41]. Introduction to Fuzzy Control, *J. of the JSTP*, **32**,361, 117/123, 1991.
[42]. Introduction to Fuzzy Measure Theory (III)., *J. of Japan Society for Fuzzy Theory and Systems*, **3**, 2, 250/262, 1991 (with T. Murofushi).
[43]. AI, Neuro and Fuzzy, *J. of Japan Society for Fuzzy Theory and Systems*, **3**, 3, 433/451, 1991 (with S. Doshita et al.).
[44]. Introduction to Fuzzy Measure Theory (IV)., *J. of Japan Society for Fuzzy Theory and Systems*, **3**, 3, 452/463, 1991 (with T. Murofushi).
[45]. Linguistic Modeling, *Mathematical Science*, **3**, 52/59, 1991 (with T. Yasukawa).

[46]. Introduction to Fuzzy Measure Theory (V)., *J. of Japan Society for Fuzzy Theory and Systems*, **3**, 4, 49/58, 1991 (with T. Murofushi).
[47]. Introduction to Fuzzy Measure Theory (VI)., ibid., **4**, 1, 81/89, 1992 (with T. Murofushi).
[48]. Introduction to Fuzzy Measure Theory (VII)., ibid., **4**, 2, 244/255, 1992 (with T. Murofushi).

Edited Books

[1]. *Industrial Applications of Fuzzy Control*, North-Holland, 1985.
[2]. *Applied Fuzzy Systems*, Ohmusha, 1989 (with T. Terano and K. Asai).
[3]. *Fuzzy Systems Theory and its Applications*, Academic Press, 1993 (with T. Terano and K. Asai).
[4]. *Fuzzy Control*, SOFT Lecture Series, Vol.5, Nikkan-Kogyo-Shinbun Press, 1993 with S. Murakami et al.).
[5]. *Industrial Applications of Fuzzy Technology in the World*, World Scientific, 1995 (with K. Hirota).

Authored Books

[1]. *Introduction to Fuzzy Engineering*, Kodansha, 1981 (with T. Terano et al.).
[2]. *Fuzzy Control*, Nikkan-Kogyo-Shinbun Press, 1988.
[3]. *Development of Fuzzy Theory: Recovery of Subjectivity in Science*, Science-sha, 1989.
[4]. *Introduction to Fuzzy Theory*, Science-sha, 1988.
[5]. *Fuzzy: A New Development of Intelligence*, Nikkan-Kogyo-Shinbun Press, 1989.
[6]. *Introduction to Fuzzy Business*, Nikkan-Kogyo-Shinbun Press, 1990 (with Y. Hasegawa).
[7]. *Fuzzy Measure*, SOFT Lecture Series, Vol.3, Nikkan-Kogyo-Shinbun Press, 1993 (with T. Murofushi).

Index

A

α-cuts, 1
Active image sensor, 21
Antitorque pedals, 18
Archimedean t-conorm, 306, 316, 331, 366
Archimedean t-norm, 2
Automatic heading control, 42
Autonomous flight, 41
Autonomous unmanned helicopter, 16

B

Belief function, 306, 328, 329
Binary fuzzy relation, 5
Borel σ-field, 9

C

Cartesian product, 2
Center-of-Gravity method, 6
Centroid method, 6
Choquet integral, 9, 279, 280, 281, 289, 290, 291, 292, 294, 295, 299, 300, 301, 306, 307, 310, 312, 313, 314, 315, 318, 320, 323, 325, 328, 329, 331
Collective module, 26
Collective pitch control, 15
Collective pitch, 18
Collision avoidance, 42
Command interpretation module, 30
Command-based-control, 15
Comonotonic, 11
Composition of relations, 5
Compositional rule of inference, 45, 46, 48, 55
Conditional proposition, 5
Consequent parameter adjustment, 181, 185, 187, 193
Consequent structure identification, 184
Continuous piecewise polynomial membership function, 254, 256
Contrast intensification, 181, 182, 189, 190, 191, 192
Control law, 6
Control theory, 77, 253
Coupling compensation, 38
Cylindrical Extension, 2

D

Decomposable measure, 306, 309, 312, 324, 326, 327
Decomposition theorem, 370
Defuzzification, 6
DeMorgan algebra, 3
DeMorgan system, 3
Design of a fuzzy controller, 30
Differential GPS system, 16, 21
Disjoint sequence, 369, 373
Duality, 305, 306, 307, 313, 316, 318, 330

E

Engine control, 17
Extension Principle, 2, 47, 57

F

Fly-by-wire controller, 16, 22
Fuzzy adjustment rule, 181, 182, 185, 186, 188, 193, 194

Fuzzy algorithm, 77, 78, 79, 83, 88
Fuzzy algorithmic control, 77, 78, 87
Fuzzy block diagrams, 254, 260
Fuzzy block, 253, 255, 258, 261, 262, 275, 276, 277, 278
Fuzzy control algorithms, 107
Fuzzy control rules, 59, 67, 68, 77, 85, 87
Fuzzy control structure, 22
Fuzzy control systems, 253, 255, 264, 272, 275
Fuzzy control, 4, 13, 67, 68, 69, 72, 73, 75, 77, 115
Fuzzy controller management, 23, 27
Fuzzy controller, 15, 59, 60, 68, 74, 75, 89, 90, 97, 102, 105
Fuzzy coupling compensation, 38
Fuzzy event, 305, 306, 307, 317, 318, 319, 320, 323, 328, 329, 330, 331
Fuzzy implication, 3, 45, 60, 89, 90, 92, 93, 95, 96, 101, 103, 105, 116, 117, 122, 123
Fuzzy integral, 10, 280, 295, 299, 305, 306, 307, 330, 331, 334, 344, 349, 350, 351, 352, 353, 355, 366, 386
Fuzzy logic, 45, 56
Fuzzy measure analysis, 331
Fuzzy measure, 7, 279, 305, 357, 365, 366, 367
Fuzzy model, 59, 115, 182, 201, 253
Fuzzy modeling, 4, 45, 56, 68
Fuzzy partition, 2
Fuzzy quantity, 1
Fuzzy reasoning, 45, 52, 89, 105
Fuzzy relation, 2
Fuzzy subset, 1, 5
Fuzzy subspaces, 117, 119
Fuzzy systems identification, 116
Fuzzy variable, 24, 95

G

Globally asymptotically stable, 266
GPS-guided navigation system, 14, 40

GPS-guided flights, 16
Grade of fuzziness, 7

H

Helicopter flight dynamics, 14
Hierarchical control system, 23
Hierarchical control, 22
Hierarchical fuzzy controller, 15
Höhle's integral, 280

I

Idempotent t-norm, 2
Identification, 115
Image-based control system, 14
Image-guided flights, 16
Implication system, 93
Inferencing, 5
Infimum, 2
Input-output map, 5
Integrated inertia sensor, 20
Intelligent unmanned helicopter, 13
Interval-valued fuzzy sets, 4

L

Lateral cyclic pitch, 18
Lateral module, 25
Lebesgue integral, 7, 280, 306, 365
Linear monotone membership functions, 56, 57
Linguistic approximation, 201
Linguistic modeling, 77, 201
Linguistic truth value, 61
Linguistic variables, 24
Longitudinal cyclic pitch, 18
Longitudinal module, 22
L-R type flat fuzzy number, 88
Lukasiewicz's infinite valued logic, 45, 59, 64
Lyapunov stability theorem, 267
Lyapunov's direct method, 253, 269, 275

M

Max-Min compositional form, 5

Index 425

Measurable function, 9, 280, 300
Measurable space, 308, 366, 369
Membership function, 2
Min and Max operators, 306, 307
Model-based design, 116, 120, 125
Model-based fuzzy controller, 253, 255, 269
Modular fuzzy controller, 15
Modus ponens, 5, 45, 47, 48, 51, 52, 53, 66
Modus tollens, 45, 48, 50, 54, 55
Monotone sequence, 7
Multi-dimensional fuzzy reasoning, 45, 46, 48, 54, 55, 59, 60
Multidimensional fuzzy relation, 182
Multi-input and single-output controller, 68
Multi-input/multi-output fuzzy model, 182, 200
Multivariable control, 131
Mutually disjoint sequences, 373

N

Navigation module, 29
Necessity coefficient, 332
Negation operators, 3
Nilpotent t-norm, 3

O

On-board fuzzy controller, 13, 15, 20
Order isomorphism, 3
Ordinary measure, 366, 369, 370, 371, 372, 373, 378
Overlapping coefficient, 332

P

Parameter design, 30
Parameter identification, 117, 119, 132, 141, 205, 224, 231
Pitch angle control, 22
Positive definite matrix, 265
Premise parameter adjustment, 181, 185, 192
Premise structure identification, 184
Probability law, 9

Probability measure, 8
Process identification, 117, 126
Pseudo-additive measure, 281, 300, 365, 367, 369

Q

Quadratic performance index, 122
Qualitative modeling, 201, 231, 247
Qualitative reasoning, 202

R

Radio wave speed meters, 20
Radon-Nikodym derivatives, 11
Radon-nikodym-like theorem, 365, 367, 379, 380, 383
Remote control devices, 21
Representation of fuzzy measure, 279
Robust control, 36
Roll angle control, 25
Rule base, 5
Rule-based control, 78, 82, 84

S

Self-organising fuzzy controller, 108
Semi-continuous function, 11
Set function, 9, 366, 367, 369
Singletons, 24
Software-based-simulator, 15
Stability analysis, 253, 254, 255, 267, 275
σ-additivity, 8
σ-decomposability, 370
σ-algebra, 67, 367
σ-finiteness, 370, 371, 372
State-variable model, 6
Strong α-cut, 1
Strong negation operator, 3
Structure design, 30
Structure identification, 117, 119, 132, 205, 331
Subjective measure, 332, 335
Successive fuzzy modeling, 182, 184, 194, 195, 197, 198, 200
Successive identification algorithm,

182, 184, 185
Sugeno λ-measure, 7
Sugeno integral, 8, 306, 307, 310, 312, 313, 316, 317, 318, 319, 322, 324, 330, 331
Sugeno negation, 3
Sugeno's fuzzy measure, 279
Supremum, 2

T

T-conorm, 3, 281
Throttle control, 17

Takagi and Sugeno's fuzzy model, 183
Tsukamoto's method, 46

U

Unary fuzzy relation, 5
Unbiasedness criterion, 184, 200
Universal approximation, 6

W

Weighted recursive least square algorithm, 181, 185
Wireless command-based-control, 15